T0392938

Redox Biology in Plasma Medicine

Plasma medicine uses non-equilibrium plasmas generated under atmospheric-pressure conditions. Therapeutical plasmas can stimulate tissue regeneration or inactivate cancer cells. This book reviews the interrelation between plasma chemistry and biochemistry complemented by discussion of the ways plasmas inactivate various pathogens. Focus is on the plasma effects on mammalian cells, subsequent consequences for cell-biological processes, and plasma applicability to specific medical therapies. Contributions illustrate the ways cold atmospheric-pressure plasma can be used as a controllable source of redox-active species and as a useful tool for research in redox biology.

Key Features

- Summarizes plasma chemistry, biochemistry, and microbiology
- Documents the ways plasmas interact with lipids, membranes, and cells
- Reviews therapeutic uses of plasmas in medicine
- Focuses on uses of plasmas in cancer treatment

Oxidative Stress and Disease

Series Editors
Helmut Sies, MD
Enrique Cadenas, MD, PhD
University of Southern California School of Pharmacy
Los Angeles, California

Lipid Oxidation in Health and Disease
Edited by Corinne M. Spickett and Henry Jay Forman

Diversity of Selenium Functions in Health and Disease
Edited by Regina Brigelius-Flohé and Helmut Sies

Mitochondria in Liver Disease
Edited by Derick Han and Neil Kaplowitz

Fetal and Early Postnatal Programming and its Influence on Adult Health
Edited by Mulchand S. Patel and Jens H. Nielsen

Biomedical Application of Nanoparticles
Edited by Bertrand Rihn

The Biology of the First 1,000 Days
Edited by Crystal D. Karakochuk, Kyly C. Whitfield, Tim J. Green, and Klaus Kraemer

Hydrogen Peroxide Metabolism in Health and Disease
Edited by Margreet C M Vissers, Mark Hampton, and Anthony J. Kettle

Glutathione
Edited by Leopold Flohé

Vitamin C: Biochemistry and Function
Edited by Margreet C M Vissers and Qi Chen

Cancer and Vitamin C
Edited by Margreet C M Vissers and Qi Chen

Mammalian Heme Peroxidases: Diverse Roles in Health and Disease
Edited by Clare Hawkins and William M. Nauseef

Proteostasis and Proteolysis
Edited by Niki Chondrogianni, Elah Pick and Anna Gioran

For more information about this series, please visit: www.crcpress.com/Oxidative-Stress-and-Disease/book-series/CRCOXISTRDIS

Redox Biology in Plasma Medicine

Edited by
Sander Bekeschus and Thomas von Woedtke

CRC Press
Taylor & Francis Group
Boca Raton London New York

CRC Press is an imprint of the
Taylor & Francis Group, an **informa** business

Designed cover image: Alice Martinet

First edition published 2025
by CRC Press
2385 NW Executive Center Drive, Suite 320, Boca Raton FL 33431

and by CRC Press
4 Park Square, Milton Park, Abingdon, Oxon, OX14 4RN

CRC Press is an imprint of Taylor & Francis Group, LLC

ISBN: 9781032356921 (hbk)
ISBN: 9781032356945 (pbk)
ISBN: 9781003328056 (ebk)

DOI: 10.1201/9781003328056

Typeset in Times
by Apex CoVantage, LLC

Contents

PART I: *Plasma Chemistry, Biochemistry, and Microbiology*

Chapter 7 Cold Atmospheric Pressure Plasma: A Novel Approach for Virus Treatment Through Reactive Oxygen and Nitrogen Species 84

Pasquale Isabelli, Matteo Gherardi, Romolo Laurita

PART II: Plasma Cell Biology and Medical Therapy

Chapter 8 The Role of Electric Fields in Plasma Treatment of Cells 97

Eric Robert

Chapter 9 Reactive Species and Redox Processes in Medical Gas Plasma-Promoted Wound Healing 109

Anke Schmidt, Sander Bekeschus

About the Editors

Sander Bekeschus was born in Berlin, Germany, in 1985. After obtaining both American and German high school degrees, he studied human biology at Greifswald University (Germany) with a major in immunology. He received a PhD from the Institute of Immunology at the same university in 2015 after two international fellowships at the Centre of Free Radical Research (Christchurch, New Zealand) and Beth Israel Deaconess Medical Center (Boston, USA). From 2016 to 2021, Bekeschus established an independent, third-party-funded junior research group at ZIK plasmatis (Leibniz Institute for Plasma Science and Technology (INP), Greifswald, Germany), which was consolidated under his leadership from 2021 through 2023. Since 2023, Sander Bekeschus has held a Professorship for Translational Plasma Research at the Clinic and Policlinic for Dermatology and Venerology (Rostock University Medical Center, Germany) and served as the Research Program Manager for Plasma Medicine at INP.

Having started his career in plasma sciences as early as 2010, his research has explored the underlying mechanisms of action of plasma and plasma-derived reactive oxygen and nitrogen species (ROS/RNS) in biology and medicine, the directed targeting of plasma parameters towards specific biomedical applications, the elucidation of the inflammatory and immunological consequences of plasma exposure, and the establishment and investigation of translational research models in Plasma Medicine to facilitate and support clinical applications of medical gas plasma technologies.

In addition, he is an executive board member of the International Society of Plasma Medicine (ISPM), the International Workshop on Plasma Cancer Treatment (IWPCT), and the German National Center for Plasma Medicine (NZPM), besides long-standing memberships at the German Society of Immunology (DGfI) and the Society for Free Radical Research Europe (SFRR-E). Bekeschus received Junior Awards Distinctions from the ISPM, INP, and SFRR-E; has authored and co-authored over 200 peer-reviewed publications; and is the initiator and speaker of Europe's first Marie-Curie Doctoral Network on Plasma Medicine, plasmACT.

Thomas von Woedtke was born in Görlitz, Germany, in 1962. He studied pharmacy at Greifswald University, Germany. He received a pharmacist diploma from Greifswald University in 1988, a doctoral degree in pharmaceutical technology in 1995, and a habilitation degree in pharmaceutical technology in 2007 from the same university. From 2008 to 2023 he was Research Program Manager for Plasma Medicine at the Leibniz Institute for Plasma Science and Technology (INP) Greifswald, and since 2020 he has been a member of the board of INP, responsible for the Research Division Health & Hygiene. He has held a professorship for plasma medicine at Greifswald University Medical Center since 2011.

His research interests include medical applications of cold atmospheric pressure plasmas, basic processes of plasma-liquid interactions, and basic research problems of plasma-cell and plasma-tissue interactions, including antimicrobial plasma effects.

He is a founding member of the International Society for Plasma Medicine (ISPM) since 2009, member of the board of directors of ISPM since 2016, founding member of the German National Center for Plasma Medicine (NZPM) since 2013, and Deputy Head of NZPM since 2015.

He received the Plasma Physics Innovation Prize of the Plasma Physics Division of the European Physical Society (EPS), together with Klaus-Dieter Weltmann, INP Greifswald, "for their pioneering work in the field of plasma medicine" in 2016, and the Plasma Medicine Award of the International Society for Plasma Medicine (ISPM) in 2021.

He has authored and co-authored over 280 peer-reviewed publications and book chapters and is co-editor of three textbooks on plasma medicine.

Contributors

Yolanda Aranda Gonzalvo
University of Minnesota
Minneapolis, Minnesota

Sander Bekeschus
ZIK *plasmatis*, Leibniz Institute for Plasma
    Science and Technology (INP)
Greifswald, Germany

Annemie Bogaerts
Research Group PLASMANT,
    Department of Chemistry
University of Antwerp
Antwerp, Belgium

Peter Bruggeman
University of Minnesota
Minneapolis, Minnesota

Cristina Canal
Universitat Politècnica de Catalunya
Barcelona, Spain

Rodin Chermat
Montreal Cancer Institute, CR-CHUM
Montreal, Quebec, Canada

Ramona Clemen
ZIK *plasmatis*, Leibniz Institute for Plasma
    Science and Technology (INP)
Greifswald, Germany

Christophe Deben
University of Antwerp
Antwerp, Belgium

Albert Espona-Noguera
Universitat Politècnica de Catalunya
Barcelona, Spain

Matteo Gherardi
University of Bologna
Bologna, Italy

Audrey Glory
Montreal Cancer Institute, CR-CHUM
Montreal, Quebec, Canada

Inès Hamouda
Universitat Politècnica de Catalunya
Barcelona, Spain

María J. Herrera Quesada
North Carolina State University
Raleigh, North Carolina

Masaru Hori
Nagoya University
Nagoya, Japan

Pasquale Isabelli
University of Bologna
Bologna, Italy

Romolo Laurita
University of Bologna
Bologna, Italy

Vandana Miller
Drexel University
Philadelphia, Pennsylvania

Maria C. Oliveira
University of Antwerp
Antwerp, Belgium

Angela Privat-Maldonado
University of Antwerp
Antwerp, Belgium

Stephan Reuter
Campus de l'Université de Montreal
Montreal, Québec, Canada

Eric Robert
Université d'Orléans
Orléans Cedex, France

João Santos Sousa
CNRS and Université Paris-Saclay
Orsay, France

Anke Schmidt
ZIK *plasmatis*, Leibniz Institute for Plasma
    Science and Technology (INP)
Greifswald, Germany

Katharina Stapelmann
North Carolina State University
Raleigh, North Carolina

Francesco Tampieri
Universitat Politècnica de Catalunya
Barcelona, Spain

Hiromasa Tanaka
Nagoya University
Nagoya, Japan

Shinya Toyokuni
Nagoya University Graduate School of Medicine
Nagoya, Japan

Thomas von Woedtke
ZIK *plasmatis*, Leibniz Institute for Plasma
    Science and Technology (INP)
Greifswald, Germany

Kristian Wende
ZIK *plasmatis*, Leibniz Institute for Plasma
    Science and Technology (INP)
Greifswald, Germany

Maksudbek Yusupov
University of Antwerp
Antwerp, Belgium

Series Editor Preface

REDOX BIOLOGY IN PLASMA MEDICINE

THE EDITORS

Sander Bekeschus is *Research Program Manager 'Plasma Medicine'* at the Leibniz Institute for Plasma Science and Technology (INP, Greifswald, Germany) and *Professor for 'Translational Plasma Medicine'* at the Clinic and Polyclinic for Dermatology and Venerology (Rostock University Medical Center, Germany). His interests are in the redox biology and immunology of plasma applications and oxidant exposure in dermatology and oncology.

Thomas von Woedtke is *Research Division Manager 'Health & Hygiene'* at INP and *Professor for 'Plasma Medicine'* at the Institute for Hygiene and Environmental Medicine (Greifswald University Medical Center, Germany). Dr. von Woedtke's research focuses on improved wound healing mediated by cold atmospheric plasma and medical applications of nonthermal atmospheric pressure plasma.

THE TOPIC: REDOX BIOLOGY IN PLASMA MEDICINE

Plasma medicine is an innovative field of research at the interface between physics and life sciences, which has experienced a substantial international upswing in recent years. Cold atmospheric pressure plasma technology generates a plethora of different mixtures of reactive oxygen and nitrogen species, serving as a novel concept in redox medicine. Biological targets being exposed to plasma-generated reactive oxygen and nitrogen species (and secondary species) trigger redox-related signaling pathways in cells and tissues.

Following an Introductory chapter on redox biology in plasma medicine by the Editors, there are two parts: (I) Plasma chemistry, biochemistry, and microbiology, and (II) Plasma cell biology and medical therapy, covered by an international team of scientists. Topics in Part (I) range from plasma chemistry as a source of reactive oxygen and nitrogen species to plasma-mediated inactivation of bacteria and viruses. Part (II) addresses several topics, from the more general cell-death pathways in plasma-treated cells to specific roles of plasma in cancer treatment, wound healing, immunogenicity, and risk assessment of plasma exposure.

We congratulate the Editors of this monograph on a highly significant and innovative research field, which brings together current knowledge on the physical, chemical, and biological bases of the effects of physical plasma.

Enrique Cadenas
Helmut Sies

1 Introduction to Redox Biology in Plasma Medicine

Thomas von Woedtke, Sander Bekeschus

1.1 COLD ATMOSPHERIC PLASMA (CAP)

Plasma medicine can be defined as the application of physical plasma directly on or in the human body for therapeutic purposes. It is a relatively new field of transdisciplinary research that has emerged over the last 15–20 years at the interface of physics, life sciences (biology, chemistry), and medicine, including clinical practice.

Physical plasma is a particular gas state often called the fourth state of matter, following solid, liquid, and gaseous. Physical plasmas are generated by supplying energy to a neutral gas up to its partial ionization, i.e., electron dissociation from atomic nuclei. This can be done in various ways, e.g., by chemical processes, heating, compression, or using laser beams. The most common method is the application of electric fields or electromagnetic radiation. This can occur under low pressure (near-vacuum), normal (atmospheric) pressure, and high pressure. Depending on the plasma-igniting parameters and above all the working power, non-thermal or thermal plasmas are generated. Plasmas for biomedical research and clinical application are configured in such a way that they work at or near body temperatures, i.e., $\leq 40°C$, at the target site during treatment, as well as in an atmospheric-pressure environment. Because distinguishing physical plasma from blood plasma is a continuous challenge in the biomedical community, several terms are in use, e.g., (medical) gas plasma, non-thermal plasma (NTP), low-temperature plasma (LTP), tissue-tolerable plasma (TTP), cold physical plasma (CPP), non-invasive physical plasma (NIPP), and others; however, cold atmospheric plasma or cold atmospheric-pressure plasma (CAP) is the most common term and abbreviation. CAP is a multi-component system containing ultraviolet, visible, and infrared radiation; electric fields; free electrons and ions; reactive oxygen and nitrogen species (ROS/RNS, also referred to as RONS); and excited gas atoms and molecules in the bulk of neutral gas atoms and molecules [1]. More details on the characteristics of physical plasma and plasma components causing biological effects are given in Chapter 2.

1.2 APPLICATIONS OF CAP IN MEDICINE

The fact that it is a reactive, multi-component system makes CAP a useful tool for various technical and industrial applications, allowing the treatment of delicate, thermo-sensitive materials. Consequently, it is used to design and optimize specific properties of various surfaces and devices by cleaning, etching, coating, activating, or functionalizing in various technical fields, including for medical devices such as implants [2–4].

A special property of CAP is that it can inactivate microorganisms and viruses without permanently influencing or destroying their surrounding structures. This property was explored for use in gentle decontamination or "sterilization" methods [5] and has received increased attention in research and technology development since the 1990s, when the availability of CAP technology improved [6–11].

This special ability of CAP to gently inactivate microorganisms inspired the idea of using it to treat contaminated or infected body regions such as skin or wounds. This was a first step toward a

DOI: 10.1201/9781003328056-1

biomedical application of CAP. The medical use of physical plasma in general had been established many years earlier, in the context of so-called electro-surgical methods such as argon plasma coagulation (APC). However, these techniques for cauterization, blood coagulation, and tissue sectioning are mainly based on thermal bio-destructible plasma effects [12, 13], whereas CAP acts in a nonthermal, physical and (bio-) chemistry-driven manner. Experimental work on CAP inactivation of microorganisms and testing of cultured mammalian cells led to the discovery that cells can be manipulated selectively without killing them at sufficiently low application of energy and/or short plasma treatment times. The first observed effects were reversible cell detachment and stimulation of cell proliferation [14–16]. Based on such experimental findings and several theoretical considerations as well as practical requirements in the medical sector, research in plasma medicine was focused initially on wound healing [17, 18]. An early hypothesis held that plasma could support wound healing in two ways: by inactivation of wound-contaminating microorganisms and by direct stimulation of tissue regeneration [19]. During the last 15 years, this hypothesis has been confirmed by a considerable number of *in vitro* experiments using microorganisms and cultivated mammalian cells, animal studies, and clinical trials. Moreover, fundamental molecular mechanisms of plasma-supported tissue regeneration have been elucidated [20–25]. Chapter 9 presents more details on plasma-supported wound healing, which is the first CAP application that has reached clinical practice. A medical guideline on "Rational therapeutic use of cold physical plasma" was published in 2022 in Germany. There are CAP devices on the market that are certified as medical devices for the purpose of treating chronic wounds and pathogen-based skin diseases. In addition to wound treatment, CAP treatment is indicated for other diseases, especially in the dermatological field, e.g., local skin disinfection, atopic eczema, acne, and others [26, 27]. Cosmetic applications are also under discussion [28].

The second major research field in plasma medicine is CAP application in cancer treatment. Investigations of plasma treatment of mammalian cells *in vitro* have shown that cells can be not only stimulated but—with higher treatment intensities—also killed by induction of regulated cell death (e.g., apoptosis). Interestingly, this is also possible with cancer cells, even though cancer cells have several strategies to limit or circumvent apoptosis [29]. These first *in vitro* findings led to high expectations predicting a "paradigm shift in cancer therapy" [30]. Many *in vitro* experiments and animal studies with a significant number of cancer cell lines have substantiated the potential of CAP in cancer treatment [31–33]. Molecular mechanisms of CAP effects on cancer cells comprise both direct killing of cancer cells by induction of regulated cell death and indirect (systemic) effects by initiation of immunogenic cell death [34, 35]. Current knowledge in the specific field of plasma-induced immunogenic cancer cell death is presented in Chapter 10, and in a promising new field of possible anticancer vaccination using plasma technology in Chapter 11. Clinical applications of CAP in cancer treatment are rare for various reasons [36]. For example, it must be unambiguously clarified whether the stimulating plasma effect that is the basis of tissue regeneration in wound healing can lead to cancer metastasis in the case of sub-effective treatment of cancer tissue [37]. Another question to be resolved is the possibility of developing resistance in cancer cells with repeated plasma treatment [38].

Because of the limited penetration depth of CAP and, consequently, an expected limitation in the treatment of bulk tumors, plasma applications in combination with established cancer treatment approaches such as surgery, radiotherapy, chemotherapy, or immunotherapy seem to be most promising from the current perspective. Aspects of plasma cancer treatment in combination with other physical modalities are presented in Chapter 12. Initial clinical experience has been gained in palliative CAP applications on advanced and ulcerated squamous cell carcinoma of the head and neck area [39, 40] and in combination with surgical tumor resection [41]. Furthermore, initial clinical case reports, as well as one randomized clinical trial on the treatment of precancerous actinic keratosis [42, 43] and one single-armed trial on cervical intraepithelial neoplasia [44], indicate additional promising areas for CAP application in the treatment of precancerous lesions.

CAP applications in dermatology and cancer treatment are currently the most prominent areas in preclinical and clinical research and practice. Other fields, including ophthalmology [45], orthopedics [46, 47], and dentistry [48–52], are also under investigation. There is hardly any clinical experience to date in these areas of application.

A primary requisite of any clinical application of CAP is proof of safety with regard to both acute side effects and long-term damage, in particular genotoxicity and mutagenicity. A significant number of *in vitro* and *in vivo* studies have been published on these topics, including long-term follow-up investigations in animals, healthy volunteers, and CAP-treated patients (see Chapter 15).

1.3 REDOX BIOLOGY IN CAP EFFECTS

Research focused on clinical applications has been accompanied by intensive fundamental research on mechanisms of the biological effects of CAP. The primary effects of CAP—inactivation of microorganisms, stimulation of tissue regeneration, and inactivation of mammalian cells, including cancer cells—have been achieved consistently with diverse plasma sources, which differ with regard to their technical setup, the applied power, or the working gases used for plasma generation (noble gases such as argon or helium, molecular gases such as oxygen, nitrogen, and air, or gas mixtures). This finding inspired the fundamental question of whether there is a uniform principle of action. The interaction of electrons and high energy states of atoms or molecules with reaction partners in both the plasma phase and its close vicinity (ambient air, liquids, surfaces) generates secondary and tertiary reactive species, which increase the complexity of the plasma "cocktail" (see Chapter 2). A common feature of all CAP sources is that they work in an ambient air environment or use air as working gas. Consequently, oxygen and nitrogen are the dominating compounds for plasma-chemical reactions that result in the formation of reactive oxygen and nitrogen species (ROS/RNS) [53].

Several early experimental studies [14, 54] supported the hypothesis that ROS/RNS play a central role in biological CAP effects. For example, it was shown that the toxic effect of CAP on cultured cells *in vitro* depended on the type of cell culture medium used. Cells showed less CAP-induced toxicity in culture media high in antioxidative capacity [55]. In other studies, the biological effects of CAP could be partially or fully counteracted by the addition of free radical scavenger substances or antioxidants such as N-acetylcysteine (NAC), glutathione, or ascorbic acid (vitamin C) [56–58]. In one of the first *in vivo* studies with human volunteers on plasma effects in healthy skin, a marked reduction of beta-carotene level in the upper layers of the Stratum corneum was identified, indicative of elevated local oxidant or free radical levels [59].

In 2012, Graves summarized some of the existing knowledge on the role of ROS/RNS in biological processes and drew a link to plasma-initiated biological effects [60]. This was a key moment in the development of plasma medicine, relating it to the well-established field of redox biology. It became evident that biologically active plasma-generated species such as superoxide, hydrogen peroxide, hydroxyl radical, singlet oxygen, nitric oxide, and others are the same as species generated by regular physiological and biochemical processes in the organism [61, 62]. This finding led to the assumption that any therapeutic application of CAP, e.g., in wound healing, is not based on the effect of "foreign" plasma-produced compounds but rather is a consequence of existing signaling pathways re-adjusted by the action of CAP-derived ROS/RNS. Thus, redox biology has proved to be the scientific basis of plasma medicine. Many biological effects caused by CAP can now be explained and interpreted using findings from redox biology [63, 64]. A central insight of this research is that CAP effects on cells follow the principle of hormesis, i.e., biphasic responses to CAP impact, with stimulatory effects observed with low CAP intensities and inhibitory or toxic effects with high intensities [63, 65]. This finding revealed clear parallels to the concept of oxidative eustress and oxidative distress introduced by Sies [66, 67] and led to the designation of plasma medicine as applied redox biology [68]. Detailed investigations of interactions of CAP or CAP-generated ROS/RNS—also in combination with physical plasma compounds such as electric fields (see Chapter 8)—with proteins and liquids led to the identification of fundamental molecular mechanisms of CAP effects on biological structures and processes. This research provided a hypothetical basis to explain the underlying mechanisms of clinical efficacy of CAP. More details are presented in Chapters 4 and 5. In addition, these insights were also the basis of the targeted identification of possible risk factors of CAP application and their systematic investigation and exclusion [69, 70] (see Chapter 15).

In contrast to the enormous volume of research in recent years on elucidating mechanisms of plasma effects on mammalian cells, including cancer cells, knowledge of the inactivation mechanisms of microorganisms by CAP is still limited. In general, microorganism inactivation is attributed to the activity of ROS/RNS via three mechanisms: (i) permeabilization or damage of the cell membrane or wall, (ii) modification or damage of intracellular proteins, and (iii) DNA damage. The antimicrobial effectivity of CAP depends on plasma processing parameters, environmental factors, and the properties of the respective microorganisms. In general, antimicrobial plasma effects are promising both for therapeutic applications such as wound healing or treatment of pathogen-associated skin diseases and for hygiene and food processing [64]. The antiviral effect of CAP application in combating pathogens, a highly topical subject, is discussed in Chapter 7.

Research has shown that liquid phases in the cellular environment play a crucial role in the transmission and/or conversion of reactive species from the plasma/gas phase to cellular structures and components [71]. Chapter 3 gives more insight into the complex chemistry resulting from CAP treatment of liquids. Detailed investigations of plasma-liquid interaction have significantly contributed to elucidating the transmission mechanisms of biological effects from plasma to biological targets. Meanwhile, hydrogel-based models are used as simple tissue models to investigate three-dimensional propagation mechanisms of plasma-generated reactive species (see Chapter 6). Moreover, plasma treatment of liquids has been found to result in at least temporary biological activity of these liquids. Several terms have become commonplace in the literature, such as plasma-treated liquids (PTL), plasma-activated liquid (PAL), plasma-activated water (PAW), plasma-activated medium (PAM), or plasma-processed water (PPW). Many studies have explored the potential application of such plasma-treated liquids in disinfection or antiseptics, food processing, agriculture, and other fields [72, 73]. Medical applications of plasma-treated liquids are mainly focused on cancer treatment, which is explained in more detail in Chapter 13. Medical applications of hydrogels as vehicles for the delivery of plasma-generated ROS/RNS are a newer field of research in plasma medicine (see Chapter 6). In general, there is an increasing focus in biomedical sciences on three-dimensional assay systems which allow to simulate parts of the spatial biology relevant *in vivo*, as reflected in Chapter 14.

1.4 CONCLUSION, OUTLOOK, AND AIM OF THIS BOOK

Within the last 15 years, an independent plasma medicine research community has been established. An International Conference on Plasma Medicine (ICPM) has been held every two years since 2007. Especially in Europe, three COST Actions (European Cooperation in Science & Technology; www.cost.eu) are perpetuating and intensifying the scientific exchange and cooperation on plasmas in medicine: MP1101—Biomedical Applications of Atmospheric Pressure Plasma Technology, 2011–2015; TD1208—Electrical discharges with liquids for future applications, 2013–2017; and CA20114—Therapeutical applications of Cold Plasmas (PlasTHER), 2021–2025.

Increased cooperation and networking between plasma medicine and redox biology are needed. For plasma medicine, it will be essential to consolidate redox biology as a scientific basis to further explain biological plasma effects. Conversely, research results and experience from plasma medicine can also contribute to specific research in redox biology. Plasma medicine and redox biology are interested in some similar questions, for example [68]:

i). How do we identify and analyze specific (CAP-derived/inflammation-derived) ROS/RNS types at their site of action?

ii). Which cell biological mechanisms are responsible for the varying sensitivity of several cell types to the impact of (CAP-derived/inflammation-derived) ROS/RNS?

iii). What is the penetration depth of different (CAP-derived/inflammation-derived) ROS/RNS types within tissues, and what is the dependence on the tissue type?

This book will contribute to bridging the gap between plasma medicine and redox biology by allowing plasma medical scientists to reflect the current knowledge about reactive species-driven chemical and biological consequences. Conversely, by presentation of the ROS/RNS-centered research focus in the plasma medical field, cold plasma science may spark interest within selected redox chemistry, redox biology, and redox medicine communities on generating, characterizing, and applying novel reactive species mixtures for various research purposes.

REFERENCES

1. K. Wende, R. Brandenburg. Cold Physical Plasma: A Short Introduction. In: H.-R. Metelmann, T. von Woedtke, K.-D. Weltmann, S. Emmert (eds). Textbook of Good Clinical Practice in Cold Plasma Therapy, Springer, Cham 2022, p. 37; 10.1007/978-3-030-87857-3_2
2. K.-D. Weltmann, J.F. Kolb, M. Holub, D. Uhrlandt, M. Šimek, K. Ostrikov, S. Hamaguchi, U. Cvelbar, M. Černák, B. Locke, A. Fridman, P. Favia, K. Becker. The Future for Plasma Science and Technology. Plasma Process. Polym. 16 (2019) e1800118; 10.1002/ppap.201800118
3. Y. Kim. Exploring Emerging Technologies with Analysis of Bibliographic Data Focused on Plasma Surface Treatment. Coatings. 11 (2021) 1291; 10.3390/coatings11111291
4. F. Förster. Atmospheric Pressure Plasma in Industrial Applications: Surface Treatment of Thermally Sensitive Polymers. Plasma Process. Polym. 19 (2022) e2100240; 10.1002/ppap.202100240
5. W.P. Menashi. Treatment of Surfaces. US Patent. 3,383,163 (1968) 383, 163.
6. M. Laroussi. Sterilization of Contaminated Matter with an Atmospheric Pressure Plasma. IEEE Trans. Plasma Sci. 24 (1996) 1188; 10.1109/27.533129
7. M. Laroussi. Low Temperature Plasma-Based Sterilization: Overview and State-of-the-Art. Plasma Process. Polym. 2 (2005) 391; 10.1002/ppap.200400078
8. V. Scholtz, J. Pazlarova, H. Souskova, J. Khun, J. Julak. Nonthermal Plasma—A Tool for Decontamination and Disinfection. Biotechnol. Adv. 33 (2015) 1108; 10.1016/j.biotechadv.2015.01.002
9. N.M. Marsit, L.E. Sidney, M.J. Branch, S.L. Wilson, A. Hopkinson. Terminal Sterilization: Conventional Methods Versus Emerging Cold Atmospheric Pressure Plasma Technology for Non-Viable Biological Tissues. Plasma Process. Polym. 14 (2017) e1600134; 10.1002/ppap.201600134
10. S. Bekeschus, P. Favia, E. Robert, T. von Woedtke. White Paper on Plasma for Medicine and Hygiene: Future in Plasma Health Sciences. Plasma Process. Polym. 16 (2019) e1800033; 10.1002/ppap.201800033
11. A. Barjasteh, Z. Dehghani, P. Lamichhane, N. Kaushik, E.H. Choi, N.K. Kaushik. Recent Progress in Applications of Non-Thermal Plasma for Water Purification, Bio-Sterilization, and Decontamination. Appl. Sci. 11 (2021) 3372; 10.3390/app11083372
12. J. Raiser, M. Zenker. Argon Plasma Coagulation for Open Surgical and Endoscopic Applications: State of the Art. J. Phys. D: Appl. Phys. 39 (2006) 3520; 10.1088/0022-3727/39/16/S10
13. A. Taheri, P. Mansoori, L.F. Sandoval, S.R. Feldman, D. Pearce, P.M. Williford. Electrosurgery Part I: Basics and Principles. J. Am. Acad. Dermatol. 70 (2014) 591.e1; 10.1016/j.jaad.2013.09.056
14. E. Stoffels, I.E. Kieft, R.E.J. Sladek. Superficial Treatment of Mammalian Cells Using Plasma Needle. J. Phys. D: Appl. Phys. 36 (2003) 2908; 10.1088/0022-3727/36/23/007
15. E. Stoffels. "Tissue Processing" with Atmospheric Plasmas. Contrib. Plasma Phys. 47 (2007) 40; 10.1002/ctpp.200710007
16. S. Kalghatgi, G. Friedman, A. Fridman, A. Morss Clyne. Endothelial Cell Proliferation is Enhanced by Low Dose Non-Thermal Plasma Through Fibroblast Growth Factor-2 Release. Ann. Biomed. Eng. 38 (2010) 748; 10.1007/s10439-009-9868-x
17. G. Fridman, G. Friedman, A. Gutsol, A.B. Shekhter, V.N. Vasilets, A. Fridman. Applied Plasma Medicine. Plasma Process. Polym. 5 (2008) 503; 10.1002/ppap.200700154
18. G. Lloyd, G. Friedman, S. Jafri, G. Schultz, A. Fridman, K. Harding. Gas Plasma: Medical Uses and Developments in Wound Care. Plasma Process. Polym. 7 (2010) 194–211; 10.1002/ppap.200900097

19. A. Kramer, N.-O. Hübner, K.-D. Weltmann, J. Lademann, A. Ekkernkamp, P. Hinz, O. Assadian. Polypragmasia in the Therapy of Infected Wounds—Conclusions Drawn from the Perspectives of Low Temperature Plasma Technology for Plasma Wound Therapy. GMS Krankenhaushyg. Interdiszip. 3 (2008) Doc13; www.egms.de/en/journals/dgkh/2008-3/dgkh000111.shtml

20. S. Arndt, A. Schmidt, S. Karrer, T. von Woedtke. Comparing Two Different Plasma Devices kINPen and Adtec SteriPlas Regarding Their Molecular and Cellular Effects on Wound Healing. Clin. Plasma Med. 9 (2018) 24; 10.1016/j.cpme.2018.01.002

21. A. Schmidt, T. von Woedtke, B. Vollmar, S. Hasse, S. Bekeschus. Nrf2 Signaling and Inflammation are Key Events in Physical Plasma-Spurred Wound Healing. Theranostics. 9 (2019) 1066; 10.7150/thno.29754

22. A. Schmidt, G. Liebelt, F. Nießner, T. von Woedtke, S. Bekeschus. Gas Plasma-Spurred Wound Healing is Accompanied by Regulation of Focal Adhesion, Matrix Remodeling, and Tissue Oxygenation. Redox Biol. 38 (2021) 101809; 10.1016/j.redox.2020.101809

23. S. Bekeschus, T. von Woedtke, S. Emmert, A. Schmidt. Medical Gas Plasma-Stimulated Wound Healing: Evidence and Mechanisms. Redox Biol. 46 (2021) 102116; 10.1016/j.redox.2021.102116

24. Y. Wu, S. Yu, X. Zhang, X. Wang, J. Zhang. The Regulatory Mechanism of Cold Plasma in Relation to Cell Activity and Its Application in Biomedical and Animal Husbandry Practices. Int. J. Mol. Sci. 24 (2023) 7160; 10.3390/ijms24087160

25. T. Bolgeo, A. Maconi, M. Gardalini, D. Gatti, R. Di Matteo, M. Lapidari, Y. Longhitano, G. Savioli, A. Piccioni, C. Zanza. The Role of Cold Atmospheric Plasma in Wound Healing Processes in Critically Ill Patients. J. Pers. Med. 13 (2023) 736; 10.3390/jpm13050736

26. T. Bernhardt, M.L. Semmler, M. Schäfer, S. Bekeschus, S. Emmert, L. Boeckmann. Plasma Medicine: Applications of Cold Atmospheric Pressure Plasma in Dermatology. Oxidative Med. Cell. Longev. 2019 (2019) 3873928; 10.1155/2019/3873928

27. L. Gan, J. Jiang, J.W. Duan, X.J.Z. Wu, S. Zhang, X.R. Duan, J.Q. Song, H.X. Chen. Cold Atmospheric Plasma Applications in Dermatology: A Systematic Review. J. Biophotonics. 14 (2021) e202000415; 10.1002/jbio.202000415

28. G. Busco, E. Robert, N. Chettouh-Hammas, J.-M. Pouvesle, C. Grillon. The Emerging Potential of Cold Atmospheric Plasma in Skin Biology. Free Radic. Biol. Med. 161 (2020) 290; 10.1016/j.freeradbiomed.2020.10.004

29. D. Hanahan, R.A. Weinberg. Hallmarks of Cancer: The Next Generation. Cell. 144 (2011) 646; 10.1016/j.cell.2011.02.013

30. M. Keidar, R. Walk, A. Shashurin, P. Srinivasan, A. Sandler, S. Dasgupta, R. Ravi, R. Guerrero-Preston, B. Trink. Cold Plasma Selectivity and the Possibility of a Paradigm Shift in Cancer Therapy. Br. J. Cancer. 105 (2011) 1295; 10.1038/bjc.2011.386

31. A. Dubuc, P. Monsarrat, F. Virard, N. Merbahi, J.-P. Sarrette, S. Laurencin-Dalicieux, S. Cousty. Use of Cold-Atmospheric Plasma in Oncology: A Concise Systematic Review. Ther. Adv. Med. Oncol. 10 (2018) 1; 10.1177/1758835918786475

32. D. Yan, A. Malyavko, Q. Wang, K. Ostrikov, J.H. Sherman, M. Keidar. Multi-Modal Biological Destruction by Cold Atmospheric Plasma: Capability and Mechanism. Biomedicines. 9 (2021) 1259; 10.3390/biomedicines9091259

33. S. Bekeschus, G. Liebelt, J. Menz, J. Berner, S. Kumar Sagwal, K. Wende, K.-D. Weltmann, L. Boeckmann, T. von Woedtke, H.-R. Metelmann, S. Emmert, A. Schmidt. Tumor Cell Metabolism Correlates with Resistance to Gas Plasma Treatment: The Evaluation of Three Dogmas. Free Radic. Biol. Med. 167 (2021) 12; 10.1016/j.freeradbiomed.2021.02.035

34. M. Khalili, L. Daniels, A. Lin, F.C. Krebs, A.E. Snook, S. Bekeschus, W.B. Bowne, V. Miller. Non-Thermal Plasma-Induced Immunogenic Cell Death in Cancer. J. Phys. D: Appl. Phys. 52 (2019) 423001; 10.1088/1361-6463/ab31c1

35. M.L. Semmler, S. Bekeschus, M. Schäfer, T. Bernhardt, T. Fischer, K. Witzke, C. Seebauer, H. Rebl, E. Grambow, B. Vollmar, J.B. Nebe, H.-R. Metelmann, T. von Woedtke, S. Emmert, L. Boeckmann. Molecular Mechanisms of the Efficacy of Cold Atmospheric Pressure Plasma (CAP) in Cancer Treatment. Cancers. 12 (2020) 269; 10.3390/cancers12020269

36. S. Bekeschus. Medical Gas Plasma Technology: Roadmap on Cancer Treatment and Immunotherapy. Redox Biol. 65 (2023) 102798; 10.1016/j.redox.2023.102798

37. S. Bekeschus, E. Freund, C. Spadola, A. Privat-Maldonado, C. Hackbarth, A. Bogaerts, A. Schmidt, K. Wende, K.-D. Weltmann, T. von Woedtke, C.-D. Heidecke, L.-I. Partecke, A. Käding. Risk Assessment of kINPen Plasma Treatment of Four Human Pancreatic Cancer Cell Lines with Respect to Metastasis. Cancers. 11 (2019b) 1237; 10.3390/cancers11091237

38. J. Berner, L. Miebach, M. Kordt, C. Seebauer, A. Schmidt, M. Lalk, B. Vollmar, H.-R. Metelmann, S. Bekeschus. Chronic Oxidative Stress Adaptation in Head and Neck Cancer Cells Generates Slow-Cyclers with Decreased Tumour Growth *in Vivo*. Brit. J. Cancer. 129 (2023) 869; 10.1038/s41416-023-02343-6

39. H.-R. Metelmann, D.S. Nedrelow, C. Seebauer, M. Schuster, T. von Woedtke, K.-D. Weltmann, S. Kindler, P.H. Metelmann, S.E. Finkelstein, D.D. Von Hoff, F. Podmelle. Head and Neck Cancer Treatment and Physical Plasma. Clin. Plasma Med. 3 (2015) 17; 10.1016/j.cpme.2015.02.001

40. H.-R. Metelmann, C. Seebauer, V. Miller, A. Fridman, G. Bauer, D.B. Graves, J.-M. Pouvesle, R. Rutkowski, M. Schuster, S. Bekeschus, K. Wende, K. Masur, S. Hasse, T. Gerling, M. Hori, H. Tanaka, E.H. Choi, K.-D. Weltmann, P.H. Metelmann, D.D. Von Hoff, T. von Woedtke. Clinical Experience with Cold Plasma in the Treatment of Locally Advanced Head and Neck Cancer. Clin. Plasma Med. 9 (2018) 6; 10.1016/j.cpme.2017.09.001

41. J. Canady, S.R.K. Murthy, T. Zhuang, S. Gitelis, A. Nissan, L. Ly, O.Z. Jones, X. Cheng, M. Adileh, A.T. Blank, M.W. Colman, K. Millikan, C. O'Donoghue, K.M. Stenson, K. Ohara, G. Schtrechman, M. Keidar, G. Basadonna. The First Cold Atmospheric Plasma Phase I Clinical Trial for the Treatment of Advanced Solid Tumors: A Novel Treatment Arm for Cancer. Cancers. 15 (2023) 3688; 10.3390/cancers15143688

42. P.C. Friedman, V. Miller, G. Fridman, A. Lin, A. Fridman. Successful Treatment of Actinic Keratosis Using Nonthermal Atmospheric Pressure Plasma: A Case Series. J. Am. Acad. Dermatol. 76 (2017) 349; 10.1016/j.jaad.2016.09.004

43. M. Wirtz, I. Stoffels, J. Dissemond, D. Schadendorf, A. Roesch. Actinic Keratoses Treated with Cold Atmospheric Plasma. JEADV. 32 (2018) e37; 10.1111/jdv.14465

44. J. Marzi, M.B. Stope, M. Henes, A. Koch, T. Wenzel, M. Holl, S.L. Layland, F. Neis, H. Bösmüller, F. Ruoff, M. Templin, B. Krämer, A. Staebler, J. Barz, D.A. Carvajal Berrio, M. Enderle, P.M. Loskill, S.Y. Brucker, K. Schenke-Layland, M. Weiss. Noninvasive Physical Plasma as Innovative and Tissue-Preserving Therapy for Women Positive for Cervical Intraepithelial Neoplasia. Cancers. 14 (2022) 1933; 10.3390/cancers14081933

45. H.H. Reitberger, M. Czugala, C. Chow, A. Mohr, A. Burkovski, A.K. Gruenert, R. Schoenebeck, T.A. Fuchsluger. Argon Cold Plasma—A Novel Tool to Treat Therapy-Resistant Corneal Infections. Am. J. Ophthalmol. 190 (2018) 150; 10.1016/j.ajo.2018.03.025.

46. L. Nguyen, P. Lu, D. Boehm, P. Bourke, B.F. Gilmore, N.J. Hickok, T.A. Freeman. Cold Atmospheric Plasma is a Viable Solution for Treating Orthopedic Infection: A Review. Biol. Chem. 400 (2019) 77; 10.1515/hsz-2018-0235

47. L. Nonnenmacher, M. Fischer, L. Haralambiev, S. Bekeschus, F. Schulze, G.I. Wassilew, J. Schoon, J.C. Reichert. Orthopaedic Applications of Cold Physical Plasma. EFORT Open Rev. 8 (2023) 409; 10.1530/EOR-22-0106

48. W.L. Hui, V. Perrotti, F. Iaculli, A. Piattelli, A. Quaranta. The Emerging Role of Cold Atmospheric Plasma in Implantology: A Review of the Literature. Nanomaterials 10 (2020) 1505; 10.3390/nano10081505

49. S. Lata, S. Chakravorty, T. Mitra, P.K. Pradhan, S. Mohanty, P. Patel, E. Jha, P.K. Panda, S.K. Verma, M. Suar. Aurora Borealis in Dentistry: The Applications of Cold Plasma in Biomedicine. Mater. Today Bio. 13 (2022) 100200; 10.1016/j.mtbio.2021.100200

50. A.C. Borges, K.G. Kostov, R.S. Pessoa, G.M.A. de Abreu, G. de M.G. Lima, L.W. Figueira, C.Y. Koga-Ito. Applications of Cold Atmospheric Pressure Plasma in Dentistry. Appl. Sci. 11 (2021) 1975; 10.3390/app11051975

51. Z. Liu, X. Du, L. Xu, Q. Shi, X. Tang, Y. Cao, K. Song. The Therapeutic Perspective of Cold Atmospheric Plasma in Periodontal Disease. Oral Dis. (2023) 1; 10.1111/odi.14547

52. A.B. Muniz, M.R. da Cruz Vegian, L.D. Pereira Leite, D. Morais da Silva, N. Vicensoto Moreira Milhan, K.G. Kostov, C.Y. Koga-Ito. Non-Thermal Atmospheric Pressure Plasma Application in Endodontics. Biomedicines. 11 (2023) 1401; 10.3390/biomedicines11051401

53. X. Lu, G.V. Naidis, M. Laroussi, S. Reuter, D.B. Graves, K. Ostrikov. Reactive Species in Non-Equilibrium Atmospheric-Pressure Plasmas: Generation, Transport, and Biological Effects. Phys. Rep. 630 (2016) 1; 10.1016/j.physrep.2016.03.003

54. M. Laroussi, F. Leipold. Evaluation of the Roles of Reactive Species, Heat, and UV Radiation in the Inactivation of Bacterial Cells by Air Plasmas at Atmospheric Pressure. Int. J. Mass Spectrom. 233 (2004) 81; 10.1016/j.ijms.2003.11.016

55. K. Wende, S. Straßenburg, B. Haertel, M. Harms, S. Holtz, A. Barton, K. Masur, T. von Woedtke, U. Lindequist. Atmospheric Pressure Plasma Jet Treatment Evokes Transient Oxidative Stress in HaCaT Keratinocytes and Influences Cell Physiology. Cell Biol. Int. 38 (2014) 412; 10.1002/cbin.10200

56. S. Kalghatgi, C.M. Kelly, E. Cerchar, B. Torabi, O. Alekseev, A. Fridman, G. Friedman, J. Azizkhan-Clifford. Effects of Non-Thermal Plasma on Mammalian Cells. PLoS One. 6 (2011) e16270; 10.1371/journal.pone.0016270

57. K.P. Arjunan, A. Morss Clyne. Hydroxyl Radical and Hydrogen Peroxide are Primarily Responsible for Dielectric Barrier Discharge Plasma-Induced Angiogenesis. Plasma Process. Polym. 8 (2011) 1154; 10.1002/ppap.201100078

58. S. Blackert, B. Haertel, K. Wende, T. von Woedtke, U. Lindequist. Influence of Non-Thermal Atmospheric Pressure Plasma on Cellular Structures and Processes in Human Keratinocytes (HaCaT). J. Dermatol. Sci. 70 (2013) 173; 10.1016/j.jdermsci.2013.01.012

59. J.W. Fluhr, S. Sassning, O. Lademann, M.E. Darvin, S. Schanzer, A. Kramer, H. Richter, W. Sterry, J. Lademann. *In Vivo* Skin Treatment with Tissue-Tolerable Plasma Influences Skin Physiology and Antioxidant Profile in Human Stratum Corneum. Exp. Dermatol. 21 (2011) 130; 10.1111/j.1600-0625.2011.01411.x

60. D.B. Graves. The Emerging Role of Reactive Oxygen and Nitrogen Species in Redox Biology and Some Implications for Plasma Applications to Medicine and Biology. J. Phys. D: Appl. Phys. 45 (2012) 263001; 10.1088/0022-3727/45/26/263001

61. W. Dröge. Free Radicals in the Physiological Control of Cell Function. Physiol. Rev. 82 (2002) 47; 10.1152/physrev.00018.2001

62. F.F. Fang. Antimicrobial Reactive Oxygen and Nitrogen Species: Concepts and Controversies. Nat. Rev. Microbiol. 2 (2004) 820; 10.1038/nrmicro1004

63. A. Privat-Maldonado, A. Schmidt, A. Lin, K.-D. Weltmann, K. Wende, A. Bogaerts, S. Bekeschus. ROS from Physical Plasmas: Redox Chemistry for Biomedical Therapy. Oxid. Med. Cell. Longev. 2019 (2019) 9062098; 10.1155/2019/9062098

64. T. von Woedtke, M. Laroussi, M. Gherardi. Foundations of Plasmas for Medical Applications. Plasma Sources Sci. Technol. 31 (2022) 054002; 10.1088/1361-6595/ac604f

65. E.J. Szili, F.J. Harding, S.-H. Hong, F. Herrmann, N.H. Voelcker, R.D. Short. The Hormesis Effect of Plasma-Elevated Intracellular ROS on HaCaT Cells. J. Phys. D: Appl. Phys. 48 (2015) 495401; 10.1088/0022-3727/48/49/495401

66. H. Sies. Hydrogen Peroxide as a Central Redox Signaling Molecule in Physiological Oxidative Stress: Oxidative Eustress. Redox Biol. 11 (2017) 613; 10.1016/j.redox.2016.12.035

67. H. Sies. On the History of Oxidative Stress: Concept and Some Aspects of Current Development. Curr. Opin. Toxicol. 7 (2018) 122; 10.1016/j.cotox.2018.01.002

68. T. von Woedtke, A. Schmidt, S. Bekeschus, K. Wende, K.-D. Weltmann. Plasma Medicine: A Field of Applied Redox Biology. *In Vivo* 33 (2019) 1011; 10.21873/invivo.11570

69. D. Boehm, P. Bourke. Safety Implications of Plasma-Induced Effects in Living Cells—A Review of *in Vitro* and *in Vivo* Findings. Biol. Chem. 400 (2019) 3; 10.1515/hsz-2018-0222

70. A. Schmidt, S. Bekeschus. How Safe is Plasma Treatment in Clinical Applications?. In: Metelmann, H.-R., von Woedtke, T., Weltmann, K.-D., Emmert, S. (eds) Textbook of Good Clinical Practice in Cold Plasma Therapy. Springer, Cham 2022, p. 99–162; 10.1007/978-3-030-87857-3_5

71. P.J. Bruggeman, M.J. Kushner, B.R. Locke, J.G.E. Gardeniers, W.G. Graham, D.B. Graves, R.C.H.M. Hofman-Caris, D. Maric, J.P. Reid, E. Ceriani, D. Fernandez Rivas, J.E. Foster, S.C. Garrick, Y. Gorbanev, S. Hamaguchi, F. Iza, H. Jablonowski, E. Klimova, J. Kolb, F. Krcma, P. Lukes, Z. Machala, I. Marinov, D. Mariotti, S. Mededovic Thagard, D. Minakata, E.C. Neyts, J. Pawlat, Z. Lj. Petrovic, R. Pflieger, S. Reuter, D.C. Schram, S. Schröter, M. Shiraiwa, B. Tarabová, P.A. Tsai, J.R.R. Verlet, T. von Woedtke, K.R. Wilson, K. Yasui, G. Zvereva. Plasma—Liquid Interactions: A Review and Roadmap. Plasma Sources Sci. Technol. 25 (2016) 053002; 10.1088/0963-0252/25/5/053002

72. Y. Gao, K. Francis, X. Zhang. Review on Formation of Cold Plasma Activated Water (PAW) and the Applications in Food and Agriculture. Food Res. Int. 157 (2022) 111246; 10.1016/j.foodres.2022.111246

73. K.S. Wong, N.S.L. Chew, M. Low, M.K. Tan. Plasma-Activated Water: Physicochemical Properties, Generation Techniques, and Applications. Processes. 11 (2023) 2213; 10.3390/pr11072213

Part I

Plasma Chemistry, Biochemistry, and Microbiology

2 Physical Plasma as Source of Reactive Oxygen and Nitrogen Species

Stephan Reuter, João Santos Sousa

2.1 PLASMA SOURCES FOR USE IN MEDICAL THERAPY

2.1.1 REACTIVE OXYGEN AND NITROGEN SPECIES GENERATION BY COLD PLASMAS

Cold plasmas generate a variety of biologically active components, namely an electric field, ultra-violet to infrared light, charged and excited species such as electrons and atomic and molecular ions, as well as chemically active species, in both radical and non-radical form. It is well established that the biological effects of plasma are in large part caused by the reactive oxygen and nitrogen species it generates [1, 2]. We need to understand that the majority of reactive species generated by cold plasmas are the same as those the human body itself generates for metabolic processes and cell-cell communication.

Cold plasmas that can be used in medicine have a gas temperature that is tissue tolerable, i.e. below 40°C. All cold plasmas are non-equilibrium plasmas, where the electrons are hot (they can have multiple tens of thousands of degrees Celsius) and the heavy particles (ions, neutral atoms and molecules) are at a low temperature (in plasmas used in medicine they are at about room temperature). The reactivity of cold plasmas is thus electron driven, and cold plasmas can generate reactive species that otherwise would typically only be generated at temperatures above 1000s of degrees Celsius.

A plasma's key chemically active species can be grouped into pure reactive oxygen species (ROS) and reactive nitrogen species (RNS). Both groups include molecules that can contain hydrogen. The term ROS originates from biology, referring to those species in which the chemical activity is associated with the oxygen part of the molecule. This is evident, for example, in oxygen-containing radicals such as the $\bullet$OH-radical,[1] where the radical's unpaired electron is located on the oxygen atom, and bonds are most likely to occur on this oxygen atom.

Reactive nitrogen species, on the other hand, are often defined as species derived from nitric oxide NO. A number of nitrogen-containing molecules that do not fall under this definition are, nevertheless, considered to be RNS due to their specific metabolic pathways.

The ability to control the plasma-generated ROS and RNS composition allows us to use plasma as a tool for treating multiple illnesses that involve redox imbalances of the human body. It is the precise control—the target-specific tailoring—of the ROS and RNS composition that suggests plasma as future tool for precision medicine and for targeted diagnostics of redox-based illnesses.

Key reactive oxygen and nitrogen species that cold plasmas generate are atomic species such as oxygen, nitrogen, and hydrogen atoms [3], as well as various metastable species, including singlet molecular oxygen or nitrogen metastable species that a) carry considerable energy; b) are present in the human body, e.g. in the respiratory chain for the case of singlet delta oxygen; and c) cannot easily be administered for therapy without the use of complex chemicals by means other than plasma. Mounting research evidence points to the high relevance of atomic oxygen as a primary

---

[1] In the rest of the chapter, the radical dot notation will be omitted for better readability.

DOI: 10.1201/9781003328056-3

plasma-generated species that is the source of a broad variety of bioactive species such as, for example, hypochlorite ClO^- or hydroperoxyl HO_2.

2.1.2 PLASMA SOURCE CONCEPTS

For use in medicine, plasma sources must operate at a tissue tolerable gas temperature, i.e., below 40°C. In plasma source engineering, a handful of plasma source concepts exist to reach these low gas temperatures. First and foremost are plasmas in which the gas temperature never exceeds tissue tolerable temperatures. Cold plasma source concepts used in plasma medicine are cold plasma jets (see Figure 2.1 a) [4] and dielectric barrier discharges (DBDs)[5], which in turn can be separated into volume barrier discharges, where the biological tissue acts as a counter electrode (see Figure 2.1 b) [6, 7], and surface barrier discharges, where the plasma ignites between a structured grounded electrode and a driven high voltage electrode, separated by a dielectric barrier (see Figure 2.1 c) [8]. The low temperature of the reactive component output is achieved by the plasma source's technical design: in cold plasmas, the electron current is limited to prevent overheating of the gas. For this purpose, DBDs have one or more dielectric layers between the electrodes. While most cold plasma jets also use a dielectric layer, in some jets the electron current is limited by the mode of power supply, either through short electric pulses down to nanosecond range, or by using radio frequency excitation. Both approaches prevent the build-up of current and thus excessive heating of the gas.

Additionally, so-called warm plasmas, such as microwave [9, 10] and spark discharges [11], are commercially available for plasma medical use. In those types of plasma sources, the non-tissue tolerable temperature of the gas stream is cooled down before actually reaching the patient (see Figure 2.1 d).

The existing plasma source concepts vary in their reactive species composition. Most notably, the reactive chemistry differs between warm and cold plasmas. If a hotter plasma is generated, and the feed gas contains air or air admixture, this plasma's chemical composition tends more towards reactive nitrogen species (RNS). For cold plasmas that contain air or air admixture, their resulting chemistry tends towards reactive oxygen species (ROS), and a major component is ozone.

Also, reactive species generation and destruction pathways differ significantly among the different cold plasma source concepts, i.e. in noble gas plasma jets versus dielectric barrier discharges in air. For example, in argon plasma jets, water dissociation occurs dominantly through argon metastable species as opposed to electron dissociation, even at molecular gas admixtures up to 1% [12]. In air discharges, the electron energy distribution function (EEDF) tends towards higher energies, little ro-vibrational excitation contributes to the chemistry, and metastable Penning processes of noble gases play no role [5] (see the following sections for the individual processes). In plasma liquid

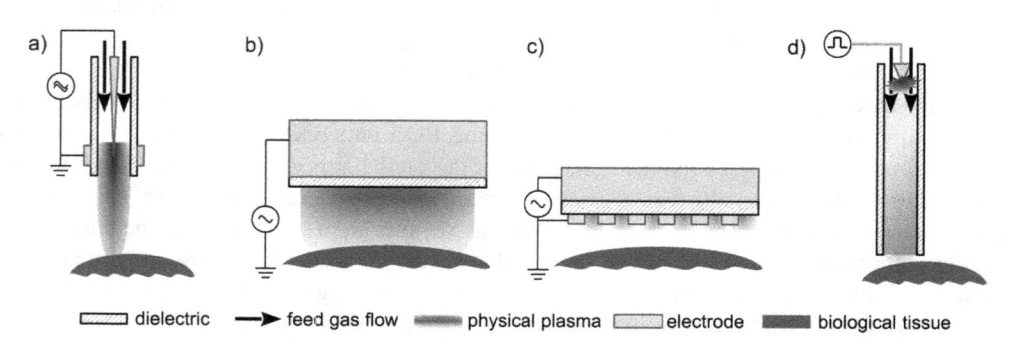

FIGURE 2.1   Four fundamentally different plasma source concepts: a) cold plasma jet; b) volume dielectric barrier discharge; c) surface dielectric barrier discharge; d) warm plasma (microwave or spark discharge) with subsequent cooling down of gas stream.

treatment, DBDs, which are operated in ambient air, have a lower hydrogen peroxide production compared to noble gas plasmas.

Another feature that distinguishes most DBDs is that they exhibit a filamentary pattern. The discharge's ionization waves are confined in a volume defined by space charges. Upon surface impact, the filamentary discharge spreads and covers a surface multiple times larger than the filament diameter [5]. For example, when applied to a wound, the discharge can create local electric fields in the surrounding skin tissue on the order of several 10s to 100s of kV/cm [13], high enough to generate electroporation [14]. Whether the discharge filaments are stationary in space or distributed stochastically on the electrode's surface depends on the discharge's polarity [5] and on the mobility of the charge carriers in the treated surface [15]. In plasma medicine, the spatial filament dynamics are important not only because of their direct electric field effects on, e.g., treated tissue, but also because they strongly influence the chemistry induced in a plasma treated liquid, such as a wound liquid [16].

2.2 REACTIVE SPECIES GENERATION AND DESTRUCTION MECHANISMS

2.2.1 Electron Collisions and Electron Energy Distribution Function

In cold plasmas, electron collisions play a significant role in the generation and destruction of reactive species. Below is a list of the key mechanisms associated with electron impact in cold plasmas:

1. Electron impact ionization: when an energetic electron collides with a neutral atom or molecule, it can transfer enough energy to the neutral species to ionize it. This results in the creation of new charged species, such as positive ions and free electrons, in the plasma.
2. Electron impact excitation: when an electron collides with a neutral species, it can transfer energy to the neutral species, exciting it to a higher electronic, rotational, or vibrational energy level. If the energy transfer is sufficient, molecular neutral species can be dissociated into their constituent atoms or molecules, in a process called electron impact dissociation.
3. Electron impact deexcitation: when an excited neutral species collides with an electron, it can transfer energy to the electron, causing it to relax to a lower energy level. This results in the destruction of the excited neutral species and the creation of new neutral species in the plasma.
4. Electron impact attachment: when a low-energy electron collides with a neutral molecule or atom, it can form a negative ion. Attachment processes can thus lead to the depletion of neutral species and the formation of negative ions. The negative ions can then react with other species in the plasma, leading to the formation of new species through dissociative attachment or associative detachment reactions.
5. Electron impact detachment: when a high-energy electron collides with a negative ion, the former causes the latter to lose an electron and become a neutral species or a positive ion. Detachment processes can also occur through thermal or photochemical processes in which high temperatures or UV radiation cause negative ions to lose electrons and become neutral species or positive ions.

Rydberg atoms are a reactive species that can play a dominant role in energy transport in plasmas. These are electronically highly excited atoms with one or more electrons in very high-energy orbitals, known as Rydberg states. In cold plasmas, Rydberg atoms can be formed through various processes, including electron impact excitation and ionization followed by attachment. Rydberg atoms can then react with other species in the plasma, leading to the formation of new species and the destruction of others (namely the Rydberg atoms themselves).

Crucial for determining the contribution of the different mechanisms of electron impact in a cold plasma is the Electron Energy Distribution Function (EEDF). The EEDF describes the distribution of electron energies within the plasma, providing information on the number of electrons at each energy level. The probability of a specific electron impact reaction occurring is determined both by the EEDF and the reaction cross-section. For electron impact ionization, the ionization cross-section is dependent on the electrons' energy. Therefore, the EEDF can determine the rate of ionization, as well as the types of ions that are produced. For example, if the EEDF is skewed towards high-energy electrons, there will be a higher rate of ionization, and the resulting ions will have higher energies. For electron impact excitation and deexcitation, the rate of these processes is also dependent on the EEDF and the reaction cross-section. Due to the different thresholds for the respective collisional processes to occur, high-energy electrons are more likely to cause excitation, while low-energy electrons are more likely to cause deexcitation. The EEDF can also affect the types of species that are produced through these processes. The same applies to attachment and detachment processes. For example, attachment processes are more likely to occur at lower electron energies and higher neutral densities, while detachment processes are more likely to occur at higher electron energies and lower negative ion densities. In a cold plasma, the EEDF is often characterized by a relatively low number of high-energy electrons and a relatively large number of low-energy electrons. These low-energy electrons can play a significant role in the overall composition and in the behavior of the plasma, as they lead to ro-vibrational excitation of molecules, which influences the reaction rates of chemical reactions in a plasma. In summary, electron-impact-driven reactions are determined by an interplay between the EEDF and the ionization, excitation and deexcitation, attachment and detachment cross-sections, and respective densities of the different neutral species in the plasma.

2.2.2 Neutral Atom Collisions (Penning Reactions)

Penning energy transfer occurs when a so-called metastable atom or molecule in a plasma collides with another atom or molecule. Just as the previously described electronic processes, Penning energy transfer is another process that leads to the generation and destruction of reactive species in a cold plasma. A metastable species is an atom or molecule that is in an excited state but has a relatively long radiative lifetime compared to other excited species, because a fast energy transfer through emission of a photon is forbidden according to the quantum energy level transition rules. Excess energy of metastable species can be transferred to other species in a cold plasma through a resonant interaction, leading to the excitation or deexcitation of those species. This so-called Penning energy transfer can also result in ionization of neutral species collision partners—this is then called Penning ionization. The energy transfer in Penning resonance occurs through an intermediate state formed when the metastable species collides with another species. The intermediate state is typically a long-lived excited state of that other species that can decay through radiative or non-radiative processes, releasing energy in the form of photons, heat, or chemical reactions. If an excited species undergoes a Penning resonance, it can also dissociate into its constituent atoms or molecules, creating new species in the plasma. Penning resonance is an important mechanism in cold plasmas, where metastable species are abundant and therefore can play a key role in controlling the plasma's chemistry. The rate of metastable, i.e. Penning, energy transfer depends on the density and energy of the metastable species themselves, as well as on the energy levels of the other species involved. If the density of metastable species is high, the rate of metastable energy transfer will also be high, resulting, for instance, in the excitation of a large number of other species. Metastable energy transfer can also play a role in the destruction of species in the plasma. An excited species, for example, can be de-excited through a collision with a metastable species.

A special type of metastable species are excimers, the term being derived from "excited dimer". An excimer is a type of molecule formed when two atoms form a bond that leads to a stable molecule, only in an excited state. Excimers have a relatively long lifetime compared to other excited species in the plasma. Since excimers carry considerable energy, they play an important role in energy transfer and reaction chemistry. Excimers can lead to both generation and destruction of species in a cold plasma, through processes such as excitation transfer and dissociation. The rate of these processes depends on the density and energy of the excimers, as well as the energy levels of the other species involved. Excimers furthermore are a source of high-energy ultraviolet and vacuum ultraviolet photons, which drive photo-chemical processes. Accordingly, excimers play a key role in the overall plasma chemistry.

Both metastable atoms and molecules including excimers can—due to their long lifetime—act as intermediates for electron driven chemistry. When the plasma's electrons create metastable atoms or excimers, these species can subsequently create highly active species that form the basis for the plasma chemical composition. This was shown, for example, for a plasma jet (kINPen) with argon as main feed gas, where excited argon atoms or excimers (Ar_2^*) play the dominant role in initiating the plasma chemistry [12]. For example, atomic oxygen, nitrogen, or hydrogen can be created from air species through impact with excited argon species (see details in section 2.4.1).

2.2.3 PHOTOINDUCED REACTIONS

Ultraviolet (UV) radiation in a cold plasma can be of significance. Photoionization and photochemistry are two mechanisms by which UV radiation can play a role in the generation and destruction of species in a cold plasma. Photoionization is the process by which UV radiation provides sufficient energy to an atom or molecule to remove one or more electrons, creating a positive ion. This results in the creation of new species in the plasma, including both positive ions and free electrons. Photoionization can, on the other hand, also play a role in the destruction of species in the plasma, as it can remove electrons from excited species, causing them to deexcite and become stable. Photochemistry comprises chemical reactions that occur as a result of the absorption of UV radiation. In a cold plasma, photochemistry can lead to the creation of new species through processes such as dissociation and recombination. For example, UV radiation can provide enough energy to break bonds and cause a species to dissociate into its constituent atoms or molecules. Photochemistry can also play a role in the destruction of species in the plasma by, for example, triggering an excited species to deexcite. The energy required to break a bond is related to the bond energy of the respective species. Stronger bonds will require more energy to break, while weaker bonds can be broken more easily by UV radiation. The amount of UV radiation present in a cold plasma will depend on the energy and distribution of the electrons in the plasma, as well as the nature of the species involved. The extent to which these UV-based mechanisms affect the plasma will depend not only on the amount of UV radiation present, but also on the strength of the bonds in the species (i.e., the nature and the energy levels of the species involved). In noble gas (argon or helium) plasma jets, photons can be transported within the noble gas channel, leading to the formation of reactive species at higher distances from the jet's nozzle. However, while photons are important for plasma generation (through photoionization), they play a comparatively minor role within the plasma's reaction chemistry.

2.2.4 ION CHEMISTRY AND WATER CLUSTERS

In a cold plasma, ions collide with neutral species and transfer their energy, leading to the formation of new species through e.g., ionization or dissociation. This process is known as ion-neutral chemistry. For example, if a positive ion collides with a neutral species, it can transfer enough energy to cause the neutral species to dissociate into its constituent atoms or molecules, creating new species in the plasma. Ion chemistry can also play a role in the destruction of species in the plasma. For example, if an excited species collides with a positive ion, the ion can cause the excited species to deexcite and become stable. This process is known as deexcitation transfer. In addition to ion-neutral chemistry and deexcitation transfer, ions can also react with other ions to form new species in the plasma. This is known as ion-ion chemistry. For example, if two positive ions collide, they can combine to form a new species in the plasma. The extent to which these processes occur will depend on the density and energy of the ions, as well as the energy levels of the other species involved.

Water clusters also influence the generation and destruction of species in a plasma and affect its overall composition and properties, through processes such as dissociation, deexcitation transfer, and electron scavenging. In cold plasmas, water clusters often form as a result of water vapor condensation. These water clusters can then act as reactants in plasma reactions, leading to the formation of new species. For instance, water clusters can dissociate into their constituent ions and molecules. Water clusters can also play a role in the destruction of species in the plasma. For example, excited species in the plasma can collide with water clusters and transfer their energy, leading to the deexcitation and stabilization of the species. In addition, water clusters can also act as electron scavengers, removing electrons from the plasma and thereby influencing the EEDF. This can alter the effect of electron impact and, thus, the generation and destruction of other species in the plasma. One feature of plasma-generated water clusters, important in medical applications, is that they enhance species transfer into liquids. This is due to the fact that the solubility of water clusters can be orders of magnitudes higher than the solubility of the cluster-forming ion.

2.2.5　Non-Equilibrium Chemistry and Ro-Vibrational Excitation

Reaction chemistry in a plasma commonly refers to the chemical reactions that do not involve electron collisions. These reactions can lead to the formation of new species and the destruction of existing species in the plasma. For example, if two neutral species collide, they can react to form a new species. Similarly, if an excited species collides with a neutral species, it can transfer its energy to the neutral species, leading to the formation of a new species. In addition, reaction chemistry can also play a role in the destruction of species. For example, if two excited species collide, they can transfer their energy to each other, leading to the deexcitation of both species and their stabilization. The extent to which reaction chemistry affects the generation and destruction of species in the plasma, through processes such as species-species reactions, energy transfer, and dissociation, will depend on the nature and energy levels of the species involved, as well as the neutral species density and the gas temperature. The main processes of non-electronic energy transfer, characteristic for non-equilibrium plasmas, are Penning energy transfer (see section 2.2.2), ro-vibrational energy transfer, and surface reactions.

Stepwise ionization refers to the process by which an atom or molecule becomes progressively ionized in multiple stages, rather than being ionized in a single, rapid event. In a cold plasma, this process can occur through the successive absorption of energy by the atom or molecule. By increasing the ionization rate of the atoms and molecules and, thus, affecting the EEDF, stepwise ionization can influence the generation and destruction of species and affect the overall composition and properties of the plasma. Stepwise ionization can lead to the formation of new species, such as positive ions and free electrons. It can also play a role in the destruction of species by removing electrons from excited species, leading to their deexcitation and stabilization. The extent to which stepwise ionization affects the generation and destruction of species in the plasma will depend on the nature and energy levels of the species involved, as well as the electron density and temperature.

Ro-vibrational energy transfer refers to the transfer of rotational and vibrational energy between species. In a cold plasma, this process can occur through collisions between excited species and neutral species, or between excited species and other excited species. Ro-vibrational energy transfer can play a role in the generation of species in the plasma by creating new excited species through energy transfer from other excited species. Ro-vibrational energy transfer can also play a role in the destruction of species by deexciting excited species through energy transfer to other species in the plasma. This can lead to the stabilization of the excited species and a return to their ground state. The extent to which ro-vibrational energy transfer affects the generation and destruction of species in the plasma will depend on the nature and energy levels of the species involved.

In general, cold plasmas boost ro-vibrational excitation of molecules, because rotational and vibrational energy levels have a low excitation energy that is resonant with low-energy electrons. Thus, the relative abundance of ro-vibrationally excited molecules is higher in cold plasmas than, e.g., in hot plasmas or equilibrium chemical reactions. Ro-vibrational excited molecules can have a much stronger reaction rate than ground state molecules. For instance, the quenching of singlet delta oxygen by ro-vibrationally excited ozone is higher by a factor of ~40 compared to ground state ozone [17, 18]. Hence, in cold plasmas, ro-vibrational excitation significantly contributes to the overall reactivity of the plasma.

Vibrational chemistry refers to the chemical reactions that occur between species due to vibrational energy transfer. In a cold plasma, this process can occur through collisions between excited species and neutral species, or between excited species and other excited species. Vibrational chemistry can play a role in the generation of species by creating new excited species through energy transfer from other excited species in the plasma. Vibrational chemistry can also play a role in the destruction of species by deexciting excited species through energy transfer to other species in the plasma. This can lead to the stabilization of the excited species and a return to their ground state. One mechanism of vibrational excitation that can lead to dissociation of molecules is the so-called vibrational ladder climbing. Vibrational ladder climbing—similar to stepwise ionization—means that a molecule is excited from one vibrational energy state to the next higher state and so forth, until an energy level from which the molecule can dissociate is reached.

Surface reactions refer to reactions that occur at the surface of a solid or liquid, such as a plasma reactor wall or a biological medium. In a cold plasma, these reactions can involve species that have been generated in the plasma and are deposited on the surface, as well as species that are adsorbed onto the surface from the surrounding gas. Surface reactions can play a role in the generation of species in the plasma by creating new species through chemical reactions at the surface. For example, a neutral species adsorbed onto the surface may react with an excited species from the plasma to form a new species in the plasma. Surface reactions can also play a role in the destruction of species in the plasma by removing species from the plasma and stabilizing them at the surface. For example, an excited species may be adsorbed onto the surface and deexcited through collisional energy transfer with other species at the surface. The extent to which vibrational chemistry and surface reactions affect the generation and destruction of species in the plasma, through processes such as energy transfer, deexcitation, and chemical reactions, will depend on the nature and energy levels of the species involved, as well as the species density and gas temperature and the properties of the surface.

2.3 MAIN CHEMICAL COMPONENTS AND THEIR PATHWAYS

2.3.1 REACTIVE OXYGEN SPECIES (ROS)

Atomic oxygen O is the single most important component in reactive oxygen species (ROS) generation in a cold plasma. Atomic oxygen is an unstable species and typically is generated in cold plasmas from molecular oxygen O_2 dissociation. Plasma generation of atomic oxygen is a result not only of the efficient electron impact dissociation of molecular oxygen (see section 2.2.1) but also of the dissociation of molecular oxygen by Penning collision (see section 2.2.2) with excited molecules such as molecular argon or nitrogen. The key generation pathway for atomic oxygen in a cold plasma is:

$$X^* + O_2 \rightarrow 2O + product \quad (R_1)$$

where X^* can be free electrons, excited noble gas species, or excited molecular nitrogen. Due to the very high reaction rate of excited molecular nitrogen, an admixture of nitrogen to an oxygen-containing plasma does not dominantly lead to the generation of reactive nitrogen species, as might be expected, but rather to formation of additional atomic oxygen in the plasma.

The presence of atomic oxygen in a cold plasma jet operated in ambient air typically results in generation of ozone O_3, according to the reaction:

$$O + O_2 + M \rightarrow O_3 + M \quad (R_2)$$

where M is a third-body reaction partner. It is the high reaction rate of R_2 that is responsible for cold plasmas typically exhibiting a dominant ozone production.

Additionally, excited oxygen species play an important role in reactive oxygen chemistry in cold plasmas. Reactions among the excited metastable molecular oxygen species $O_2\left(b^1\Sigma\right)$ and $O_2\left(a^1\Delta\right)$, and the metastable atomic oxygen species $O(^1D)$ and $O(^1S)$ with ozone and atomic oxygen, are circular, transforming one species into the other, according to the following repeating reaction scheme [18]:

$$\left.\begin{aligned} O_3 + h\nu &\rightarrow O(^1D) + O_2\left(a^1\Delta\right) \\ &\rightarrow O(^3P) + O_2\left(X^3\Sigma\right) \end{aligned}\right\} \quad (R_3)$$

$$\left.\begin{aligned} O(^1D) + O_2 &\rightarrow O(^3P) + O_2\left(b^1\Sigma\right) \\ &\rightarrow O(^3P) + O_2\left(a^1\Delta\right) \\ &\rightarrow O(^3P) + O_2\left(X^3\Sigma\right) \end{aligned}\right\} \quad (R_4)$$

$$\left.\begin{aligned} O_2\left(b^1\Sigma\right) + O_3 &\rightarrow O(^3P) + 2O_2\left(X^3\Sigma\right) \\ &\rightarrow O\left(a^1\Delta\right) + O_3 \\ &\rightarrow O_2\left(X^3\Sigma\right) + O_3 \end{aligned}\right\} \quad (R_5)$$

Depending on the source of energy input into reactions (R_3 to R_5), one of the four excited oxygen species becomes dominant, which leads to significant altering of the subsequent reaction pathways.

The final group of reactive oxygen species relevant to reaction chemistry in cold plasmas is ionic species. The negative atomic oxygen ion O^- and the superoxide anion O_2^- are the main source of ionic ROS generation pathways in the plasma. Both oxygen ions (O^- and O_2^-) dominantly originate from electron impact ionization of neutral oxygen species. In connection with humidity, oxygen ions are responsible for water cluster formation in the plasma (see section 2.2.4.).

2.3.2 HYDROGEN OXYGEN COMPOUNDS

Plasma generated reactive hydrogen oxygen compounds and reactive hydrogen nitrogen compounds (see section 2.3.3.) originate from humidity in either the plasma feed gas or the ambient gas. Predominantly through electron impact, water molecules are dissociated into OH-radicals and hydrogen ions, according to reaction R_6 (see e.g. [19]). To a lesser degree, water molecules are dissociated by excited molecular nitrogen, by excited noble gas species, or by oxygen atoms: the water molecules' dissociation into OH and products follows reaction R_7. Apart from reactions R_6 and R_7, reactions R_8 and R_9 are also a source of hydroxyl radical OH in the plasma through reaction of hydrogen atoms with metastable molecular oxygen $O_2\left(a^1\Delta\right)$ or hydrogen superoxide HO_2.

$$H_2O + e^- \rightarrow OH + H^- \quad (R_6)$$

$$H_2O + X^* \rightarrow OH + H + products \quad (R_7)$$

$$H + O_2\left(a^1\Delta\right) \rightarrow OH + O \quad (R_8) \tag{20}$$

$$H + HO_2 \rightarrow OH + O \quad (R_9) \tag{21}$$

The hydroxyl radical OH is a primary species or hub species—a term that will be explained in greater detail in section 2.4.2—in a humid plasma gas chemistry. In plasma medicine, the relatively stable hydrogen peroxide molecule H_2O_2 plays an important role [22, 23]. The self-recombination reaction of OH is the dominant generation pathway for hydrogen peroxide:

$$OH + OH \rightarrow H_2O_2 \quad (R_{10})$$

Hydrogen peroxide can furthermore be created through the reaction of superoxide anion and the positive hydrogen ion (hydron):

$$2O_2^- + 2H^+ \rightarrow O_2 + H_2O_2 \quad (R_{11})$$

In a long and slow process, hydrogen peroxide eventually decays into O_2 and water according to

$$H_2O_2 \rightarrow O_2 + 2H_2O \quad (R_{12})$$

2.3.3 Reactive Nitrogen Species (RNS)

Reactive nitrogen species (RNS) are molecules originating from nitric oxide NO. For the generation of RNS, the strong bond of molecular nitrogen's two atoms needs to be broken. Generally, high temperatures—typical for spark, microwave, or arc plasmas—help to break the nitrogen bond. At high temperatures, nitric oxide NO is formed through the extended Zeldovich mechanism (e.g. [24]) where reaction

$$N_2 + O \rightarrow NO + N \quad (R_{13})$$

is followed by

$$N + O_2 \rightarrow NO + O \quad (R_{14})$$

and

$$N + OH \rightarrow NO + H \quad (R_{15})$$

At room temperatures, however, R_{13} and R_{14} become insignificant, and molecular and atomic oxygen are consumed by R_2, leading to ozone formation.

Cold plasmas generate nitric oxide NO mainly through the reaction of excited molecular nitrogen with oxygen atoms:

$$N_2^* + O \rightarrow NO + N \quad (R_{16}) \tag{25}$$

NO at low temperatures in the presence of oxidizing species is not stable and oxidizes quickly according to reactions

$$NO + O + M \rightarrow NO_2 + M \quad (R_{17})$$

$$NO + HO_2 \rightarrow NO_2 + OH \quad (R_{18})$$

and

$$NO + O_3 \rightarrow NO_2 + O_2 \quad (R_{19})$$

After longer interaction times, recombination of nitrogen dioxide and ozone leads to the generation of nitrate radicals and oxygen molecules:

$$NO_2 + O_3 \rightarrow NO_3 + O_2 \quad (R_{20}) \tag{26}$$

Recombination of nitrogen dioxide and nitrate radicals leads to formation of N_2O_5 according to

$$NO_2 + NO_3 + M \rightarrow N_2O_5 + M \quad (R_{21})$$

In plasma-liquid interaction, for every N_2O_5 molecule that is dissolved, two nitric acid molecules HNO_3 are formed [27]. The equilibrium concentration of a gaseous species in water is proportional to its so-called Henry constant. For nitrous and nitric acid (HNO_2 and HNO_3), the Henry constants are three to five orders of magnitude higher than those of NO_2 and NO_3. Thus, in plasma-liquid interaction, nitrous and nitric acid contribute dominantly to RNS transport from the plasma/gas phases to the liquid phase. Nitrous and nitric acid are generated by reaction with the hydroxyl radical OH of NO and NO_2, respectively. The reaction of NO_2 with OH leads to the formation of nitrous acid in 90% of the cases and peroxynitrous acid (HOONO) in 10% of the cases. Peroxynitrous acid has been identified as an important precursor of the biological activity of plasmas (see Chapter 3).

2.3.4 OTHER REACTIVE SPECIES

Plasma medical research mostly focuses on ROS and RNS. However, these are not the only reactive species that cold plasmas can generate. One of the precursor molecules not discussed above that is present in air discharges, in plasma jets operated in ambient air, or as feed gas impurity in plasma jets, is carbon dioxide CO_2. The presence of CO_2 in cold plasmas can lead to the formation of highly reactive carbon monoxide CO through, e.g., dissociation by singlet delta oxygen. Carbon monoxide has a strong biological impact, and its role in plasma medicine may be as significant as that of RONS discussed in the previous sections [28].

Since interaction of plasmas with biological liquids such as blood or wound liquid plays a key role in plasma medicine (see Chapter 3), salts that are present in these liquids also form important reaction partners for species generated by cold plasmas. The reaction of salts such as sodium chloride NaCl with plasma generated atomic oxygen species leads to biologically highly active species. The interaction of excited atomic oxygen $O(^1D)$ with chloride Cl^- forms hypochlorite ClO^- :

$$O(^1D) + Cl^- \rightarrow ClO^- \quad (R_{22})$$

The interaction of $O(^1D)$ with water leads to the creation of hydroxyl cations OH^+ and, subsequently, the reaction of OH^+ with Cl^- leads to the formation of hypochlorous acid $HOCl$ [29, 30]. Both, ClO^- and $HOCl$ are strong oxidizers.

To put it in a nutshell, it is the cold plasma generated reactive species that are key to the efficacy of plasmas in medical therapy. Reactive oxygen and nitrogen species involving oxygen, nitrogen, and hydrogen play an especially fundamental role, not least because they are the same species that the human body itself produces for cell-cell communication and metabolism (see chapters 9 to 12). The reactions described in the sections above, among others, allow modification and control of the reactive species composition in cold plasmas by controlling the precursor molecules or the generation and destruction of reactants. What sets plasma apart from other medical therapies is its ability to supply atomic species such as oxygen, nitrogen, or hydrogen atoms. With implications that go beyond plasma medicine, these highly reactive species in cold plasmas allow us to influence chemical reaction chains in unconventional ways, opening a path to tailor chemistry [30, 31].

In the following section, strategies for reactivity control of cold plasmas are described.

2.4 CONTROLLING THE REACTIVE SPECIES COMPOSITION

2.4.1 CONTROL OVER SPECIES GENERATION THROUGH INFLUENCE OF PLASMA PARAMETERS

Control over reactive species in cold plasmas can be achieved through tuning the plasma processes of species generation and destruction (see Figure 2.2). One of the key ways to tailor the reactive species composition in a cold plasma is by influencing the generation of species through their ionization and excitation pathways (such as electronic, rotational, or vibrational excitation) and the resulting chemistry. By controlling the generation pathways of species, it is possible to influence the number and type of reactive species present in the plasma, as well as their energy levels and reactivity. Pathways are commonly manipulated by controlling electron energy in the plasma. The electron

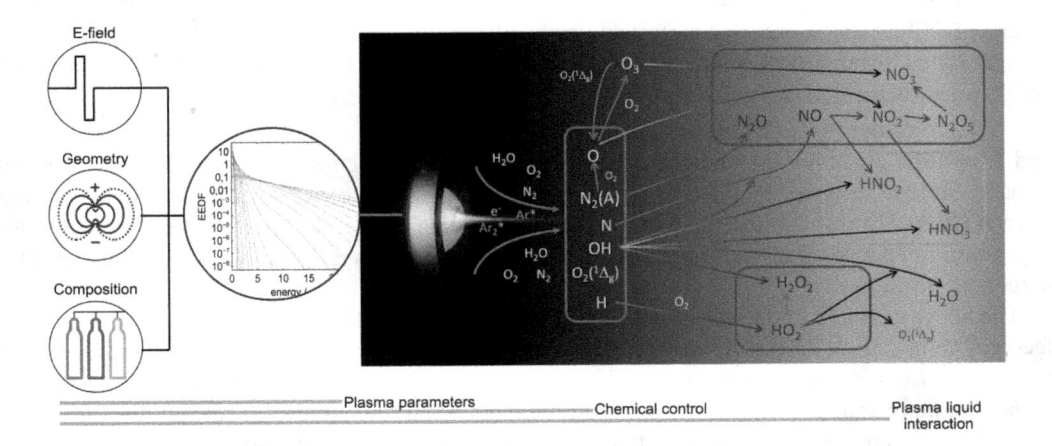

FIGURE 2.2 A plasma's reactivity is controlled in three steps (the picture shows a photo of a cold argon plasma jet ©2023 [32]): 1) control of electron energy distribution function (EEDF) through plasma parameters, defined by feed gas nature and composition, electric field distribution and strength, and plasma source geometry and materials. 2) Chemical control, through feed-gas admixtures and through influencing the ambient gas composition in the interaction zone between the plasma jet and its surrounding. 3) In plasma-liquid systems, e.g., when a plasma jet interacts with biological liquids, the plasma reactivity and the plasma-liquid interaction can be controlled by designing the gas and the liquid flows, by adjusting the liquid composition, and by adapting the plasma-liquid interaction time.

energy and concentration determine the degree of ionization and the types of ions and excited species that are generated. Furthermore, by controlling the electron density as well as the properties of the treated surface, it is possible to define the extent of vibrational chemistry and surface reactions, and the formation of new species through these processes. Additionally, reactive species in a cold plasma can be controlled through their destruction processes, such as dissociation and neutralization or quenching. Also, by removing or stabilizing reactive species, it is possible to control the number, type, and reactivity of reactive species present in the plasma.

As explained in section 2.2.1, the EEDF plays a crucial role in the generation of reactive species in a cold plasma. The EEDF describes the distribution of electron energies in a plasma and provides information about the electron populations and their energy levels. The EEDF influences the ionization and excitation processes in the plasma, which in turn impact the generation of reactive species. For example, high energy electrons can ionize neutral species and excite ions and neutral species, leading to the formation of new species through processes such as dissociation and ion-molecule reactions. On the other hand, low energy electrons can neutralize ions and prevent the formation of new species. The shape of the EEDF can also impact the generation of reactive species. For example, a skewed EEDF with a high-energy tail can lead to a higher degree of ionization and excitation, and therefore a higher production of reactive species. On the other hand, a flatter EEDF with lower energy electrons can lead to a lower production of reactive species.

In a cold plasma, both the electron impact and the Penning processes can influence the generation and destruction mechanisms of species, and their interactions can be complex and influence the properties and behavior of the plasma in significant ways. For example, electrons can influence the degree of ionization and excitation in the plasma, which in turn can impact the Penning processes and the formation of new species through this process. Similarly, the Penning processes can impact the ionization and excitation processes in the plasma, influencing the generation of new species through these processes. The energy distribution of electrons and excited species can also impact the production of new species through these processes. Whether certain reactions occur dominantly through Penning collisions or through electron impact depends on the reactions' cross sections as well as on the densities of heavy particles and electrons.

In a study of an argon plasma jet with different air impurities (0.1 permille, 1 permille, and 1 percent) [12], it was shown that metastable molecular oxygen $O_2\left(a^1\Delta\right)$ and $O_2\left(b^1\Sigma\right)$ are predominantly excited through electron impact, while atomic hydrogen and nitrogen, as well as hydroxyl OH are dominantly created through argon metastable impact reactions. Besides, electronic excitation of nitrogen molecules $N_2\left(a^3\Sigma\right)$ occurs dominantly through argon metastable impact excitation at 0.1 and 1 permille air impurity, whereas at higher air impurity, it occurs dominantly through electron impact. Similar behavior was found for the generation of atomic oxygen and metastable atomic oxygen $O(^1D)$. However, while the ratio of electronic excitation increases with increasing air impurity, argon impact generation of excited or ground state atomic oxygen remains dominant even at 1 percent of air impurity.

The above example of the role of gas impurities in an argon plasma highlights the importance of feed gas composition for controlling the reactive species generation pathways. The gas composition influences whether species are generated through electronic or Penning processes. Penning processes are more abundant in plasmas operated with noble gas and molecular admixtures than in air discharges. It is important to mention that the highest influence of impurities on plasma chemistry can be observed when changing from almost no gas impurities to very little [33, 34]. Thus, to have a stable, repeatable, and controllable reactive species composition, it can in some cases be advisable to have a controlled amount of impurities (e.g., humidity or air) in the plasma.

2.4.2   Chemical Pathway Analysis to Determine Key Species in Reaction Chemistry

The reactive species generation pathways are dominantly determined through the plasma parameters as described in the previous section. A strong control over the reactive species composition can be imposed by controlling the destruction pathways in the plasma reaction chemistry. First, the chemical reaction pathways in the plasma must be analyzed. In the following paragraphs, reactive species control is illustrated using the examples of ozone versus nitric oxide generation in cold plasmas and increasing singlet delta oxygen concentration in a cold plasma. Subsequently, reaction pathway analysis is described as a fundamental tool for reactive species control in cold plasmas.

The generation and destruction mechanisms of the educts and intermediate reactants (that is, reactive species that are the basis for further desired or undesired reaction pathways) are both relevant to a plasma's reactive species composition. The significance of the destruction processes is shown in a study of the control of ROS versus RNS production in a cold plasma jet [35, 36]. At room temperature, the production of nitric oxide NO, an important precursor for NO_x species, occurs through reactions of atomic nitrogen N with the hydroxyl radical OH (R_{15}). Reactions of N_2 with atomic oxygen O (R_{13}) as well as atomic nitrogen N with O_2 (R_{14}) are relevant only at higher temperatures, which explains why microwave or spark discharges produce more nitric oxide dominant chemistry compared to cold plasma jets. In cold plasmas, atomic oxygen will dominantly lead to ozone formation through reaction with molecular oxygen O_2 (R_2). The information gained from a chemical pathway analysis can be applied to, e.g., change a plasma jet's ozone-based chemistry to an NO_x-based chemistry through the following steps:

1. Adding water to the feed gas quenches atomic oxygen (through reaction with OH or HO_2, forming molecular oxygen and H or OH, respectively) and forms NO (according to R_{15}) or HNO_x;
2. To further suppress ozone formation, reaction R_2 – the reaction of O_2 and O – needs to be prohibited in the plasma jet effluent. This can be achieved by shielding the plasma jet's effluent from ambient oxygen with a nitrogen gas curtain (see also section 2.5.2).

A second example of reactive species composition control through chemical pathways is the increase of singlet delta molecular oxygen $O_2(a^1\Delta)$ concentration in the plasma. As in the previous case, atomic oxygen needs to be quenched. However, admixture of water would be detrimental, because water quenches not only atomic oxygen but also singlet delta oxygen itself (e.g., by reaction R_7 and R_8). The concentration of singlet delta oxygen can instead be increased by limiting its quenching reaction with atomic oxygen through addition of nitric oxide NO [37]. NO removes atomic oxygen through reaction R_{17}, forming nitrogen dioxide NO_2. O_3 is also a main quencher of $O_2(a^1\Delta)$. Adding NO into the gas mixture, even at rather low concentrations (10s-100s of ppm), can cause a huge decrease in the O_3 concentration through reaction R_{19}, and, concomitantly, a large increase in the $O_2(a^1\Delta)$ concentration [38]. Indeed, as the O_3 generation results almost exclusively from the conversion of atomic oxygen through reaction R_2, the reduction of the atomic oxygen concentration induced by NO admixture leads to an even larger decrease in the O_3 concentration. Since NO is not lost in the process but rather is recycled in collisions

$$O + NO_2 \rightarrow NO + O_2 \quad (R_{23})$$

the effects of NO admixture are catalytic. However, NO and NO_2 also quench $O_2(a^1\Delta)$, and, thus, an optimum NO admixture can be found to maximize the $O_2(a^1\Delta)$ concentration [39].

Elucidating a balance between destruction and production mechanisms that is optimal to increase the concentration of a specific species requires a detailed analysis of the chemical processes in the plasma. A reaction analysis yields the relevance of certain pathways over others [40, 41]. In many cases, reaction products depend on relatively few "hub species", meaning reactive species that are key to specific reaction pathways. These hub species can be identified by an analytical approach based on graph theory as shown by Murakami et al. [42]. In graph theory, the species in the chemical formula and the interactions between these species are represented as nodes and edges, respectively. Figure 2.3 a) shows two exemplary chemical reactions occurring in a helium/oxygen plasma. These

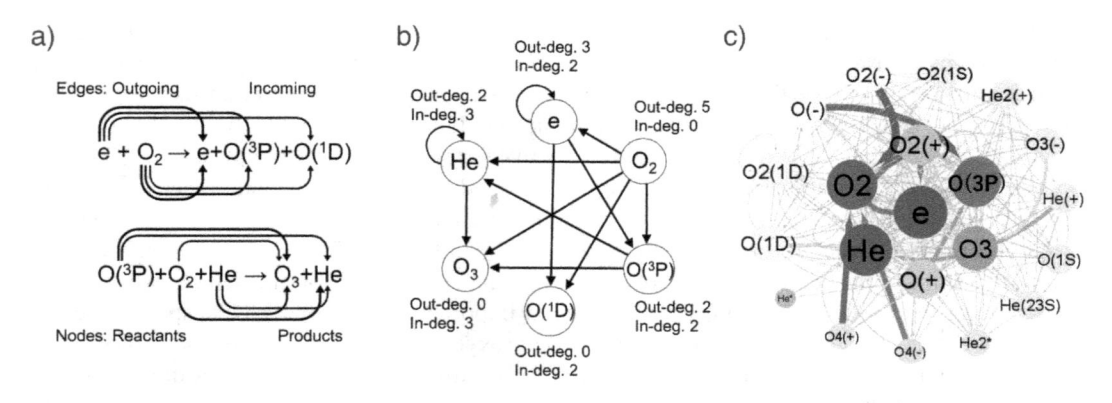

FIGURE 2.3 Chemical reactions (including plasma chemical reactions involving electrons) can be analyzed in a reaction network represented by edges and nodes (a). Graph diagrams can represent the respective relations of the reactive species (b), which can help in reducing the reactions required to describe a chemical situation, and to identify relevant hub species (c) that are elemental to a majority of the reaction pathways [40, 41].

two chemical reactions can be depicted according to the rules in graph theory: outgoing and incoming edges are defined according to the direction of reaction from reactants to products. The chemical formulas of Figure 2.3 a) can thus be encoded into the corresponding directed graph diagram according to the out-degree/in-degree counts, as shown in Figure 2.3 b). The term "degree" indicates the total number of edges connecting one species with other species. For directed graph diagrams, the in-degree is the number of incoming edges (right side of the reaction) and the out-degree is the number of outgoing edges (left side of the reaction). Encoding all chemical formulas of a plasma reaction set yields the network-like diagram shown in Figure 2.3 c). Here, the circle size corresponds to the sum of the in- and out-degrees, that is the total number of edges going out from or coming in to the species of that circle, linked to the total number of reactions that the species takes part in. The edge thickness corresponds to the sum of the edges that link two species, that is, how many reactions lead to transforming the first species into the second. As visualized in Figure 2.3 c), the hub species, electrons, helium He, and molecular oxygen O_2 are at the center of the network diagram. The hub species generate a variety of reactive species in the plasma.

Graph theory and corresponding network analysis are helpful tools for reducing a plasma's set of chemical reactions to a number of reactions that can be implemented in reasonably fast computational chemical models [33, 40, 41]. A network analysis furthermore allows identification of key species in desired or undesired reaction pathways. This information can be used to control the reactive species composition of a cold plasma. A reaction network analysis is not restricted to the gas-phase chemistry of plasmas but can also be extended to biochemical processes in biological systems treated with plasma. This versatility of reaction network analysis makes it a powerful tool in plasma medicine [43].

In conclusion, the plasma reactive species composition can be tailored through an analysis of the chemical reaction network of a plasma and subsequent knowledge-based control of destruction and generation pathways in the plasma chemistry. In the next and final section of Chapter 2, technological examples and implementations of plasma tailoring are described.

2.5 TECHNOLOGICAL IMPLEMENTATION FOR REACTIVE SPECIES CONTROL

2.5.1 CONTROL OF GAS COMPOSITION IN PLASMA JETS

Since plasma jets for medical use and therapy typically are (and need to be) operated in ambient air, ambient air has significant influence on the reactions occurring in the active effluent, the region where the plasma jet's ionization wave enters the surrounding air. Indeed, the ratio of nitrogen to oxygen diffusing into the active effluent is defined by the surrounding air composition. Furthermore, ambient humidity can differ drastically, depending on the surrounding conditions. It has been shown that in a noble gas plasma jet, however, the feed gas humidity is significantly more influential than ambient humidity [44].

Given the influence of ambient air, a widely followed approach to improving control over the plasma's reactive species output is to envelop the active effluent of a plasma jet in a shielding gas curtain [36, 45]. This has two distinct and crucially important effects: firstly, a well-designed gas curtain prohibits the influx of air species into the active effluent; secondly, by specifically designing the gas curtain composition, the plasma's reactive species composition can be tailored. By applying the principles described in section 2.4.2, the reactive species composition in a cold plasma jet can be changed from ROS dominated to RNS dominated. The ability to define the plasma's reactive species composition provides a vital tool for plasma medical research, as it helps to identify the ROS and the RNS effect in, e.g., plasma-based wound healing and plasma-assisted cancer therapy.

In summary, adding a shielding gas curtain to the plasma jet is a vital step towards the ultimate goal of tailoring plasma treatment. Tailored plasma treatment will open the path to individual therapeutic application of plasma and can prove a useful tool for precision medicine.

2.5.2 Reactivity Control Through the Applied Electric Field

In plasma, the electric field drives the electrons, which in turn drive the plasma-chemistry. This highlights the relevance of the electric field as means to influence the plasma chemistry. The plasma chemistry can be influenced not only through the amplitude of the electric field, but also by the electrode geometry and configuration, as well as by the electric field gradient in time and space. Furthermore, the plasma chemistry depends on the mode of electric field application: as constant applied voltage, single or repetitively pulsed, or in a periodic manner as high-, radio-, or microwave frequency.

The electrode geometry and direction of the applied electric field are important. A comparison of linear electric field (i.e., an electric field that has a strong component in the direction of the plasma jet expansion) and cross electric field (where the electric field is mainly perpendicular to the plasma jet expansion) shows the distinct effect of the electric field direction on the reactivity of the plasma [46]. This comparison was recently repeated with the focus on plasma-liquid interaction, similarly showing that an electric field in the direction of the plasma flow creates a higher reactive species concentration—especially of electron impact derived species—within a treated liquid sample [47].

By shaping the electric field, or by combining high and low fields, the reactive species composition can be modified. By using multiple excitation frequencies, e.g., the overall EEDF can be shaped, which has a strong influence over the plasma chemistry. Different electric fields can be combined through the plasma source geometry. An example of this is micro-cathode sustained discharge (MCSD) [48], whereby a small, submillimeter-sized hollow cathode creates a comparatively dense plasma ($n_e > 10^{16}\,cm^{-3}$), which is then extracted from the hollow cathode to a much larger plasma volume by a biased third electrode. Between the micro-hollow cathode and the extraction electrode, a low electron energy glow discharge develops. This low electron energy discharge promotes, e.g., the generation of singlet delta oxygen $O_2(a^1\Delta)$, as well as other species of high relevance in plasma medicine [39, 49, 50].

Electric field control has a further effect: plasma ionization along long tubes is driven through short electric pulses, whereby the high temporal gradient of the electric pulse propels the plasma. Channeling the plasma generated reactive species in or through long tubes is one of the approaches that were pursued early on in plasma medical research [51]. The motivation of researchers to develop plasmas in long tubes is not least to be able to apply plasma in endoscopic procedures.

In summary, the reactive species composition of plasmas used in medicine can be controlled through the processes described in this chapter. Many of these means of control have already found their way into commercial technical implementations of plasma sources and are in use in plasma medicine. Future studies of the effect of controlled reactive species composition in medical applications of plasma will fully unleash the potential of plasma-based therapy.

REFERENCES

1. Graves, D.B., *The Emerging Role of Reactive Oxygen and Nitrogen Species in Redox Biology and Some Implications for Plasma Applications to Medicine and Biology.* Journal of Physics D: Applied Physics, 2012. **45**: p. 263001.
2. Hasse, S., T. Meder, E. Freund, T. von Woedtke, and S. Bekeschus, *Plasma Treatment Limits Human Melanoma Spheroid Growth and Metastasis Independent of the Ambient Gas Composition.* Cancers (Basel), 2020. **12**: p. 2570.
3. Gorbanev, Y., J. Golda, V. Gathen, and A. Bogaerts, *Applications of the COST Plasma Jet: More Than a Reference Standard.* Plasma, 2019. **2**: p. 316–327.
4. Reuter, S., T. von Woedtke, and K.-D. Weltmann, *The kINPen—A Review on Physics and Chemistry of the Atmospheric Pressure Plasma Jet and Its Applications.* Journal of Physics D: Applied Physics, 2018. **51**: p. 233001.
5. Brandenburg, R., *Dielectric Barrier Discharges: Progress on Plasma Sources and on the Understanding of Regimes and Single Filaments.* Plasma Sources Science and Technology, 2017. **26**: p. 053001.
6. Fridman, G., A. Shereshevsky, M.M. Jost, A.D. Brooks, A. Fridman, A. Gutsol, V. Vasilets, and G. Friedman, *Floating Electrode Dielectric Barrier Discharge Plasma in Air Promoting Apoptotic Behavior in Melanoma Skin Cancer Cell Lines.* Plasma Chemistry and Plasma Processing, 2007. **27**: p. 163–176.
7. Helmke, A., D. Hoffmeister, N. Mertens, S. Emmert, J. Schuette, and W. Vioel, *The Acidification of Lipid Film Surfaces by Non-Thermal DBD at Atmospheric Pressure in Air.* New Journal of Physics, 2009. **11**: p. 115025.
8. Oehmigen, K., M. Hähnel, R. Brandenburg, C. Wilke, K.D. Weltmann, and T. von Woedtke, *The Role of Acidification for Antimicrobial Activity of Atmospheric Pressure Plasma in Liquids.* Plasma Processes and Polymers, 2010. **7**: p. 250–257.
9. Shimizu, T., B. Steffes, R. Pompl, F. Jamitzky, W. Bunk, K. Ramrath, M. Georgi, W. Stolz, H.-U. Schmidt, T. Urayama, S. Fujii, and G.E. Morfill, *Characterization of Microwave Plasma Torch for Decontamination.* Plasma Processes and Polymers, 2008. **5**: p. 577–582.
10. Arndt, S., A. Schmidt, S. Karrer, and T. von Woedtke, *Comparing Two Different Plasma Devices kIN-Pen and Adtec SteriPlas Regarding Their Molecular and Cellular Effects on Wound Healing.* Clinical Plasma Medicine, 2018. **9**: p. 24–33.
11. Butenko, A.V., A.B. Shekhter, A.V. Pekshev, A.B. Vagapov, A.L. Fayzullin, N.B. Serejnikova, N.A. Sharapov, V.A. Zaborova, and V.N. Vasilets, *Review of Clinical Applications of Nitric Oxide-Containing Air-Plasma Gas Flow Generated by Plason Device.* Clinical Plasma Medicine, 2020. **19–20**: p. 100112.
12. Schmidt-Bleker, A., J. Winter, A. Bösel, S. Reuter, and K.-D. Weltmann, *On the Plasma Chemistry of a Cold Atmospheric Argon Plasma Jet with Shielding Gas Device.* Plasma Sources Science and Technology, 2016. **25**: p. 015005.
13. Babaeva, N.Y., W. Tian, and M.J. Kushner, *The Interaction between Plasma Filaments in Dielectric Barrier Discharges and Liquid Covered Wounds: Electric Fields Delivered to Model Platelets and Cells.* Journal of Physics D: Applied Physics, 2014. **47**: p. 235201.
14. Pakhomov, A.G., J.F. Kolb, J.A. White, R.P. Joshi, S. Xiao, and K.H. Schoenbach, *Long-Lasting Plasma Membrane Permeabilization in Mammalian Cells by Nanosecond Pulsed Electric Field (nsPEF).* Bioelectromagnetics, 2007. **28**: p. 655–663.
15. Bronold, F.X., K. Rasek, and H. Fehske, *Electron Microphysics at Plasma—Solid Interfaces.* Journal of Applied Physics, 2020. **128**: p. 180908.

16. Tian, W., and M.J. Kushner, *Long-Term Effects of Multiply Pulsed Dielectric Barrier Discharges in Air on Thin Water Layers Over Tissue: Stationary and Random Streamers.* Journal of Physics D: Applied Physics, 2015. **48**: p. 494002.

17. Kurylo, M.J., W. Braun, A. Kaldor, S.M. Freund, and R.P. Wayne, *Infra-Red Laser Enhanced Reactions: Chemistry of Vibrationally Excited O3 with NO and O2(1Δ).* Journal of Photochemistry, 1974. **3**: p. 71–87.

18. Azyazov, V.N., P. Mikheyev, D. Postell, and M.C. Heaven, *O2(a1Δ) Quenching in the O/O2/O3 System.* Chemical Physics Letters, 2009. **482**: p. 56–61.

19. Bissonnette-Dulude, J., S. Coulombe, T. Gervais, and S. Reuter, *Coupling the COST Reference Plasma Jet to a Microfluidic Device: A New Diagnostic Tool for Plasma-Liquid Interactions.* Plasma Sources Science and Technology, 2023. **32**: p. 055003.

20. Cupitt, L.T., G.A. Takacs, and G.P. Glass, *Reaction of Hydrogen Atoms and O2(1?g).* International Journal of Chemical Kinetics, 1982. **14**: p. 487–497.

21. Hack, W., and H. Kurzke, *Kinetic Study of the Elementary Chemical Reaction Atomic Hydrogen(2S1/2) + Molecular Oxygen(1.DELTA.g). Fwdarw. Hydroxyl(2.pi.) + Atomic Oxygen(3P) in the Gas Phase.* The Journal of Physical Chemistry, 2002. **90**: p. 1900–1906.

22. Girard, P.M., A. Arbabian, M. Fleury, G. Bauville, V. Puech, M. Dutreix, and J.S. Sousa, *Synergistic Effect of H2O2 and NO2 in Cell Death Induced by Cold Atmospheric He Plasma.* Scientific Reports, 2016. **6**: p. 29098.

23. Winter, J., H. Tresp, M.U. Hammer, S. Iseni, S. Kupsch, A. Schmidt-Bleker, K. Wende, M. Dünnbier, K. Masur, K.D. Weltmann, and S. Reuter, *Tracking Plasma Generated H2O2 from Gas into Liquid Phase and Revealing Its Dominant Impact on Human Skin Cells.* Journal of Physics D: Applied Physics, 2014. **47**: p. 285401.

24. Lavoie, G.A., J.B. Heywood, and J.C. Keck, *Experimental and Theoretical Study of Nitric Oxide Formation in Internal Combustion Engines.* Combustion Science and Technology, 1970. **1**: p. 313–326.

25. van Gaens, W., S. Iseni, A. Schmidt-Bleker, K.D. Weltmann, S. Reuter, and A. Bogaerts, *Numerical Analysis of the Effect of Nitrogen and Oxygen Admixtures on the Chemistry of an Argon Plasma Jet Operating at Atmospheric Pressure.* New Journal of Physics, 2015. **17**: p. 033003.

26. Graham, R.A., and H.S. Johnston, *The Photochemistry of the Nitrate Radical and the Kinetics of the Nitrogen Pentoxide-Ozone System.* The Journal of Physical Chemistry, 2002. **82**: p. 254–268.

27. Feng, Y., and J.E. Penner, *Global Modeling of Nitrate and Ammonium: Interaction of Aerosols and Tropospheric Chemistry.* Journal of Geophysical Research, 2007. **112**: p. 1–24.

28. Carbone, E., and C. Douat, *Carbon Monoxide in Plasma and Agriculture: Just a Foe or a Potential Friend?* Plasma Medicine, 2018. **8**: p. 93–120.

29. Xu, S., V. Jirasek, and P. Lukes, *Elucidation of Molecular-Level Mechanisms of Oxygen Atom Reactions with Chlorine Ion in NaCl Solutions using Molecular Dynamics Simulations Combined with Density Functional Theory.* ChemistrySelect, 2023. **8**: p. e202203937 (1–7).

30. Gorbanev, Y., J. Van der Paal, W. Van Boxem, S. Dewilde, and A. Bogaerts, *Reaction of Chloride Anion with Atomic Oxygen in Aqueous Solutions: Can Cold Plasma Help in Chemistry Research?* Physical Chemistry Chemical Physics, 2019. **21**: p. 4117–4121.

31. Elg, D.T., I.W. Yang, and D.B. Graves, *Production of TEMPO by O Atoms in Atmospheric Pressure Non-Thermal Plasma—Liquid Interactions.* Journal of Physics D: Applied Physics, 2017. **50**: p. 475201.

32. [1.8.2023]; Available from: https://plasma.polymtl.ca.

33. Murakami, T., K. Niemi, T. Gans, D. O'Connell, and W.G. Graham, *Afterglow Chemistry of Atmospheric-Pressure Helium—Oxygen Plasmas with Humid Air Impurity.* Plasma Sources Science and Technology, 2014. **23**: p. 025005.

34. Winter, J., K. Wende, K. Masur, S. Iseni, M. Dunnbier, M.U. Hammer, H. Tresp, K.D. Weltmann, and S. Reuter, *Feed Gas Humidity: A Vital Parameter Affecting a Cold Atmospheric-Pressure Plasma Jet and Plasma-Treated Human Skin Cells.* Journal of Physics D: Applied Physics, 2013. **46**: p. 295401.

35. Pipa, A.V., S. Reuter, R. Foest, and K.D. Weltmann, *Controlling the NO Production of an Atmospheric Pressure Plasma Jet.* Journal of Physics D-Applied Physics, 2012. **45**: p. 085201.

36. Schmidt-Bleker, A., R. Bansemer, S. Reuter, and K.-D. Weltmann, *How to Produce an NOx- Instead of Ox-Based Chemistry with a Cold Atmospheric Plasma Jet.* Plasma Processes and Polymers, 2016. **13**: p. 1120–1127.

37. Sousa, J.S., and V. Puech, *Diagnostics of Reactive Oxygen Species Produced by Microplasmas.* Journal of Physics D: Applied Physics, 2013. **46**: p. 464005.

38. Sousa, J.S., G. Bauville, B. Lacour, V. Puech, and M. Touzeau, *Atmospheric Pressure Generation of O2(a1Δg) by Microplasmas.* The European Physical Journal Applied Physics, 2009. **47**: p. 22807.

39. Sousa, J.S., G. Bauville, and V. Puech, *Arrays of Microplasmas for the Controlled Production of Tunable High Fluxes of Reactive Oxygen Species at Atmospheric Pressure.* Plasma Sources Science and Technology, 2013. **22**: p. 035012.

40. Murakami, T., and O. Sakai, *Rescaling the Complex Network of Low-Temperature Plasma Chemistry Through Graph-Theoretical Analysis.* Plasma Sources Science and Technology, 2020. **29**: p. 115018.

41. Sakai, O., S. Kawaguchi, and T. Murakami, *Complexity Visualization, Dataset Acquisition, and Machine-Learning Perspectives for Low-Temperature Plasma: A Review.* Japanese Journal of Applied Physics, 2022. **61**.

42. Murakami, T., and O. Sakai, *Rescaling the Complex Network of Low-Temperature Plasma Chemistry Through Graph-Theoretical Analysis.* Plasma Sources Science and Technology, 2020. **29**.

43. Murakami, T., *Numerical Modelling of the Effects of Cold Atmospheric Plasma on Mitochondrial Redox Homeostasis and Energy Metabolism.* Scientific Reports, 2019. **9**: p. 17138.

44. Reuter, S., J. Winter, S. Iseni, A. Schmidt-Bleker, M. Dunnbier, K. Masur, K. Wende, and K.-D. Weltmann, *The Influence of Feed Gas Humidity Versus Ambient Humidity on Atmospheric Pressure Plasma Jet-Effluent Chemistry and Skin Cell Viability.* IEEE Transactions on Plasma Science, 2015. **43**: p. 3185–3192.

45. Iseni, S., A. Schmidt-Bleker, J. Winter, K.D. Weltmann, and S. Reuter, *Atmospheric Pressure Streamer Follows the Turbulent Argon Air Boundary in a MHz Argon Plasma Jet Investigated by OH-Tracer PLIF Spectroscopy.* Journal of Physics D: Applied Physics, 2014. **47**: p. 152001.

46. Walsh, J.L., F. Iza, N.B. Janson, V.J. Law, and M.G. Kong, *Three Distinct Modes in a Cold Atmospheric Pressure Plasma Jet.* Journal of Physics D: Applied Physics, 2010. **43**: p. 075201.

47. Xu, H., C. Chen, D. Liu, D. Xu, Z. Liu, X. Wang, and M.G. Kong, *Contrasting Characteristics of Aqueous Reactive Species Induced by Cross-Field and Linear-Field Plasma Jets.* Journal of Physics D: Applied Physics, 2017. **50**: p. 245201.

48. Sousa, J.S., G. Bauville, B. Lacour, V. Puech, M. Touzeau, and L.C. Pitchford, *O[sub 2](a[sup 1]Δ[sub g]) Production at Atmospheric Pressure by Microdischarge.* Applied Physics Letters, 2008. **93**: p. 011502.

49. Sousa, J.S., G. Bauville, B. Lacour, V. Puech, M. Touzeau, and J.-L. Ravanat, *DNA Oxidation by Singlet Delta Oxygen Produced by Atmospheric Pressure Microdischarges.* Applied Physics Letters, 2010. **97**: p. 141502.

50. Sousa, J.S., P.M. Girard, E. Sage, J.L. Ravanat, and V. Puech, *DNA oxidation by reactive oxygen species produced by atmospheric pressure microplasmas,* in *Plasma for Bio-Decontamination, Medicine and Food Security, NATO Science for Peace and Security Series A: Chemistry and Biology,* Z.M.e.a. (eds.). 2012, Springer Science+Business Media B.V. p. 107–119.

51. Robert, E., E. Barbosa, S. Dozias, M. Vandamme, C. Cachoncinlle, R. Viladrosa, and J.M. Pouvesle, *Experimental Study of a Compact Nanosecond Plasma Gun.* Plasma Processes and Polymers, 2009: p. 795–802.

3 Plasma Chemistry in Aqueous Solutions

Peter Bruggeman, Yolanda Aranda Gonzalvo

3.1 INTRODUCTION

As described in Chapter 2, gas-phase plasmas are a rich source of reactive oxygen and nitrogen species (RONS). When plasmas are used for biological and medical applications, in most cases an aqueous liquid medium is present between the plasma and the living matter, such as a wound fluid covering a wound surface or the presence of blood during surgical applications of plasmas. Similarly, when biological matter is treated during *in-vitro* studies, the living matter is often contained in physiological solutions. The presence and composition of a liquid medium can drastically impact the resulting plasma-produced RONS delivered to the biological substrate. In addition, several treatment approaches are based on the application of plasma-treated liquids, solutions or media which have shown to have decontamination capabilities and can be applied for cancer treatment [1] (see also Chapter 13). The large number of plasma sources that have been developed and studied for plasma redox biology applications, as discussed in Chapter 2, leads to a broad range of RONS. Nonetheless, only a subsection of those RONS have been shown to date to have a direct impact on biological processes. The different properties of the plasma-produced gas-phase species, including widely varying lifetimes and Henry law's constants, lead to a large variation in their ability to transfer into the liquid phase. These large variations in transport properties can have a significant impact on the subsequent plasma-enabled liquid phase reactivity. Hence, understanding the interaction of plasmas with liquids and the resulting RONS is a critical step towards a detailed understanding of plasma redox biology. This chapter summarizes our current understanding of plasma chemistry in an aqueous solution relevant to plasma redox biology.

While plasmas can be directly produced in liquids [2], we will focus on gas-phase plasmas in contact with liquids, as used in most applications of plasma redox biology. In the next section, we introduce the basics of plasma-liquid interactions and the key impact of the transport of gas-phase reactive species through solutions. In these interactions, the treatment modality can have a major impact on both the RONS production efficacy and the RONS delivered to the liquid, which are discussed in a subsequent section. This discussion is followed by a summary of some of the major chemical mechanisms that have been reported to enable decontamination and plasma medicine or that play a key role in plasma redox biology. We conclude this chapter with a brief overview of the state of the art in the quantification of RONS in solution and relevant modeling.

3.2 BASICS OF PLASMA-LIQUID INTERACTIONS

The interaction of plasmas with aqueous solutions enables the production of both short- and long-lived reactive species either due to photolysis directly in the liquid phase or due to transfer of ionic, electron or neutral species from the gas-phase plasma into the solution [3]. These species can react near the plasma-liquid interface or penetrate into the liquid and initiate secondary chemical processes, ultimately leading to the generation of long-lived reactive species. For example, the injection of plasma-generated gas-phase electrons into the liquid phase generates solvated (or hydrated) electrons (e_{aq}^-), which are strongly reducing species with a typical lifetime less than 5×10^{-4} s [4].

DOI:10.1201/9781003328056-4

This short lifetime usually enables the penetration of solvated electrons by only tens of nanometers into the liquid before they recombine into long-lived species [5]. Another abundant species produced during the interaction of plasmas with aqueous solutions is the OH radical, which is a strong oxidizing species and readily oxidizes many hydrocarbons [6] leading to typical penetration depths on the order of 1 μm or less. The OH radical is ubiquitous in the presence of aqueous solutions even when using dry gases, as evaporation from the aqueous solution will cause a water-vapor layer near the plasma-liquid interface, which can easily allow for the generation of OH radicals through, for example, electron-induced dissociation of water or hydration reactions of H_2O^+ with H_2O forming H_3O^+[3]. Plasmas interacting with liquid water, are also a copious source of hydrogen peroxide (H_2O_2), a species that is readily dissolved in the liquid due to its high Henry's law constant. The lower reactivity of H_2O_2 enables H_2O_2 typically to transport into the bulk liquid and accumulate over time in the solution.

While plasmas are extensively used as (vacuum) UV ((V)UV) light sources, the direct impact of photolysis, which is particularly important in noble gases that can produce excimer radiation, is often limited to near the plasma-liquid interface because of the small penetration depth of the VUV, and can be seen as a source of solvated electrons, hydrogen atoms and OH radicals near the plasma-liquid interface [3]. In many cases, near-UV and visible emission are not present at sufficient fluxes to make a major impact in bulk solutions; hence, photo-induced chemistry is not focused on in this section. However, a study combining plasmas with an LED source has shown possible photon-driven reactions [7].

While short-lived species typically play a key role in liquid-induced chemistry, their role can be more limited when treating biological substrates covered by a liquid. In this case, short-lived species will react with chemical compounds in the liquid before reaching the biological substrate. Hence, the substrate will be mainly exposed to secondary longer-lived reaction products of the plasma-produced primary radicals. A striking example is the important role of solvated electrons in the synthesis of nanoparticles. The solvated electrons, in spite of their limited penetration depth, enable the reduction of metal precursor ions near the plasma-liquid interface, and the generated gradient in metal ions will drive the transport of the metal ion precursor to the plasma-liquid interface [8]. While such short-lived species might still play a role when treating planktonic bacteria or viruses in a cuvette filled with solution, the treatment of a biological substrate covered with a liquid layer of finite thickness does not allow the penetration of most short-lived radical species. In the case of solvated electrons, they might, for example, attach to dissolved O_2 and form superoxide (O_2^-), which has a longer lifetime (and penetration depth), or be converted in other RONS before reaching the substrate.

3.3   PLASMA TREATMENT MODALITIES

Depending on the plasma source configuration and the distance between the solution and the plasma, three distinct treatment modalities can be distinguished [9]: direct, remote and indirect treatments. A schematic of each of these three treatment modalities is shown in Figure 3.1. For a direct treatment, the solution acts as an electrode, and the electrical current of the plasma flows through the solution. In this modality, there is typically an abundant production of short-lived plasma-produced reactive species, such as radicals, electrons and ions to the solution, resulting in fluxes of short-lived species significantly higher as compared to the two other treatment modalities. As the liquid is in direct contact with the plasma, a high electric field will be present at the plasma-liquid interface, which might have a finite penetration depth beyond the interface. However, in cases where biological substrates are covered by a solution with a large electrical conductivity (such as saline) with a finite thickness, the penetration depth of the electric field might remain small and not reach the underlying biological substrate. As explained above, even in direct plasma treatments, plasma-produced radicals are not always interacting with the biological substrate if a liquid layer with finite thickness is present on top of the substrate.

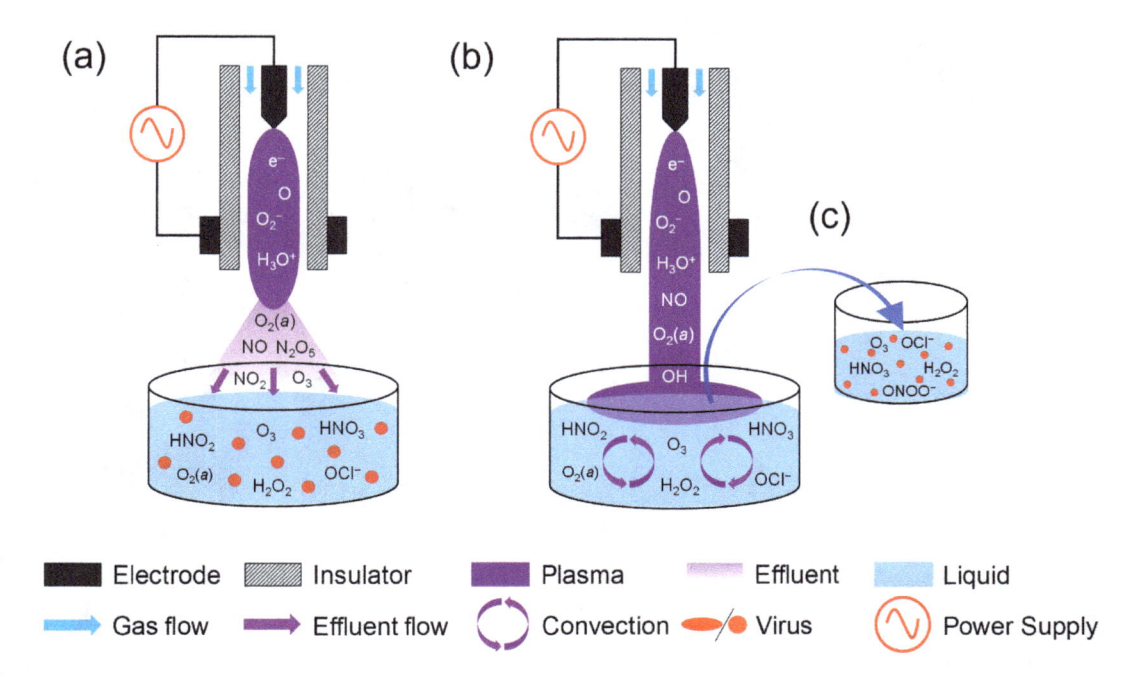

FIGURE 3.1 Schematic representation of a plasma jet treatment of an aqueous solution in the three different treatment modalities: (a) remote, (b) direct and (c) indirect treatment. In this example, the biological substrate a suspension of virus particles in solution. The figures are taken from [9].

During remote plasma treatments, the plasma is generated at a finite distance from the solution and the plasma effluent is convectively transported to the solution. In this case, the plasma effluent contains reactive species with a lifetime typically longer than the convective transport time-scales. For most plasma jets, as described in Chapter 2, this will lead to gas residence times on the order of milliseconds to seconds. The lifetime of most radicals and ionic species is significantly shorter, leading to small or even negligible fluxes of such species compared to direct treatment. In such cases, the liquid-phase chemistry is driven by the transfer of long-lived species (> 10 s of ms) such as O_3, H_2O_2 and HNO_2. While the dominant species in the plasma effluent will often be significantly less reactive than plasma-produced radicals, remote treatment modalities have been shown to be highly effective for decontamination applications [10]. Furthermore, it has been established that plasma-treated solutions can exhibit bactericidal effects for several minutes up to days after the plasma treatment [11]. This finding gave rise to the indirect treatment modality, which refers to the treatment of biological substrates with solutions pre-treated with plasma. These solutions are often referred to in the literature as plasma-activated medium or plasma-activated liquid (PAM or PAL) [12, 13].

3.4 SELECTED KEY CHEMICAL MECHANISMS

Plasmas used for redox biology are often generated in (humid) air or in a noble gas such as argon or helium, but even in the latter case, the feed gas typically contains air (or just O_2 or N_2) and humidity. The gas-phase chemistry in such mixtures can contain 100 different species and involve 1000+ reactions [14]. This complexity in the gas phase is also reflected in an equally broad range of RONS in the liquid phase. Despite the complexity of plasma-induced liquid-phase chemistry, this section highlights a few key reactive species that have been identified in recent years. An overview of these reactions is shown in Figure 3.2.

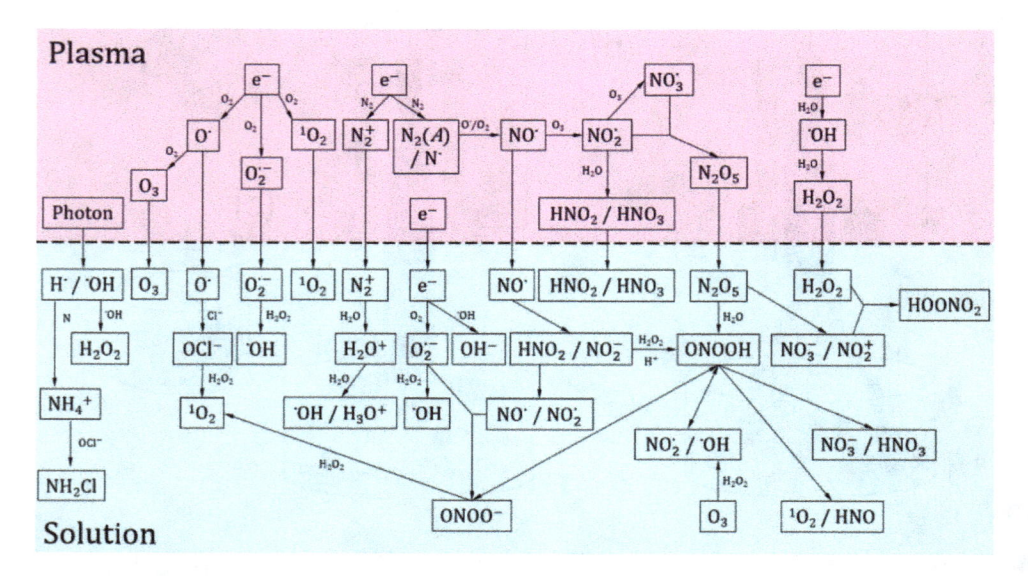

FIGURE 3.2 Schematic representation of selected key plasma produced RONS reported in literature to be responsible for plasma redox biology. Reaction pathways are provided both in the gas and liquid phase. The horizontal dashed line represents the plasma-liquid interface. This image is a modified version of the schematic reported in [9].

3.4.1 OXYGEN CHEMISTRY: OZONE, SINGLET DELTA OXYGEN AND SUPEROXIDE

Plasmas containing oxygen produce a rich chemistry with a variety of different reactive oxygen species (ROS) that can have a profound biological impact. Ozone is produced by commercial ozonizers for water disinfection at large scale and has been extensively studied for its potential for virus and bacteria inactivation in both water and gas phase [15]. Compared to many other RONS, O_3 has a moderate Henry's Law constant, and mixing is typically needed to enhance its mass transfer into solution [16]. While significant O_3 - induced oxidation of biomolecules has been observed, several studies have indicated an important role for more short-lived ROS in plasma redox biology. This includes atomic oxygen (O), singlet delta oxygen ($O_2(a^1\Delta_g)$) and superoxide (O_2^-).

O radicals in solution phase have not been extensively studied beyond the field of plasma science and technology, as they are uniquely produced by plasmas. Nonetheless, several studies suggest that O radicals can be a potent oxidizer of organic compounds in aqueous solutions and can behave very similarly to OH radicals discussed in the next section [17]. Nonetheless, O radicals are a key enabler of plasma-induced chlorine chemistry when plasmas interact with Cl-containing solutions such as, for example, saline, through the formation of the hypochlorite ion (OCl^-) with known bactericidal effects [18]. We discuss chlorine chemistry in more detail below.

Singlet delta oxygen, $O_2(a^1\Delta_g)$, is the first electronically excited state of molecular oxygen with an energy of 0.98 eV larger than the ground state [19]. It is abundantly produced by electron impact in non-equilibrium plasmas and has a radiative lifetime of ~ 75 min and low collisional quenching probabilities, enabling high densities of $O_2(a^1\Delta_g)$ in gas-phase plasmas. Singlet delta oxygen can be an important active agent in photodynamic therapy [20] and has even been suggested to be responsible for the observed selectivity of plasma treatment of cancer cells or tumors [21]. In addition to the transfer of gas-phase $O_2(a^1\Delta_g)$ into solution, $O_2(a^1\Delta_g)$ can also be produced through radical reactions in the liquid phase [22]. Despite the challenge of detecting $O_2(a^1\Delta_g)$ in plasma-treated liquids, colorimetric methods have been able to correlate the presence of $O_2(a^1\Delta_g)$ with the inactivation of virus and trace back oxidation-reaction products of amino acids comprising the capsid protein of feline calicivirus to singlet delta oxygen-mediated chemistry [23].

Superoxide is well established as a bioactive molecule [24, 25], and several studies have indicated that biocidal effects can be suppressed by superoxide dismutase (SOD), a highly specific scavenger of O_2^- [23]. The negative ion is abundantly produced in the gas or liquid phase, due to the attachment of electrons to O_2. Nonetheless, the ion density is typically much smaller than the dominant neutral ROS density in low-temperature plasmas, and the impact of short-lived species including O and singlet delta oxygen either directly or through secondary reaction products should not be disregarded a priori in the presence of O_2^-.

3.4.2 HYDROGEN PEROXIDE

As mentioned earlier, OH radicals are abundantly produced during plasma-liquid interactions. While OH radicals are powerful oxidizers, their short lifetime significantly limits their penetration into solution, and in most cases OH formed in the bulk of solution through secondary reactions of long-lived plasma-produced species is probably more impactful in the treatment of biological substrates covered by a liquid layer. Nonetheless, the recombination product of OH, hydrogen peroxide (H_2O_2), is a well-known disinfectant, and many commercial H_2O_2-based decontamination solutions and systems have been tested against different bacteria and viruses [26]. Plasmas containing water vapor can be an abundant source of H_2O_2 and have been suggested to enable plasma-induced biological effects particularly for plasma-induced treatment of cancer cells and with mammalian cells [18]. Nonetheless, in many cases, the inactivation of virus or bacteria was observed at concentrations lower than needed for H_2O_2 only, indicating that while H_2O_2 could contribute to plasma-induced biological effects, the inactivation is mainly the product of a different chemistry, which might include secondary products formed upon reaction with H_2O_2 [23]. Examples of secondary reaction products of H_2O_2 reacting with reactive nitrogen species (RNS) are discussed in the next section.

3.4.3 NO, ACIDIFIED NITRATES, AND PEROXYNITRITE CHEMISTRY

Many studies have shown that the bactericidal effect of plasma-treated solutions is facilitated by an acidic environment (typically pH values of 3–4). These effective conditions have been attributed to RNS such as acidified nitrites and peroxynitrite chemistry [27]. Air plasmas can be an abundant source of NO, particularly when operated at higher power densities and gas temperatures. In the presence of O_2 (and ozone), NO leads to the generation of NO_2, N_2O and N_2O_5. If water vapor is present, HNO_2 and HNO_3 are also formed.

Nitrous acid (HNO_2) is not stable at low pH and can lead to the formation of radicals in solution:

$$2HNO_2 \rightarrow NO + NO_2 + H_2O \quad (R_1)$$

The resulting NO and NO_2 radicals possess strong cell-toxic properties. This chemistry at low pH, which could also involve the formation of the nitrosonium ion (NO^+), is referred to as acidified nitrites and will ultimately convert NO_2^- into stable NO_3^- [27].

The interaction of air plasmas with water typically leads to the abundant production of HNO_3, which in the absence of a buffer will cause acidification of the solution. While HNO_3 is not highly biologically active, its ability to reduce the pH of solutions can have a significant impact. A substantial literature has shown the important role of peroxynitrite chemistry, which is strongly pH dependent. Both the anionic ($O=NOO^-$, peroxynitrite anion) and the protonated forms ($O=NOOH$, peroxynitrous acid) (pKa = 6.8) can participate in oxidation reactions. While the formation of the peroxynitrite anion is typically enabled by radical reactions, air-containing plasmas can create a reaction environment that allows for the formation of peroxynitrous acid through long-lived species in the bulk solution [27]

$$O_2^- + NO \rightarrow O=NOO^- \quad (R_2)$$

$$OH + NO_2 \rightarrow O=NOO^- + H^+ \quad (R_3)$$

$$NO_2^- + H_2O_2 + H^+ \rightarrow O=NOOH + H_2O \quad (R_4)$$

Peroxynitrite has been extensively studied in the biology community. Its formation in such conditions often involves the pathway through O_2^-, and the formation of peroxynitrite through H_2O_2 and NO_2^- seems to be more typical in plasma-treated solutions. Peroxynitrite can either directly or indirectly react with biomolecules through radicals forming in the following disproportionation reaction:

$$O=NOOH \leftrightarrow OH + NO_2 \quad (R_5)$$

While plasmas at elevated temperatures can produce significant amounts of NO, which has been shown to have many potentially beneficial uses in biological processes, the transfer of NO from the liquid phase to the solution is often limited due to the relatively low Henry's law constant compared to oxidized forms of NO (N_2O, NO_2 and N_2O_5) and certainly HNO_x. This low Henry's law constant has recently led to studies that show an important role for N_2O_5 in plasma-enabled decontamination [28]. While the N_2O_5 chemistry is less studied, researchers have proposed that N_2O_5 dissolves in solution and dissociates as follows [28]:

$$N_2O_5 \rightarrow [NO_2^+ \, NO_3^-] \rightarrow NO_2^+ + NO_3^- \quad (R_6)$$

The resulting nitronium ion (NO_2^+) is a reactive cation that can enable nitration of organic compounds or react with H_2O_2 to produce $HOONO_2$ (peroxynitric acid), which has known bactericidal properties [29, 30]), as follows [30]:

$$NO_2^+ + H_2O_2 \rightarrow HOONO_2 + H^+ \quad (R_7)$$

While our understanding of the detailed reaction mechanism of N_2O_5 might benefit from further study, multiple pathways can be present which could account for the reactivity of N_2O_5 with biomolecules in the liquid phase.

3.4.4 CHLORINE CHEMISTRY

Body fluids are electrolyte solutions containing ions such as Cl^-, Na^+ and K^+. Physiological solutions (substitutes for tissue fluids) typically contain 0.9% NaCl, and cells, tissue or microorganisms used for *in-vitro* experiments are often placed in solutions such as a phosphate-buffered saline (PBS) or media containing high concentrations of Cl^- ions. Hence, during both *in-vivo* and *in-vitro* plasma experiments, Cl^- ions will be exposed to plasma-produced RONS, which can lead to the generation of a rich chlorine chemistry. Sodium hypochlorite (NaOCl), commonly referred to as bleach, is a well-known disinfectant. In addition, hypochlorite compounds (HOCl, OCl^-, pKa = 7.54 [31]) are important in microbiology and medicine [3, 32].

As mentioned above, O radicals are a key enabler of plasma-induced chlorine chemistry when plasmas interact with Cl^--containing solutions through the following reaction [18]:

$$O + Cl^- \rightarrow OCl^- \quad (R_8)$$

This reaction was proposed in 2015 to explain the impact of plasma on mammalian cells in Cl^--containing solutions; the production of OCl^- has been measured, and recently the reaction was also studied computationally [33].

A more detailed study has shown that the plasma treatment of saline leads to the formation of the following species: oxoacids or their conjugate bases, hypochlorites $HOCl/OCl^-$, chlorites ClO_2^-, chlorates ClO_3^- and chlorine dioxide, ClO_2 [34]. Rather than a reduction of pH, as is found in air plasmas due to the production of HNO_3 in the liquid, the pH of plasma-treated saline (with oxygen plasma) under conditions in which these species are formed typically becomes alkaline.

The complexity of the chlorine chemistry is significantly impacted by its strong pH dependence and by the reaction of OCl^- with other plasma-produced RONS:

$$OCl^- + H_2O_2 \rightarrow H_2O_2 + Cl^- + O_2(a^1\Delta_g) \quad (R_9)$$

$$NO_2^- + HOCl \rightarrow NO_3^- + Cl^- + H^+ \quad (R_{10})$$

A unique aspect of plasma-liquid interactions is that both oxidation and reduction reactions can be enabled, and potentially even simultaneously [4]. While the above chemistry relies mainly on oxidation processes, it has been shown that N_2 can be reduced to form NH_3 at the plasma-liquid interface [35]. Interestingly, ammonia chemistry has also been identified as a possible bactericidal pathway induced by N_2 plasmas not containing O_2 [36]. The reaction with hypochlorite can lead to the formation of chloramine (NH_2Cl), according to the conventional reaction in which the compound is made:

$$NH_3 + NaOCl \rightarrow NH_2Cl + NaOH \quad (R_{11})$$

Chloramine is also a known disinfection used in water treatment. In addition, reactions of RONS with biomolecules can form reactive hydrocarbon molecules that potentially induce further chemical changes to biological substrates.

3.5 QUANTIFICATION METHODS OF PLASMA-INDUCED LIQUID-PHASE CHEMISTRY

The level of detail required in the quantification of plasma-produced RONS depends on the goal of the experiment. In the case of medical applications, it is often suggested that the biological effect relates to a 'dose', which to date is not well defined, but it also might be related to the overall redox activity of the combined RONS, rather than to individual levels of RONS species. Hence, it could suffice to measure the redox potential in the plasma-treated solution with standard electrode systems from electrochemistry [37] or by using redox indicators, i.e., colorimetric reagents that show a color change at a specific electrode potential [38]. Nonetheless, if the goal of the study is to understand the fundamental interactions of plasma-induced aqueous RONS with biological substrates, a detailed analysis of individual species in the liquid will be required.

Many methods have been developed and implemented in recent decades to probe the composition of solutions based on optical absorption spectroscopy, electron paramagnetic resonance spectroscopy and mass spectrometry [39, 40, 41]. Nonetheless, most of these techniques have been developed to measure long-lived RONS or only allow the measurement of radical species indirectly through turn-on colorimetric and/or spin-trapping methods. Both colorimetric and spin-trapping methods use stable chemical probes that are transformed through reactions with radical species. It is often assumed that the amount of colorimetric or spin-trapping product directly relates to the number of short-lived intermediates present, although in plasma-treated solutions this assumption needs to be considered with caution due to (1) the non-trivial mixture of short-lived and long-lived aqueous reactive species, which leads to inherent unselective chemical reactions or potentially even decomposition of the probe molecules, and (2) gradients in RONS present in the plasma-treated solution that could significantly impact the amount of products being formed through transport-limited reactions.

In many cases, the direct measurement of species is still preferred. For example, broadband UV/Vis absorption spectroscopy has been extensively used for the characterization of OCl^-, NO_3^-, NO_2^-, H_2O_2 and dissolved oxygen species [42, 43]. While many short-lived species have well-documented absorption spectra [44], absorption measurement of radicals remains challenging as they are often present only in the immediate vicinity of the plasma-liquid interface at sufficient densities, and their bulk liquid densities are low. Furthermore, absorption spectra in liquids are broad and often overlap, particularly in the UV part of the spectrum. This leads to challenges in the deconvolution of the absorption spectra of complex mixtures of reactive species.

Reaction pathways can be investigated non-quantitatively by adding specific scavengers to solutions, a common method in biological investigations. The addition of a scavenger leads to a depletion of the targeted RONS; hence, this approach can block specific chemical pathways. Scavenger studies complemented with positive control measurements, in which one adds the targeted RONS to the biological substrate and assesses the chemical's impact on the biological substrate, are a powerful approach to checking reaction mechanisms. Examples of scavengers include catalase for H_2O_2, superoxide dismutase (SOD) for O_2^-, histidine for singlet oxygen and ascorbic acid for peroxynitrous acid [3]. However, scavengers are typically non-selective, particularly in the presence of radicals. We therefore recommend considering different scavengers for the same targeted RONS, if possible, to assess the consistency of the obtained result. An overview of common detection techniques and scavengers for plasma-generated RONS in solutions can be found in [3].

3.6 MAJOR PLASMA-INDUCED LIQUID-PHASE MODELS

There are comprehensive models of non-equilibrium plasmas and of liquid-phase chemistry, although only a few models exist that merge both aspects [3]. In general, such models rely on reaction mechanisms developed in different fields. Studies have reported on 0-D global models as well as 1-D and multi-dimensional models. While 0-D global models allow us to capture very complex chemical mechanisms with thousands of reactions if needed, they assume a well-mixed reactor that does not allow us to accurately capture transport, which can be particularly important for short-lived RONS. Advanced global and multi-dimensional models of plasmas interacting with liquid have been developed by several researchers and can not only compute the detailed transport, which is critical in many plasma-liquid interaction studies, but also incorporate the complex coupling of the plasma phase with the liquid phase [45].

While many computational studies focus on plasma-induced liquid-phase chemistry, only a few focus on conditions relevant for biological applications. Several studies investigated the chemistry of RONS in thin layers of solution covering biological substrates [46, 47]. Such studies can assess the attenuation of plasma-produced RONS fluxes through the liquid layer onto the biological substrate and conversion reaction products. Models investigating larger liquid volumes (including convective transport) have been reported, although they often include a simplified representation of the plasma and/or gas-phase reactions [48]. Most of the computational studies consider simplified liquids, including saline solutions [49], as such models require detailed knowledge of the reaction mechanisms, including rate coefficients for each elementary reaction. Reaction sets typically do not incorporate reactions of RONS with biomolecules as many reaction rates of such reactions are not reported. While such studies have been instrumental in guiding insights in fundamental plasma-induced liquid-phase chemistry, detailed comparisons with experimental results remain limited, and many of these models do not provide quantitative information but should be used mainly to assess mechanisms. Nonetheless, developments to include biomolecules in reaction sets are starting to emerge [50].

3.7 CONCLUSION

Plasma redox biology has established itself as a mature field in the last two decades. While the underpinning mechanisms of the biological impact of plasma in the presence of aqueous solutions are complex, an understanding of the important key species is emerging. The chemistry induced by plasma-generated RONS on biomolecules remains an active area of research and is further discussed in chapters 4 and 5. However, we still lack the quantitative understanding that would enable us to determine a 'dose' concept. This is mainly due to the large range of gas-phase species and reactions in a humid air environment and the subsequent myriad of liquid-phase reactions resulting in a large range of transient and long-lived reaction products. Incomplete liquid-phase reaction sets and significant challenges to performing selective measurements in the resulting highly reactive liquid-phase mixtures have forced us to rely on indirect measurements of radicals or focus on long-lived reaction products, which are often not the direct cause of the plasma-induced biological effect. The strong dependence of the biological impact on plasma gas-phase chemistry and treatment modality further increases the difficulty of obtaining a generally valid dose concept, particularly given the large number of different plasma source designs investigated for plasma redox biology.

ACKNOWLEDGEMENTS

PJB acknowledges support from the National Science Foundation (PHY—2020695).

REFERENCES

[1] D. Yan, J.H. Sherman, X. Cheng, E. Ratovitski, J. Canady and M. Keidar, Controlling Plasma Stimulated Media in Cancer Treatment Application, Appl. Phys. Lett. (2014) **105** (22) 22410. https://doi.org/10.1063/1.4902875

[2] P. Bruggeman and C. Leys, Non-Thermal Plasmas in and in Contact with Liquids, *J. Phys. D: Appl. Phys.* (2009) **42** (5) 053001. https://dx.doi.org/10.1088/0022-3727/42/5/053001

[3] P.J. Bruggeman, M.J. Kushner, B.R. Locke, J.G.E. Gardeniers, W.G. Graham, D.B. Graves, R.C.H.M. Hofman-Caris, D. Maric, J.P. Reid, E. Ceriani, D. Fernandez Rivas, J.E. Foster, S.C. Garrick, Y. Gorbanev, S. Hamaguchi, F. Iza, H. Jablonowski, E. Klimova, J. Kolb, F. Krcma, P. Lukes, Z. Machala, I. Marinov, D. Mariotti, S. Mededovic Thagard, D. Minakata, E.C. Neyts, J. Pawlat, Z. Lj. Petrovic, R. Pflieger, S. Reuter, D.C. Schram, S. Schröter, M. Shiraiwa, B. Tarabová, P.A. Tsai, J.R.R. Verlet, T. von Woedtke, K.R. Wilson, K. Yasui and G. Zvereva, Plasma—Liquid Interactions: A Review and Roadmap, *Plasma Sources Sci. Technol.* (2016) **25** (5) 053002. https://dx.doi.org/10.1088/0963-0252/25/5/053002

[4] P.J. Bruggeman, R.R. Frontiera, U.R. Kortshagen, M.J. Kushner, S. Linic, G.C. Schatz, H. Andaraarachchi, S. Exarhos, L.O. Jones, C.M. Mueller, C.C. Rich, C. Xu, Y. Yue and Y. Zhang, Plasma-Driven Solution Electrolysis, *J. Appl. Phys.* (2021) **129** (20) 200902. https://doi.org/10.1063/5.0044261

[5] P. Rumbach, D.M. Bartels, R.M. Sankaran and D.B. Go, The Solvation of Electrons by an Atmospheric-Pressure Plasma, *Nat. Commun.* (2015) **19** (6) 7248. https://doi.org/10.1038/ncomms8248

[6] M. Sahni and B.R. Locke, Quantification of Hydroxyl Radicals Produced in Aqueous Phase Pulsed Electrical Discharge Reactors, *Ind. Eng. Chem. Res.* (2006) **45** (17) 5819–5825. https://doi.org/10.1021/ie0601504

[7] M.J. Pavlovich, Y. Sakiyama, D.S. Clark and D.B. Graves, Antimicrobial Synergy Between Ambient-Gas Plasma and UVA Treatment of Aqueous Solution, *Plasma Process. Polym.* (2013) **10** (12) 1051–1060. https://doi.org/10.1002/ppap.201300065

[8] D. Mariotti, J. Patel, V. Švrček and P. Maguire, Plasma—Liquid Interactions at Atmospheric Pressure for Nanomaterials Synthesis and Surface Engineering. *Plasma Processes Polym.* (2012) **9** (11–12) 1074–1085. https://doi.org/10.1002/ppap.201200007

[9] H. Mohamed, G. Nayak, N. Rendine, B. Wigdahl, F.C. Krebs, P.J. Bruggeman and V. Miller, Non-Thermal Plasma as a Novel Strategy for Treating or Preventing Viral Infection and Associated Disease, *Front. Phys.* (2021) **9** 683118. https://doi.org/10.3389/fphy.2021.683118

[10] G. Nayak, H.A. Aboubakr, S.M. Goyal and P.J. Bruggeman, Reactive Species Responsible for the Inactivation of Feline Calicivirus by a Two-Dimensional Array of Integrated Coaxial Microhollow Dielectric Barrier Discharges in Air, *Plasma Process. Polym.* (2018) **15** (1) 1700119. https://doi.org/10.1002/ppap.201700119

[11] M.J Traylor, M.J. Pavlovich, S. Karim, P. Hait, Y. Sakiyama, D.S. Clark and D.B. Graves, Long-Term Antibacterial Efficacy of Air Plasma-Activated Water, *J. Phys. D: Appl. Phys.* (2011) **44** (47) 472001. https://doi.org/10.1088/0022-3727/44/47/472001

[12] J.C. Harley, N. Suchowerska, L.J. Rogers, H. Lo, E.S. Choong and D.R. McKenzie, Plasma Activated Liquid Synergistically Enhances Response to Radiation for Improved Cancer Therapy, *Plasma Process. Polym.* (2022) **19** (9) 2200026. https://doi.org/10.1002/ppap.202200026

[13] D. Yan, J.H. Sherman and M. Keidar, Cold Atmospheric Plasma, a Novel Promising Anti-Cancer Treatment Modality, *Oncotarget* (2017) **8** 15977–15995. www.oncotarget.com/article/13304/text/

[14] W. Van Gaens and A. Bogaerts, Kinetic Modelling for an Atmospheric Pressure Argon Plasma Jet in Humid Air, *J. Phys. D: Appl. Phys.* (2013) **46** (27) 275201. https://doi.org/10.1088/0022-3727/46/27/275201

[15] J.C. Hoff. Inactivation of Microbial Agents by Chemical Disinfectants, U. S. Environmental Protection Agency, Cincinnati, OH (1986). EPA/600/2–86/067

[16] R. Sander, Compilation of Henry's Law Constants (Version 4.0) Forwater as Solvent, *Atmos. Chem. Phys.* (2015) **15** 4399–4498. www.atmos-chem-phys.net/15/4399/2015/, https://doi.org/10.5194/acp-15-4399-2015

[17] J. Benedikt, M. Mokhtar Hefny, A. Shaw, B.R. Buckley, F. Iza, S. Schäkermann and J.E. Bandow, The Fate of Plasma-Generated Oxygen Atoms in Aqueous Solutions: Non-Equilibrium Atmospheric Pressure Plasmas as an Efficient Source of Atomic $O_{(aq)}$, *Phys. Chem. Chem. Phys.* (2018) **20** (17) 12037–12042. https://doi.org/10.1039/c8cp00197a

[18] K. Wende, P. Williams, J. Dalluge, W. Van Gaens, H. Aboubakr, J. Bischof, T. von Woedtke, S.M. Goyal, K.-D. Weltmann, A. Bogaerts, K. Masur and P.J. Bruggeman, Identification of the Biologically Active Liquid Chemistry Induced by a Nonthermal Atmospheric Pressure Plasma Jet, *Biointerphases* (2015) **10** (2) 029518. https://doi.org/10.1116/1.4919710

[19] A.A. Ionin, I.V. Kochetov, A.P. Napartovich and N.N. Yuryshev, Physics and Engineering of Singlet Delta Oxygen Production in Low-Temperature Plasma, *J. Phys. D: Appl. Phys.* (2007) **40** R25. https://doi.org/10.1088/0022-3727/40/2/R01

[20] K.R. Weishaupt, C.J. Gomer and T.J. Dougherty, Identification of Singlet Oxygen as the Cytotoxic Agent in Photoinactivation of a Murine Tumor, *Cancer Res.* (1976) **36** 2326–2329.

[21] G. Bauer and D.B. Graves, Mechanisms of Selective Antitumor Action of Cold Atmospheric Plasma-Derived Reactive Oxygen and Nitrogen Species, *Plasma Process. Polym.* (2016) **13** (12) 1157–1178. https://doi.org/10.1002/ppap.201600089

[22] M. Riethmüller, N. Burger and G. Bauer, Singlet Oxygen Treatment of Tumor Cells Triggers Extracellular Singlet Oxygen Generation, Catalase Inactivation and Reactivation of Intercellular Apoptosis-Inducing Signaling, *Redox Biol.* (2015) **6** 157–168. https://doi.org/10.1016/j.redox.2015.07.006

[23] H.A. Aboubakr, U. Gangal, M.M Youssef, S.M. Goyal and P.J. Bruggeman, Inactivation of Virus in Solution by Cold Atmospheric Pressure Plasma: Identification of Chemical Inactivation Pathways, *J. Phys. D: Appl. Phys.* (2016) **49** (20) 204001. https://doi.org/10.1088/0022-3727/49/20/204001

[24] J.Z. Byczkowski and T. Gessner, Biological Role of Superoxide Ion-Radical, *Int. J. Biochem.* (1988) **20** (6) 569–580. https://doi.org/10.1016/0020-711x(88)90095-x

[25] J.J. Van Hemmen and W.J.A. Meuling, Inactivation of Escherichia Coli by Superoxide Radicals and Their Dismutation Products, *Arch. Biochem. Biophys.* (1977) **182** (2) 743–748. https://doi.org/10.1016/0003-9861(77)90556-2

[26] K.J. Sammarro Silva and L.P. Sabogal-Paz, A 10-Year Critical Review on Hydrogen Peroxide as a Disinfectant: Could It be an Alternative for Household Water Treatment?, *Water Supply* (2022) **2** (12) 8527–8539. https://doi.org/10.2166/ws.2022.384

[27] P. Lukes, E. Dolezalova, I. Sisrova and M. Clupek, Aqueous-Phase Chemistry and Bactericidal Effects from an Air Discharge Plasma in Contact with Water: Evidence for the Formation of Peroxynitrite Through a Pseudo-Second-Order Post-Discharge Reaction of H_2O_2 and HNO_2, *Plasma Sources Sci. Technol.* (2014) **23** (1) 015019. https://doi.org/10.1088/0963-0252/23/1/015019

[28] Y. Kimura, K. Takashima, S. Sasaki and T. Kaneko, Investigation on dinitrogen pentoxide roles on air plasma effluent exposure to liquid water solution, *J. Phys. D: Appl. Phys.* (2018) **52** (6) 064003. https://dx.doi.org/10.1088/1361-6463/aaf15a

[29] S. Ikawa, A. Tani, Y. Nakashima and K. Kitano, Physicochemical Properties of Bactericidal Plasma-Treated Water, *J. Phys. D: Appl. Phys.* (2018) **49** 425401.

[30] E. Wiberg, N. Wiberg and A. Holleman, Inorganic Chemistry, Academic Press, New York (2005).

[31] J.C. Morris, The Acid Ionization Constant of HOCl from 5 to 35°, *J. Phys. Chem.* (1966) **70** 3798–3805.

[32] O. Firuzi, P. Mladenka, R. Petrucci, G. Marrosu and L. Saso, Hypochlorite Scavenging Activity of Flavonoids, *J. Pharm. Pharmacol.* (2004) **56** 801.

[33] S. Xu, V. Jirasek, P. Lukes. Elucidation of Molecular-level Mechanisms of Oxygen Atom Reactions with Chlorine Ion in NaCl Solutions using Molecular Dynamics Simulations Combined with Density Functional Theory, *Chem. Select* (2023) **8** (23) e202203937. https://doi.org/10.1002/slct.202203937

[34] V. Jirásek and P. Lukeš, Formation of Reactive Chlorine Species in Saline Solution Treated by Non-Equilibrium Atmospheric Pressure He/O2 Plasma Jet, *Plasma Sources Sci. Technol.* (2019) **28** 035015. https://doi.org/10.1088/1361-6595/ab0930

[35] R. Hawtof, S. Ghosh, E. Guarr, C. Xu, R.M. Sankaran and J.N. Renner, Catalyst-Free, Highly Selective Synthesis of Ammonia from Nitrogen and Water by a Plasma Electrolytic System, *Sci. Adv.* (2019) **5** (1) aat5778. https://doi.org/10.1126/sciadv.aat5778

[36] S. Maheux, D. Duday, T. Belmonte, C. Penny, H.-M. Cauchie, F. Clément and P. Choquet, Formation of Ammonium in Saline Solution Treated by Nanosecond Pulsed Cold Atmospheric Microplasma: A Route to Fast Inactivation of E. Coli Bacteria, *RSC Adv.* (2015) **5** 42135–42140.

[37] J. Schüring, H.D. Schulz, W.R. Fischer, J. Böttcher and W.H.M. Duijnisveld, Redox: Fundamentals, Processes and Applications, Springer Berlin, Heidelberg (2000) https://doi.org/10.1007/978-3-662-04080-5

[38] R. Bourbonnais, D. Leech and M.G. Paice, Electrochemical Analysis of the Interactions of Laccase Mediators with Lignin Model Compounds, *Biochim. Biophys. Acta* (1998) **1379** (3) 381–390. https://doi.org/10.1016/s0304-4165(97)00117-7

[39] R.E. Ardrey, Liquid Chromatography—Mass Spectrometry: An Introduction, John Wiley & Sons Ltd., Chichester, England (2003).

[40] R.F. Pupo Nogueira, M.C. Oliveira and W.C. Paterlini, Simple and Fast Spectrophotometric Determination of H_2O_2 in Photo-Fenton Reactions Using Metavanadate, *Talanta* (2005) **66** (1) 86–91. https://doi.org/10.1016/j.talanta.2004.10.001

[41] H. Tresp, M.U. Hammer, J. Winter, K.-D. Weltmann and S. Reuter, Quantitative Detection of Plasma-Generated Radicals in Liquids by Electron Paramagnetic Resonance Spectroscopy, *J. Phys. D: Appl. Phys.* (2013) **46** (43) 435401. https://doi.org/10.1088/0022-3727/46/43/435401

[42] V.S.S.K. Kondeti, C.Q. Phan, K. Wende, H. Jablonowski, U. Gangal, J.L. Granick, R.C. Hunter and P.J. Bruggeman, Long-Lived and Short-Lived Reactive Species Produced by a Cold Atmospheric Pressure Plasma Jet for the Inactivation of Pseudomonas Aeruginosa and Staphylococcus Aureus, *Free Radic. Biol. Med.* (2018) **124** 275–287. https://doi.org/10.1016/j.freeradbiomed.2018.05.083

[43] B. He, Y. Ma, X. Gong, Z. Long, J. Li, Q. Xiong, H. Liu, Q. Chen, X. Zhang, S. Yang, and Q.H. Liu, Simultaneous Quantification of Aqueous Peroxide, Nitrate, and Nitrite during the Plasma—Liquid Interactions by Derivative Absorption Spectrophotometry, *J. Phys. D: Appl. Phys.* (2017) **50** (44) 445207. https://doi.org/10.1088/1361-6463/aa8819

[44] R.E. Buehler, J. Staehelin and J. Hoigne, Ozone Decomposition in Water Studied by Pulsed Radiolysis. 1 HO_2/O^{-2} and HO_3/O^{-3} as Intermediates, *J. Phys. Chem.* (1984) **88** (22) 5450. https://doi.org/10.1021/j150666a600

[45] W. Tian, A.M. Lietz, and M.J. Kushner, The Consequences of Air Flow on the Distribution of Aqueous Species During Dielectric Barrier Discharge Treatment of Thin Water Layers, *Plasma Sources Sci. Technol.* (2016) **25** 055020.

[46] S.A. Norberg, W. Tian, E. Johnsen and M.J. Kushner, Atmospheric Pressure Plasma Jets Interacting with Liquid Covered Tissue: Touching and Not-Touching the Liquid, *J. Phys. D: Appl. Phys.* (2014) **44** (44) 475203. https://doi.org/10.1088/0022-3727/47/47/475203

[47] C. Chen, D.X. Liu, Z.C. Liu, A.J. Yang, H.L. Chen, G. Shama and M.G. Kong, A Model of Plasma-Biofilm and Plasma-Tissue Interactions at Ambient Pressure, *Plasma Chem. Plasma Process.* (2014) **34** 403–441. https://doi.org/10.1007/s11090-014-9545-1

[48] C.C.W. Verlackt, W. Van Boxem and A. Bogaerts, Transport and Accumulation of Plasma Generated Species in Aqueous Solution, *Phys. Chem. Chem. Phys.* (2018) **20** (10) 6845–6859. http://doi.org/10.1039/C7CP07593F

[49] Z.C. Liu, L. Guo, D.X. Liu, M.Z. Rong, H.L. Chen and M.G. Kong, Chemical Kinetics and Reactive Species in Normal Saline Activated by a Surface Air Discharge: Chemical Kinetics and Reactive Species in Normal Saline, *Plasma Process. Polym.* (2017) **14** (4–5) 1600113. https://doi.org/10.1002/ppap.201600113

[50] J. Polito, M.J. Herrera Quesada, K. Stapelmann and M.J. Kushner, Reaction Mechanism for Atmospheric Pressure Plasma Treatment of Cysteine in Solution, *J. Phys. D: Appl. Phys.* (2023) **56** (39) 395205. https://doi.org/10.1088/1361-6463/ace196

4 Plasma Effects on Lipids and Proteins
Insights From Non-Reactive Molecular Dynamics Simulations

Maria C. Oliveira, Maksudbek Yusupov, Annemie Bogaerts

4.1 INTRODUCTION

Cold atmospheric plasma (CAP) has emerged as a promising approach to killing cancer cells through oxidative stress, i.e., an overproduction of reactive oxygen and nitrogen species (RONS). Examples of RONS are hydrogen peroxide (H_2O_2), hydroperoxyl radical ($HO_2^{\bullet}$), hydroxyl radical ($HO^{\bullet}$), superoxide radical ($O_2^{\bullet-}$), singlet oxygen ($^1O_2$), nitrogen dioxide ($^{\bullet}NO_2$), nitric oxide radical ($^{\bullet}NO$), and peroxynitrous acid (ONOOH) [1]. When interacting with the surface of cancer cells, RONS may cause the oxidation of membrane lipids and proteins, inducing structural and dynamic changes in the membrane. Consequently, pore formation and membrane rupture may occur, leading to cell death [2,3]. Protein oxidation also has detrimental effects on protein structure and function and contributes to cell death [4,5].

Understanding the interaction of RONS with membrane lipids and proteins, as well as its subsequent effects, would help us to improve the efficacy of CAP treatment and other treatments based on oxidative stress. One promising approach to elucidating these interactions is computer modeling. In this chapter, we will first give an overview of a widely used simulation tool called molecular dynamics (MD) for studying biomolecules at the atomic and molecular levels. Subsequently, we will present typical experimental results obtained from non-reactive MD simulations for membrane lipid and protein model systems. Finally, we will summarize and discuss some of the future challenges of modeling for plasma-cancer treatment.

4.2 MOLECULAR DYNAMICS (MD) SIMULATIONS

MD simulations can sometimes provide detailed information that is inaccessible by current experimental methods, such as information on the RONS concentration required to permeate cancer cells, or the synergistic effects between RONS and other biomolecules present in or around cancer cells.

MD simulations describe the time-dependent behavior of a molecular system, based on the principles of classical mechanics, and provide information about the trajectory of all atoms in the system in accordance with their displacement at different times. To describe the motion of the atoms, it is necessary to know the forces acting on them. First, the initial positions and velocities of all atoms must be defined, and the potential energy (V) and force (F) of each pair of atoms calculated. The potential energy of a system can be described by bonded interactions, i.e., bond, angle, and torsion (proper and improper dihedral angle) potentials, and non-bonded interactions, i.e., Coulomb and Lennard-Jones potentials. The potential energy constitutes a force field, and its

DOI: 10.1201/9781003328056-5

parameters are typically obtained by fitting the atomic or molecular properties of small molecules against calculated quantum-mechanical or experimental data [6]. Many types of force fields are available in the literature, and each of them has specific areas of application and corresponding parameters [7]. Therefore, the choice of force field depends on the system to be studied, the type of information required, and the details of the simulation method. An example of a potential energy expression is:

$$V = \sum_{bonds} \frac{1}{2} k_{ij} \left(r_{ij} - r_0 \right)^2 + \sum_{angles} \frac{1}{2} k_{ijk} \left(\theta_{ijk} - \theta_0 \right)^2 + \sum_{\substack{proper \\ dihedrals}} k_{ijkl} \left[1 + \cos \left(n_{ijkl} \phi_{ijkl} - \phi_0 \right) \right] +$$

$$+ \sum_{\substack{improper \\ dihedrals}} \frac{1}{2} k_{ijkl} \left(\xi_{ijkl} - \xi_0 \right)^2 + \sum_{i<j} \frac{q_i q_j}{4 \pi \varepsilon_0 r_{ij}} + \frac{1}{2} \sum_{i<j} 4 \varepsilon_{ij} \left[\left(\frac{\sigma_{ij}}{r_{ij}} \right)^{12} - \left(\frac{\sigma_{ij}}{r_{ij}} \right)^6 \right]$$

(Eq. 1)

where k_{ij}, k_{ijk}, and k_{ijkl} are the force constants for the harmonic bond, angle, and dihedral potentials, respectively; r_{ij} is the distance between atoms i and j; r_0, θ_0, φ_0, and ξ_0 are the reference bond length, angle, proper dihedral angle, and improper dihedral angle, respectively; θ_{ijk} is the angle between atoms i, j, and k; n_{ijkl} is the multiplicity that indicates the number of times the bond is rotated through 360°; φ_{ijkl} and ξ_{ijkl} are the proper and improper dihedral angles, respectively; ε_0 is the dielectric permittivity in vacuum; q_i and q_j are the partial charges of atoms i and j; ε_{ij} is a measure of how strongly the atoms i and j attract each other; and σ_{ij} is the distance at which the intermolecular potential between atoms i and j is zero.

As the potential energy depends only on the position of the atoms ($\vec{r}_i$), the force of each pair of atoms ($\vec{F}_{ji}$) and the resulting force applied on each atom ($\vec{F}_i$) are given by:

$$\vec{F}_{ji} = - \frac{\partial V \left(\vec{r}_i \right)}{\partial \vec{r}_i} = - \left(\frac{\partial V}{\partial x_i}, \frac{\partial V}{\partial y_i}, \frac{\partial V}{\partial z_i} \right)$$

(Eq. 2)

$$\vec{F}_i = \sum_{j \neq i} \vec{F}_{ji}$$

If we know the force acting on each atom ($\vec{F}_i$), we can calculate the acceleration of each atom ($\vec{a}_i$), applying the Newton's equation of motion:

$$\vec{a}_i = \frac{\vec{F}_i}{m_i}$$

(Eq. 3)

where m_i is the mass of the atom i. Finally, if we integrate the acceleration of each atom over time, we can determine the next velocity of that atom, and if we integrate the velocity over time, we can determine the next position of that atom:

$$\int \vec{a}_i dt = \vec{v}_i \left(t \right)$$

(Eq. 4)

$$\int \vec{v}_i dt = \vec{r}_i(t) \tag{Eq. 5}$$

In summary, an MD simulation mainly consists of an iterative process of obtaining a structure closer to the equilibrated system and, at the end, calculating the macro- and microscopic properties of the system. Just like quantum mechanics simulations, MD simulations have several limitations. The major limitation is the classical treatment of the system, which makes it impossible to access electronic properties of atoms to consider chemical reactions without describing at least part of the system quantum mechanically. However, this limitation does not cause problems in studying the consequences of oxidation in membrane lipids and proteins if one is interested in the dynamics of the system and conformational changes, rather than in the bond-breaking and formation processes during RONS-lipids/proteins interaction. Other methods can be employed to study these processes, such as reactive MD simulations [8,9]. However, in this chapter, we will focus only on the results of non-reactive MD simulations.

4.3 TYPICAL NON-REACTIVE MD SIMULATION RESULTS

In addition to being formed by CAP, RONS are products of normal cellular metabolism, generated within the mitochondria and cytoplasm [10], and since they are reactive species, they will react with other molecules at or near their site of production. At low/moderate concentrations, they are involved in physiological processes and cellular responses such as the defense against infectious agents, the functioning of several cellular signaling pathways, and the induction of mitogenic responses [11]. On the other hand, elevated levels of RONS may lead to injurious oxidative stress and trigger damage to membrane lipids, proteins, and DNA, inhibiting their normal function and contributing to several diseases, such as cancer, rheumatoid arthritis, diabetes, neurological disorders (Alzheimer and Parkinson), and ageing [12].

Under oxidative stress conditions, RONS are able to attack membrane lipids and proteins, inducing lipid oxidation [13] and protein oxidation [14]. Interestingly, evidence suggests a role for lipid oxidation products in the control of cell proliferation and induction of differentiation, maturation, and apoptosis [15,16]. Likewise, protein oxidation products affect the structure, function, and activity of proteins [17,18]. The extent of damage from these events will depend on the duration and intensity of the exposure to the increased RONS environment [19]. Thus, RONS-induced lipid oxidation and protein oxidation are promising anti-cancer therapies based on oxidative stress caused by, for instance, CAP. Below, we discuss typical MD simulation results to investigate the effect of RONS-induced oxidative stress on membrane lipids and proteins for the killing of cancer cells.

4.3.1 LIPID OXIDATION

The major structural lipids in eukaryotic membranes are the phospholipids. They have a nearly cylindrical molecular geometry composed of saturated and *cis*-unsaturated acyl chains, which can become oxidized by specific enzymatic pathways (involving, e.g. lipoxygenase, myeloperoxidase, and NADPH oxidase), or by non-enzymatic pathways involving RONS [20]. The non-enzymatic lipid oxidation induced by RONS can occur in three phases: initiation, propagation, and termination [21], as shown in Figure 4.1. The initiation phase occurs by reaction between RONS and a phospholipid molecule (LH), generating a highly stabilized mesomeric carbon-centered lipid radical (L•) by abstraction of a hydrogen atom from the CH_2 group adjacent to the double bond (Figure 4.1). The propagation phase is cycled through L• in the presence of molecular oxygen (O_2) to generate lipid peroxyl radicals (LOO•), which can react with a phospholipid molecule producing lipid hydroperoxide (LOOH). Many propagation reactions can occur with each initiation reaction. Depending on the degree and location of unsaturation of the lipid undergoing oxidation, a regio- and stereochemical equilibrium mixture of LOO• is formed during propagation, i.e., isomerization of cis double bonds to trans [22]. Furthermore, transition metal ions, such as Fe^{2+} and Cu^+, are able to catalyze the reaction of the LOOH to generate lipid alkoxyl radicals (LO•) in the so-called Fenton reaction [23]. The LO• can react with another LH to produce a lipid alcohol (LOH) and L•; i.e., more oxidized lipids are produced (Figure 4.1). In the termination phase, two LOO• react to generate a linear tetroxide intermediate that decomposes by the Russell mechanism [24] to yield singlet oxygen ($^1O_2$) and non-radical products: lipid ketone (LO) and lipid alcohol (LOH). Additionally, LO• may undergo β-Scission [25] and produce lipid aldehydes (L-ALD) (Figure 4.1).

Computer simulations have demonstrated that lipid oxidation has many effects on membrane properties, inducing structural changes related to the area per lipid, lipid order, bilayer thickness, and bilayer hydration profile [26,27]. For instance, MD simulations have shown that lipid oxidation increases membrane permeability [28,29] and induces pore formation and bilayer disintegration [30,31]. Since cell membranes are barriers that control the movement of substances/molecules into and out of cells and the flow of information between cells (e.g. recognition/sending of signaling molecules from/to other cells), and that are also involved in the capture and release of energy (e.g. photosynthesis and oxidative phosphorylation) [32], any changes in membrane permeability and structure may jeopardize cell function and induce cell death.

Oliveira et al. studied the mechanical instability of oxidized membranes, stretching them until pore formation, in order to access atomistic information about the ease of forming a pore. The results demonstrated that lipid oxidation facilitates pore formation under mechanical stress conditions, contributing to membrane leakage [33]. Previous MD studies also showed that the degree of oxidation strongly influences the time needed for the opening of a pore: a higher degree of oxidation induces spontaneous pore formation in the tens of nanoseconds (10^{-9} seconds) timescale, whereas a lower degree of oxidation induces spontaneous pore formation in the timescale of seconds [34].

Antioxidants have been widely used to protect membranes from free radical-initiated oxidation [12,35]. For example, the antioxidant alpha-tocopherol (vitamin E) reduces pore formation in oxidized membranes, even at low concentrations. At high alpha-tocopherol concentrations, no pores were observed [36]. The authors proposed that alpha-tocopherol traps the polar groups of the oxidized lipids

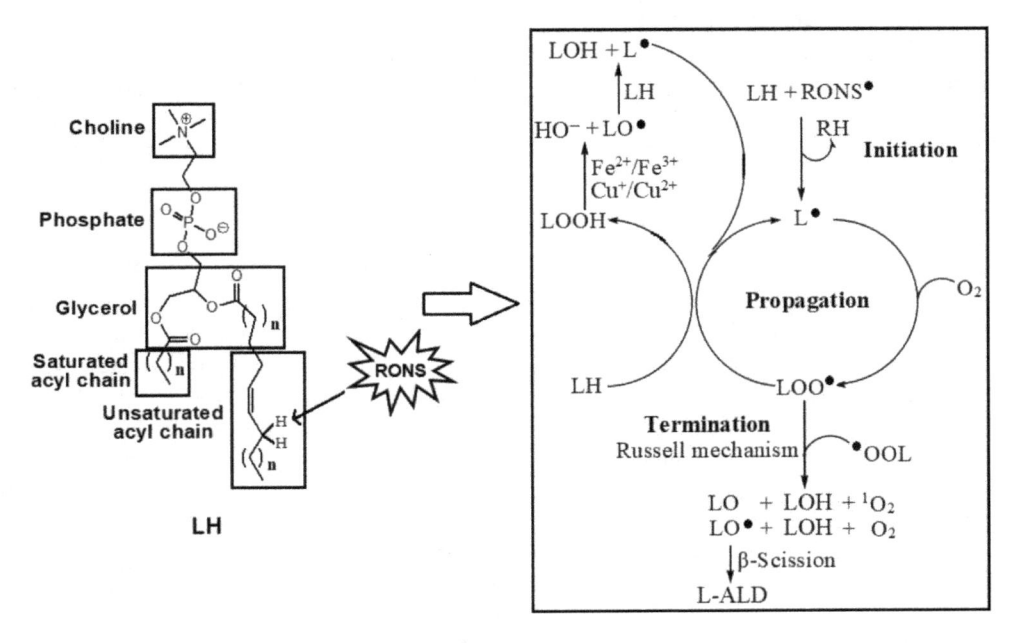

FIGURE 4.1 (left) Schematic representation of the structure of a phospholipid molecule (LH) and the attack of RONS at the hydrogen atom adjacent to the double bond, inducing the formation of lipid oxidation products (right): lipid hydroperoxide (LOOH), lipid alcohol (LOH), lipid ketone (LO), and lipid aldehyde (L-ALD).

at the membrane-water interface, decreasing the probability for the oxidized lipids to initiate pore formation [36]. Thus, alpha-tocopherol not only protects the membrane from oxidation but also helps to stabilize the membrane after lipid oxidation. Additionally, cholesterol also protects the membrane from disruption due to lipid oxidation by interacting with the oxidized chains, filling void spaces and preventing the lipid oxidation-induced damage [37,38]. In another study, Owen et al. used MD simulations to evaluate the effect of cholesterol on oxidized membranes with different concentrations of lipid oxidation products and found that the presence of cholesterol reduces the membrane permeability for water molecules: with increasing lipid oxidation, the number of hydrogen bonds between cholesterol and the acyl chains of oxidized lipids was higher than the number of contacts between water molecules and the acyl chains of oxidized lipids [39]. This finding suggests that cholesterol protects the membrane from pore formation even at high concentrations of lipid oxidation products.

Interestingly, the increased membrane permeability has been correlated with the presence of lipid aldehydes. Runas and Malmstadt observed that even low levels of lipid aldehydes (about 2.5% of the total lipids) were enough to increase the membrane permeability by around two orders of magnitude [40]. Likewise, lipid hydroperoxides also increased membrane permeability, but complete pore formation was observed only in the membrane consisting of lipid aldehydes [41]. Membranes can also organize lipids and proteins into several distinct regions or domains, which serve as rafts for the transport of selected molecules or as relay stations in intracellular signaling [42]. MD simulations demonstrated that the presence of a few lipid aldehydes at the domain interfaces of membranes increases the membrane permeability [43]. These results highlight lipid aldehydes as the major lipid oxidation products that increase membrane permeability and induce pore formation. However, we cannot rule out the effect of other lipid oxidation products: the MD simulations also showed that lipid alcohols decrease the free energy barrier for water permeation [44]. The possible mechanism by which lipid aldehydes increase membrane permeability is linked to their truncated acyl chains: the density of lipid aldehydes in the membrane interior is lower than that of lipid hydroperoxides and alcohols; thus, lipid aldehydes can be engaged in hydrogen bonds with water molecules, creating transient water bridges that may progress into pores. This hypothesis was verified by Runas et al., who found that the addition of cleaved tail groups in oxidized membranes into lipid aldehydes decreased membrane permeability, suggesting that shorter acyl chains increase membrane permeability [45].

Although these MD simulation results highlight the detrimental effects of lipid oxidation on membrane properties, they also suggest the possibility of using RONS-induced lipid oxidation as an anti-cancer therapy in, for instance, CAP treatment. In the next section, we discuss MD simulation results based on CAP experiments to kill cancer cells.

4.3.2 COLD ATMOSPHERIC PLASMA-INDUCED LIPID OXIDATION

Based on the damaging effects of RONS on membrane lipids, proteins, and DNA, CAP treatment has shown powerful capabilities to kill tumor cells both *in vitro* and *in vivo*, by inducing pro-apoptotic factors [46,47]. The possible selectivity of CAP in killing cancer cells is probably due to the noticeable increase in intracellular RONS in cancer cells as compared to normal cells exposed to the same CAP treatment [48]. In principle, cancer cells are probably more permeable to RONS than normal cells, because some of them present fewer cholesterol molecules in their membrane than do normal cells [49]. Cholesterol is embedded in the spaces between phospholipid molecules, accounting for more than 20% of the lipid components of the cell membrane and helping to maintain its fluidity, order, and permeability [50]. Hence, a reduced cholesterol content in cancer cell membranes makes them more prone to RONS permeation. The intracellular RONS level in cancer cells may also be higher than in normal cells due to intrinsic factors, e.g., aberrant metabolism, mitochondrial dysfunction, and loss of functional anti-cancer proteins [51]. Thus, RONS generated during CAP treatment can maximize the selective inactivation of cancer cells with less or no damage to normal cells.

In light of the potential anti-cancer effect of CAP on cancer cells, Yusupov et al. performed MD simulations and experiments using CAP as an external RONS source and found that lipid oxidation increases membrane fluidity [52], which indicates that RONS penetrate into the cell interior easily (e.g., through pores) [3,31,53]. Indeed, lipid oxidation leads to an overall decrease in the free energy barriers for RONS permeation, especially for oxidized membranes into lipid aldehydes: $HO_2^\bullet$ and $HO^\bullet$ radicals are more prone to permeate the membrane than are H_2O_2 molecules [31,53]. Conversely, oxidized membranes of the stratum corneum into lipid hydroperoxides were more permeable to hydrophobic RONS, such as $^\bullet NO$, $^\bullet NO_2$, and N_2O_4 molecules, than hydrophilic RONS, such as $HO_2^\bullet$ and $HO^\bullet$ radicals and H_2O_2 molecules [54]. If RONS permeate the membrane bilayer easily, they are more prone to oxidizing lipids and embedded proteins there [55], as well as to accessing the cell interior easily, thereby increasing damage to DNA and other biomolecules.

The permeation of RONS depends on many intrinsic factors, such as lipid structure and membrane composition. For instance, experimental and MD studies have showed that membranes containing unsaturated lipids are significantly more vulnerable to RONS permeation than are those containing saturated lipids [56]. However, the addition of cholesterol to membranes containing unsaturated lipids makes the membrane less vulnerable to RONS permeation [56]. This result is expected, since RONS are susceptible to reacting with the hydrogen atoms adjacent to the double bonds in lipid tails (see Figure 4.1 above), and cholesterol decreases the membrane fluidity [50]. Regarding the underlying mechanism of RONS permeation across oxidized membranes, another study showed that the addition of lipid aldehydes in membranes promotes H_2O_2 permeation by increasing the partition coefficient of the membrane via intra-molecular configurational change of oxidized lipids [57]. On the other hand, the addition of oxidized lipids with carboxyl groups (i.e., another truncated acyl chain) promotes H_2O_2 permeation by the formation of pores via inter-molecular configurational change of oxidized lipids [57].

Apart from producing RONS, CAP also generates strong electric fields that enhance the penetration of RONS. Synergistic effects between RONS and electric fields have been reported, whereby lipid oxidation facilitated pore formation induced by electric fields (electroporation) [58,59]. For instance, Yusupov et al. [31] and Yadav et al. [60] demonstrated that a higher electric-field strength leads to a shorter time required to induce pore formation, and that this effect is stronger for oxidized membranes. More recently, an MD study evaluated the effect of lipid aldehydes on the permeability of membranes exposed to pulsed electric fields, i.e., electroporation. The authors found that a small "spot" of lipid aldehydes is enough to spontaneously form a large pore (up to ~7 nm diameter) [61]. This finding indicates that the oxidation of only a small area of a cell membrane is sufficient to induce the formation of pores wide enough to transport large molecules such as RONS and drugs. Moreover, the authors also showed that the membrane fluidity and location of the "oxidative spots" can significantly affect the pore's lifespan [61].

All of these studies are important in determining how lipid oxidation affects the membrane permeability, and which CAP-generated RONS are more prone to permeating cell membranes. CAP-generated RONS permeate cell membranes, causing lipid oxidation and increasing membrane permeability, which may lead to membrane rupture and consequently death of cancer cells (Figure 4.2). However, further

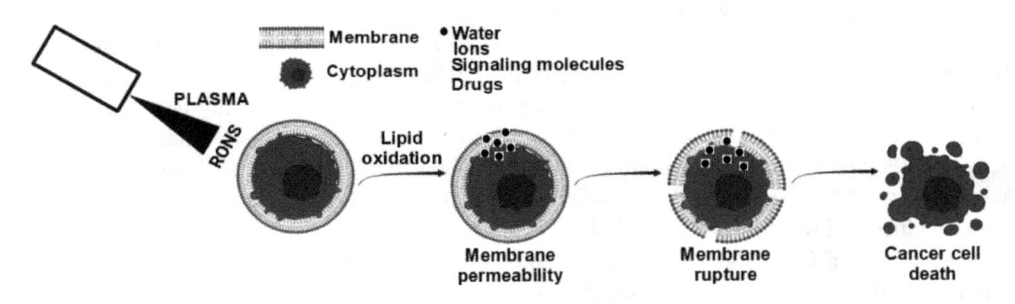

FIGURE 4.2 Schematic representation of how plasma-generated RONS may induce cancer cell death through lipid oxidation.

detailed investigations are needed to gain insights into the cooperative role of RONS-RONS, RONS-lipids, and RONS-proteins interactions in intracellular signaling processes.

4.3.3 PROTEIN OXIDATION

As discussed previously, RONS can oxidize not only lipids, but also embedded proteins in the membrane. About 50% of the membrane's surface of real cells is composed of proteins [62]. Due to their high concentration and high rate constants for reaction with multiple oxidants [4], proteins are the major target for RONS-mediated oxidation, marking them for proteolytic degradation [63]. Most protein damage is irreparable and has deleterious consequences on protein structure and function, contributing to cellular and tissue dysfunction (e.g. apoptosis, necrosis, and altered cell signaling) and a range of human pathologies [64]. The degree of protein damage caused by RONS depends on many factors, including the concentration of the target protein, the rate constant of the reaction between RONS and the target protein, the location of the target protein relative to the site of RONS formation, the occurrence of secondary damaging events (i.e., chain reactions and damage transfer processes), the occurrence of RONS-scavenging reactions, and repair reactions [4].

RONS-mediated oxidation can occur at both the protein backbone and the amino acid side-chains. The factors that might favor or disfavor a reaction at a particular position depends on: the stability of the RONS (tertiary > secondary > primary, and species with potential for electron delocalization); statistical factors (i.e., the number of each particular type of C—H bond/sites of addition); and possible steric and charge interactions [4]. When RONS react with the backbone of proteins, they induce chemical fragmentation of these proteins [65] or the formation of large aggregates of covalently cross-linked proteins (e.g., disulfide bonds (—S—S—)) and non-covalently cross-linked proteins (hydrophobic and electrostatic interactions) [66]. When RONS react with amino acid side-chains, this results in a huge variety of products, including hydroperoxides, alcohols, and carbonyl compounds (aldehydes and carboxylic acids) [67]. Among amino acids, cysteine (Cys) and methionine (Met) are particularly susceptible to oxidation by RONS because of their sulfhydryl (—SH) and thioether (—SCH$_3$) groups, respectively [4,68]. In general, amino acids that are located in buried positions of proteins are less easily oxidized by exogenous RONS than those at the surface [69,70]. Proteins may also be oxidized via mechanisms of secondary products generated by the interaction between RONS and lipids [71].

Jas and Kuczera performed MD simulations to study the effects of C-terminal Met side-chain oxidation on the properties of calmodulin, a small acidic protein that is involved in transmitting the Ca^{2+} signal to a variety of target proteins [70]. They found that Met side-chain oxidation decreased the free energy for the unfolded calmodulin state, i.e., the protein thermal stability. Moreover, oxidation at position 144 enhanced the peptide binding and weakened the Ca^{2+} binding, whereas oxidation at position 145 weakened the peptide binding and enhanced the affinity for Ca^{2+} [70]. Thus, Met residues oxidation should have a destabilizing effect on calmodulin, affecting protein thermal stability, Ca^{2+} affinity, and binding of a target peptide. Subsequent MD simulations demonstrated that Met oxidation may destabilize the binding affinity between the A1 and A2 domains of the blood protein von Willebrand factor, shutting down the auto-inhibitory mechanism that prevents aggregation of platelets to the site of vascular injury [72]. Interestingly, combined MD simulations and experiments demonstrated that Met residues in granulocyte colony-stimulating factor, a therapeutic protein useful in treating neutropenia, have significant and distinct rates of oxidation by H$_2$O$_2$ as a function of pH, and induce unique conformational changes in the protein [73]. These results indicate that oxidation of Met residues affects protein function and structure, and that both the oxidation rate of Met residues and the conformational changes of the protein are pH-dependent.

The oxidation of Cys residues has also been heavily studied both experimentally and with MD simulations [74–76]. For instance, Bock et al. used MD simulations to investigate the effect of oxidation on the formation of a disulfide bond between two Cys residues in the neuronal protein SNAP-25B [74]. They found that Cys residue oxidation destabilizes the SNARE complex, a coiled-coil of proteins composed of SNAP-25B, suggesting that an oxidation-induced conformational change may be a chemomechanical regulator of the SNARE complex formation [74]. Studies have also shown that oxidation of tyrosine residues can induce changes in the structure and stability of proteins, such as protein flexibility, hydrogen bond networks, electrostatic potential at the protein surface, and global energy of protein/molecule–protein interactions [77,78].

In conclusion, protein oxidation has a large impact on conformation of proteins, in both intra- and inter-molecular interactions. The MD simulation results are in qualitative agreement with experimental results, indicating that MD simulations are helpful analytical tools for understanding changes in biological function associated with oxidation of proteins. In the next section, we discuss MD results of protein oxidation based on CAP experiments.

4.3.4 COLD ATMOSPHERIC PLASMA-INDUCED PROTEIN OXIDATION

To determine the underlying mechanisms and effects of protein oxidation on signaling processes and cancer-cell death, experimental and computational studies have investigated the effect of CAP treatment on protein modification/degradation. For instance, a combined experimental and simulation study investigated the effect of plasma-generated RONS on the structure of cytoglobin, a redox-sensitive protein that protects the cell from oxidative stress by acting as a RONS-scavenger (tumor suppression) [79]. The authors found that the treatment of cytoglobin with CAP leads to oxidation of the two surface-exposed Cys residues of cytoglobin (Cys_{38} and Cys_{83}) and subsequently to the formation of inter- and intra-molecular disulfide bonds. This in turn affects the cytoglobin 3D structure, thereby opening access to the heme group through gate functioning of the histidine residue His_{117} [79]. Therefore, plasma-generated-RONS can indeed induce chemical modifications of proteins at specific sites, and the oxidation of extracellular Cys residues leads to distinct conformations compared to those observed under non-oxidative conditions.

Oxidation of extracellular Cys residues indeed plays an important role in the closure of protein channels. Ghasemitarei et al. demonstrated that oxidation of the extracellular Cys_{327} of the xC^- antiporter, a transmembrane protein responsible for translocation of cystine (an amino acid formed by disulfide bonds between two Cys residues) from the extracellular milieu to the cell interior, inhibits extracellular cystine uptake and thereby reduces cellular protection against intracellular oxidative stress, since the transport of cystine into the cell is important [80]. Thus, CAP can also induce cystine starvation in cancer cells through inhibition of the xC^- antiporter, which can eventually lead to cancer-cell death.

Another study evaluated the effect of plasma-generated-RONS on the signaling protein human epidermal growth factor (hEGF) and its receptor hEGFR, both of which are involved in several types of cancer and in wound healing, by considering different oxidation degrees of specific amino acid residues of hEGF [81]. The results revealed that a low degree of oxidation (used to mimic shorter CAP treatment times) did not cause significant structural changes in hEGF and had a diminished impact on the hEGF–hEGFR interaction, which is favorable in wound healing. On the other hand, a higher oxidation degree (used to mimic longer CAP treatment times) induced conformational changes and reduced the binding free energy between hEGF and hEGFR, which may inhibit cell growth or proliferation of cancer cells, potentially leading to apoptosis or even necrosis [81].

CAP experiments combined with MD simulations also demonstrated that the oxidation of the cell surface adhesion receptor protein CD44 and hyaluronan, both expressed in cancer cells, reduces the CD44–HA binding affinity, which can hinder cancer progression [82]. Protein oxidation has also been reported to induce protein denaturation and distortion after CAP treatment [83,84], demonstrating that proteins become unstable due to amino acid oxidation and are therefore more likely to influence cell function.

Interestingly, protein oxidation may also increase the efficacy of tumor suppressor proteins. In a study by Attri et al., the oxidation of the murine double minute 2 (Mdm2) protein, which is known to inhibit activation of the anti-cancer protein p53 by formation of the Mdm2–p53 complex, resulted in high flexibility of the Mdm2 that might disturb the binding with p53 [85]. This promising result suggests that the oxidation of Mdm2 might decrease the binding energy of the Mdm2–p53 complex, resulting in the availability of p53 to increase CAP-induced cancer cell death by intracellular accumulation of RONS. Likewise, protein oxidation also reduced the binding energy between CD47, a transmembrane protein with immunosuppressive functions that is overexpressed in cancer cells, and its receptor SIRPα. Consequently, this reduction in binding energy weakened the immunosuppressive signals on tumor cells, making them more susceptible to being killed by plasma-generated RONS [86]. Therefore, disruption of the protein-protein interactions involved in cancer progression is also a target for CAP-induced cancer cell death.

CAP has also anti-microbial potential, thanks to the interaction between RONS and microorganisms (viruses, bacteria, and fungi) [87,88]. In principle, RONS trigger chain reactions to disrupt the metabolism of the microorganisms, resulting in cell membrane damage, leakage of intracellular macromolecules, enzyme inactivation, and DNA fragmentation [89,90]. For instance, the oxidation of Cys residues of the C-terminal receptor-binding domain (RBD) of the SARS-CoV-2 S-glycoprotein (a structural protein responsible for the attachment and entry of coronavirus to host cells) reduced the non-bonded interaction energy between RBD and the receptor protein GRP78, inhibiting the formation of the RBD–GRP78 complex. Thus, RONS-induced RBD oxidation may inhibit the entry of SARS-CoV-2 virus to the cell [91]. RONS-induced protein oxidation was also able to destabilize/modify the MHT protein in thermophilic bacteria, i.e., bacterial protein that is resistant to temperature and chemical denaturant [92]. This finding shed light on the effect of RONS on proteins for inactivating thermophilic bacteria, which are very difficult to destabilize and thus a common problem in the food industry.

In general, plasma-generated RONS induce structural modifications in proteins, which may result in significant cytotoxic effects on cancer cells (Figure 4.3), viruses, bacteria, and fungi. However, more investigations are required, e.g., to evaluate the effects of RONS-protein interaction under the presence of other biomolecules and proteins.

4.4　CONCLUSION AND FUTURE CHALLENGES

The fluidity, functionality, and intrinsic properties of membrane lipids and proteins make them a potential target for plasma-based oxidative therapies such as CAP. The damage caused by lipid and protein oxidation has a large impact on the structure and function of lipids and proteins and may increase the intracellular RONS accumulation to kill cancer cells through oxidative stress. This damage includes an increase in membrane permeability, pore formation/membrane rupture, conformational changes in proteins, and channel opening/closing of proteins. Plasma-lipids and

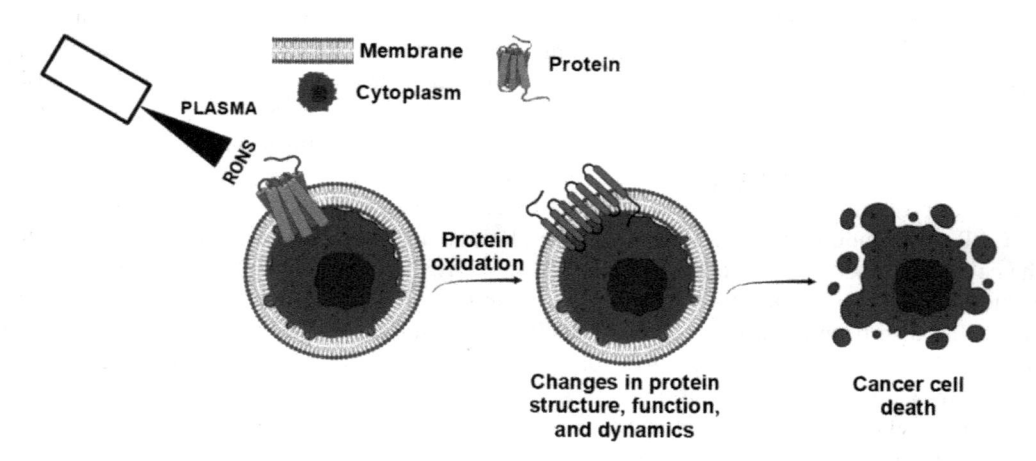

FIGURE 4.3　Schematic representation of how plasma-generated RONS may induce cancer cell death through protein oxidation.

plasma-proteins interactions are key determinants of the cell's fate after plasma treatment. Thus, it is of great interest to improve the understanding of how plasma interacts with membrane lipids and proteins.

RONS may also induce the nitration of lipids and proteins. For instance, the tumor-suppressive protein cytoglobin undergoes nitration of the heme group with longer CAP treatment time, through the binding of $\bullet NO_2$ to the vinyl groups [79]. Tyrosine residues of transmembrane proteins and lipoproteins are also susceptible to nitration mediated by $\bullet NO$ and $\bullet NO_2$. These modifications induce significant changes in protein structure and function [93]. Besides proteins, lipids have also been linked to undergo nitration, affecting lipid organization [94] and inducing a series of downstream cellular effects [95]. Interestingly, the MD simulations of Oliveira et al. demonstrated that the permeability of nitrated membranes is higher than that of oxidized membranes: the water permeability for nitrated lipids increased by three-fold compared to oxidized lipids [96]. Later, the authors found that the free energy barrier for RONS permeation is generally lower across nitrated membranes than across oxidized membranes, and hydrophilic RONS, such as $HO_2\bullet$ and ONOOH, are more prone to penetrating into nitrated membranes [97]. These results highlight nitrated proteins and lipids as novel potential targets for improving our knowledge about RONS-induced cancer cell death. However, there are few studies of nitrated proteins and lipids, and more in-depth research is needed to investigate these products.

To summarize, plasma-based cancer therapy has proven to be a powerful tool for inducing cell death based on lipid and protein oxidation. Moreover, the formation of secondary products, such as nitrated lipids and proteins, may increase the efficacy of the treatment. However, RONS-lipids and RONS-proteins interactions are only a small part of the complex process of cancer cell death; we must also consider the effect of RONS-RONS, protein-protein, protein-lipid, and lipid-lipid interactions, as well as the presence of other biomolecules around or within the cancer cell. Understanding the underlying mechanisms of these processes would improve the efficacy of not only CAP but also other pro-oxidant therapies, such as photodynamic therapy [98] and sonoporation [99].

REFERENCES

[1] D. B. Graves, *The emerging role of reactive oxygen and nitrogen species in redox biology and some implications for plasma applications to medicine and biology*, J. Phys. D: Appl. Phys. 45, 263001 (2012).

[2] R. Itri, et al., *Membrane changes under oxidative stress: The impact of oxidized lipids*, Biophys. Rev. 6, 47–61 (2014).

[3] A. Bogaerts, et al., *Plasma for cancer treatment: How can RONS penetrate through the cell membrane? Answers from computer modeling*, Front. Chem. Sci. Eng. 13, 253–263 (2019).

[4] M. J. Davies, *The oxidative environment and protein damage*, Biochim. Biophys. Acta-Proteins Proteomics 1703, 93–109 (2005).

[5] R. Clemen and S. Bekeschus, *Oxidatively modified proteins: Cause and control of diseases*, Appl. Sci. 10, 6419 (2020).

[6] Y. Duan, et al., *A point-charge force field for molecular mechanics simulations of proteins based on condensed-phase quantum mechanical calculations*, J. Comput. Chem. 24, 1999–2012 (2003).

[7] A. D. Mackerell Jr, *Empirical force fields for biological macromolecules: Overview and issues*, J. Comput. Chem. 25, 1584–1604 (2004).

[8] K. Chenoweth, et al., *ReaxFF reactive force field for molecular dynamics simulations of hydrocarbon oxidation*, J. Phys. Chem. A. 112, 1040–1053 (2008).

[9] A. Bogaerts, et al., *Reactive molecular dynamics simulations for a better insight in plasma medicine*, Plasma Proc. Polym. 11, 1156–1168 (2014).

[10] E. Cadenas and K. J. Davies, *Mitochondrial free radical generation, oxidative stress, and aging*, Free Radic. Biol. Med. 29, 222–230 (2000).

[11] W. Dröge, *Free radicals in the physiological control of cell function*, Physiol. Rev. 82, 47–95 (2002).

[12] M. Valko, et al., *Free radicals and antioxidants in normal physiological functions and human disease*, Int. J. Biochem. Cell Biol. 39, 44–84 (2007).

[13] S. T. Kodali, et al., *Chapter 5: Oxidative lipidomics: Analysis of oxidized lipids and lipid peroxidation in biological systems with relevance to health and disease*, in: L. J. Berliner and N. L. Parinandi (Ed.), Measuring Oxidants and Oxidative Stress in Biological Systems, Springer, 2020.

[14] T. Grune, et al., *Chapter 1: Oxidative stress and protein oxidation*, in: T. Grune, B. Catalgol, T. Jung and V. N. Uversky (Ed.), Protein Oxidation and Aging, John Wiley & Sons, Inc., 2012, pp. 1–214.

[15] J. M. Mates and F. M. Sanchez-Jimenez, *Role of reactive oxygen species in apoptosis: Implications for cancer therapy*, Int. J. Biochem. Cell Biol. 32, 157–170 (2000).

[16] P. Cejas, et al., *Implications of oxidative stress and cell membrane lipid peroxidation in human cancer (Spain)*, Cancer Causes Control 15, 707–719 (2004).

[17] D. I. Pattison, et al., *Photo-oxidation of proteins*, Photochem. Photobiol. Sci. 11, 38–53 (2012).

[18] M. J. Davies, *Protein oxidation and peroxidation*, Biochem. J. 473, 805–825 (2016).

[19] Q. Kong, et al., *A threshold concept for cancer therapy*, Med. Hypoth. 55, 29–35 (2000).

[20] G. O. Fruhwirth, et al., *Oxidized phospholipids: From molecular properties to disease*, Biochim. Biophys. Acta-Mol. Basis Dis. 1772, 718–736 (2007).

[21] T. A. Dix and J. Aikens, *Mechanisms and biological relevance of lipid peroxidation initiation*, Chem. Res. Toxicol. 6, 2–18 (1993).

[22] K. M. Schaich, *Lipid oxidation: New perspectives on an old reaction*, in: F. Shahidi (Ed.), Bailey's Industrial Oil and Fat Products, Wiley, 2005, pp. 269–355.

[23] H. B. Dunford, *Oxidations of iron (II)/(III) by hydrogen peroxide: From aquo to enzyme*, Coord. Chem. Rev. 233–234, 311–318 (2002).

[24] G. A. Russell, *Deuterium-isotope effects in the autoxidation of aralkyl hydrocarbons: Mechanism of the interaction of peroxy radicals*, J. Am. Chem. Soc. 79, 3871–3877 (1957).

[25] H. W. Gardner, *Oxygen radical chemistry of polyunsaturated fatty acids*, Free Radic. Biol. Med. 7, 65–86 (1989).

[26] L. Beranova, et al., *Oxidation changes physical properties of phospholipid bilayers: Fluorescence spectroscopy and molecular simulations*, Langmuir 26, 6140–6144 (2010).

[27] A. J. P. Neto and R. M. Cordeiro, *Molecular simulations of the effects of phospholipid and cholesterol peroxidation on lipid membrane properties*, Biochim. Biophys. Acta-Biomembr. 1858, 2191–2198 (2016).

[28] J. Wong-ekkabut, et al., *Effect of lipid peroxidation on the properties of lipid bilayers: A molecular dynamics study*, Biophys. J. 93, 4225–4236 (2007).

[29] J. Van der Paal, et al., *Effect of lipid peroxidation on membrane permeability of cancer and normal cells subjected to oxidative stress*, Chem. Sci. 7, 489–498 (2016).

[30] L. Cwiklik and P. Jungwirth, *Massive oxidation of phospholipid membranes leads to pore creation and bilayer disintegration*, Chem. Phys. Lett. 486, 99–103 (2010).

[31] M. Yusupov, et al., *Synergistic effect of electric field and lipid oxidation on the permeability of cell membranes*, Biochim. Biophys. Acta-Gen. Sub. 1861, 839–847 (2017).

[32] H. Watson, *Biological membranes*, in: P. K. Robinson (Ed.), Volume 59: Understanding Biochemistry: Enzymes and Membranes, Portland Press, 2015, pp. 43–69.

[33] M. C. Oliveira, et al., *Molecular dynamics simulations of mechanical stress on oxidized membranes*, Biophys. Chem. 254, 106266 (2019).

[34] M. Lis, et al., *The effect of lipidoxidation on the water permeability of phospholipids bilayers*, Phys. Chem. Chem. Phys. 13, 17555–17563 (2011).

[35] M. Parcheta, et al., *Recent developments in effective antioxidants: The structure and antioxidant properties*, Materials 14, 1984 (2021).

[36] P. Boonnoy, et al., *Alpha-tocopherol inhibits pore formation in oxidized bilayers*, Phys. Chem. Chem. Phys. 19, 5699–5704 (2017).

[37] M. Štefl, et al., *Comprehensive portrait of cholesterol containing oxidized membrane*, Biochim. Biophys. Acta-Biomembr. 1838, 1769–1776 (2014).

[38] H. Khandelia, et al., *Pairing of cholesterol with oxidized phospholipid species in lipid bilayers*, Soft Matter 10, 639–647 (2014).

[39] M. C. Owen, et al., *Cholesterol protects the oxidized lipid bilayer from water injury: An all-atom molecular dynamics study*, J. Membr. Biol. 251, 521–534 (2018).

[40] K. A. Runas and N. Malmstadt, *Low levels of lipid oxidation radically increase the passive permeability of lipid bilayers*, Soft Matter 11, 499–505 (2015).

[41] P. Boonnoy, et al., *Bilayer deformation, pores, and micellation induced by oxidized lipids*, J. Phys. Chem. Lett. 6, 4884–4888 (2015).

[42] K. Simons and D. Toomre, *Lipid rafts and signal transduction*, Nat. Rev. Mol. Cell Biol. 1, 31–39 (2000).

[43] M. C. Oliveira, et al., *Lipid oxidation: Role of membrane phase-separated domains*, J. Chem. Inf. Model. 61, 2857–2868 (2021).

[44] I. O. L. Bacellar, et al., *Photosensitized membrane permeabilization requires contact-dependent reactions between photosensitizer and lipids*, J. Am. Chem. Soc. 140, 9606–9615, 2018.

[45] K. A. Runas, et al., *Addition of cleaved tail fragments during lipid oxidation stabilizes membrane permeability behavior*, Langmuir 32, 779–786 (2016).

[46] A. Lin, et al., *Non-thermal plasma as a unique delivery system of short-lived reactive oxygen and nitrogen species for immunogenic cell death in melanoma cells*, Adv. Sci. 6, 1802062 (2019).

[47] T-H. Chung, et al., *Cell electropermeabilisation enhancement by non-thermal-plasma-treated PBS*, Cancers 12, 219 (2020).

[48] S. J. Kim, et al., *Production of intracellular reactive oxygen species and change of cell viability induced by atmospheric pressure plasma in normal and cancer cells*, Appl. Phys. Lett. 103, 153705 (2013).

[49] J. Van der Paal, et al., *Hampering effect of cholesterol on the permeation of reactive oxygen species through phospholipids bilayer: Possible explanation for plasma cancer selectivity*, Sci. Rep. 7, 39526 (2017).

[50] W. K. Subczynski, et al., *High cholesterol/low cholesterol: Effects in biological membranes: A review*, Cell Biochem. Biophys. 75, 369–385 (2017).

[51] R. A. Cairns, et al., *Regulation of cancer cell metabolism*, Nat. Rev. Cancer 11, 85–95 (2011).

[52] M. Yusupov, et al., *Effect of head group and lipid tail oxidation in the cell membrane revealed through integrated simulations and experiments*, Sci. Rep. 7, 5761 (2017).

[53] J. Razzokov, et al., *Atomic scale understanding of the permeation of plasma species across native and oxidized membranes*, J. Phys. D: Appl. Phys. 51, 365203 (2018).

[54] J. Duan, et al., *The penetration of reactive oxygen and nitrogen species across the stratum corneum*, Plasma Proc. Polym. 17, e2000005 (2020).

[55] M. Yusupov, et al., *Transport of reactive oxygen and nitrogen species across aquaporin: A molecular level picture*, Oxid. Med. Cellul. Long. 2019, 2930504 (2019).

[56] J. Van der Paal, et al., *How membrane lipids influence plasma delivery of reactive oxygen species into cells and subsequent DNA damage: An experimental and computational study*, Phys. Chem. Chem. Phys. 21, 19327–19341 (2019).

[57] Y. Ouchi, et al., *Role of oxidized lipids in permeation of H_2O_2 through a lipid membrane: Molecular mechanism of an inhibitor to promoter switch*, Sci. Rep. 9, 12497 (2019).

[58] P. T. Vernier, et al., *Electroporating fields target oxidatively damaged areas in the cell membrane*, PLoS One 4, e-7966 (2009).

[59] T. Kaneko, et al., *Improvement of cell membrane permeability using a cell-solution electrode for generating atmospheric-pressure plasma*, Biointerphases 10, 029521 (2015).

[60] D. K. Yadav, et al., *Electric-field-induced electroporation and permeation of reactive oxygen species across a skin membrane*, J. Biomol. Struct. Dyn. 39, 1343–1353 (2020).

[61] D. Wiczew, et al., *Molecular dynamics simulations of the effects of lipid oxidation on the permeability of cell membranes*, Bioelectrochemistry 141, 107869 (2021).

[62] A. D. Dupuy and D. M. Engelman, *Protein area occupancy at the center of the red blood cell membrane*, Proc. Natl. Acad. Sci. USA 105, 2848–2852 (2008).

[63] T. Grune, et al., *Degradation of oxidized proteins in mammalian cells*, FASEB J. 11, 526–534 (1997).

[64] A. Garcia-Garcia, et al., *Biomarkers of protein oxidation in human disease*, Curr. Mol. Med. 12, 681–697 (2012).

[65] W. M. Garrison, *Reaction mechanisms in the radiolysis of peptides, polypeptides, and proteins*, Chem. Rev. 87, 381–398 (1987).

[66] K. J. Davies, et al., *Protein damage and degradation by oxygen radicals*, J. Biol. Chem. 262, 9914–9920 (1987).

[67] C. L. Hawkins and M. J. Davies, *Generation and propagation of radical reactions on proteins*, Biochim. Biophys. Acta-Bioenerg. 1504, 196–219 (2001).

[68] G. Xu and M. R. Chance, *Hydroxyl radical-mediated modification of proteins as probes for structural proteomics*, Chem. Rev. 107, 3514–3543 (2007).

[69] S. W. Griffiths and C. L. Cooney, *Relationship between protein structure and methionine oxidation in recombinant human αl-antitrypsin*, Biochemistry 41, 6245–6252 (2002).

[70] G. S. Jas and K. Kuczera, *Free-energy simulations of the oxidation of c-terminal methionines in calmodulin*, Proteins 48, 257–268 (2002).

[71] A. Gęgotek and E. Skrzydlewska, *Biological effect of protein modifications by lipid peroxidation products*, Chem. Phys. Lipids 221, 46–52 (2019).

[72] R. Tsai and G. Interlandi, *Oxidation shuts down an auto-inhibitory mechanism of von Willebrand factor*, Proteins 89, 731–741 (2021).

[73] J-W. Chu, et al., *Molecular dynamics simulations and oxidation rates of methionine residues of granulocyte colony-stimulating factor at different pH values*, Biochemistry 43, 1019–1029 (2004).

[74] L. V. Bock, et al., *Chemomechanical regulation of SNARE proteins studied with molecular dynamics simulations*, Biophys. J. 99, 1221–1230 (2010).

[75] P. Marinelli, et al., *A single cysteine post-translational oxidation suffices to compromise globular proteins kinetic stability and promote amyloid formation*, Redox Biol. 14, 566–575 (2018).

[76] J. B. Behring, et al., *Spatial and temporal alterations in protein structure by EGF regulate cryptic cysteine oxidation*, Sci. Signal. 13, eaay7315 (2020).

[77] S. Mazier, et al., *Radiation damage to a DNA-binding protein: Combined circular dichroism and molecular dynamics simulation analysis*, Radiat. Res. 170, 604–612 (2008).

[78] S. Aci-Sèche, et al., *Comparing native and irradiated E: Coli lactose repressor-operator complex by molecular dynamics simulation*, Eur. Biophys. J. 39, 1375–1384 (2010).

[79] J. De Backer, et al., *The effect of reactive oxygen and nitrogen species on the structure of cytoglobin: A potential tumor suppressor*, Redox Biol. 19, 1–10 (2018).

[80] M. Ghasemitarei, et al., *Effect of oxidative stress on cystine transportation by xC^- antiporter*, Arch. Biochem. Biophys. 674, 108114 (2019).

[81] M. Yusupov, et al., *Impact of plasma oxidation on structural features of human epidermal growth factor*, Plasma Proc. Polym. 15, e1800022 (2018).

[82] M. Yusupov, et al., *Oxidative damage to hyaluronan-CD44 interactions as an underlying mechanism of action of oxidative stress-inducing cancer therapy*, Redox Biol. 43, 101968 (2021).

[83] P. Attri, et al., *Structural modification of NADPH oxidase activator (Noxa 1) by oxidative stress: An experimental and computational study*, Int. J. Biol. Macromol. 163, 2405–2414 (2020).

[84] P. Attri, et al., *Plasma treatment causes structural modifications in lysozyme, and increases cytotoxicity towards cancer cells*, Int. J. Biol. Macromol. 182, 1724–1736 (2021).

[85] P. Attri, et al., *Impact of reactive oxygen and nitrogen species produced by plasma on Mdm2—p53 complex*, Int. J. Mol. Sci. 22, 9585 (2021).

[86] A. Lin, et al., *Oxidation of innate immune checkpoint CD47 on cancer cells with non-thermal plasma*, Cancers 13, 579 (2021).

[87] G. Daeschlein, et al., *Skin and wound decontamination of multidrug-resistant bacteria by cold atmospheric plasma coagulation*, JDDG 13, 143–149 (2015).

[88] B. Boekema, et al., *Antibacterial and safety tests of a flexible cold atmospheric plasma device for the stimulation of wound healing*, Appl. Microbiol. Biotechnol. 105, 2057–2070 (2021).

[89] H. Lu, et al., *Bacterial inactivation by high-voltage atmospheric cold plasma: Influence of process parameters and effects on cell leakage and DNA*, J. Appl. Microbiol. 116, 784–794 (2014).

[90] X. Y. Liao, et al., *Nonthermal plasma induces the viable-but-nonculturable state in staphylococcus aureus via metabolic suppression and the oxidative stress response*, Appl. Environ. Microbiol. 86, e02216–e02219 (2020).

[91] M. Ghasemitarei, et al., *Effect of cysteine oxidation in SARS-CoV-2 receptor-binding domain on its interaction with two cell receptors: Insights from atomistic simulations*, J. Chem. Inf. Model. 62, 129–141 (2022).

[92] P. Attri, et al., *CAP modifies the structure of a model protein from thermophilic bacteria: Mechanisms of CAP-mediated inactivation*, Sci. Rep. 8, 10218 (2018).

[93] R. Radi, *Protein tyrosine nitration: Biochemical mechanisms and structural basis of functional effects*, Acc. Chem. Res. 46, 550–559 (2013).

[94] J. Franz, et al., *Nitrated fatty acids modulate the physical properties of model membranes and the structure of transmembrane proteins*, Chem. Eur. J. 23, 9690–9697 (2017).

[95] S. Duarte, et al., *Insight into the cellular effects of nitrated phospholipids: Evidence for pleiotropic mechanisms of action*, Free Radical Biol. Med. 144, 192–202 (2019).

[96] M. C. Oliveira, et al., *How do nitrated lipids affect the properties of phospholipid membranes?* Arch. Biochem. Biophys. 695, 108548 (2020).

[97] M. C. Oliveira, et al., *Unraveling the permeation of reactive species across nitrated membranes by computer simulations*, Comput. Biol. Med. 136, 104768 (2021).

[98] L. Huang, et al., *Photodynamic therapy for hypoxic tumors: Advances and perspectives*, Coord. Chem. Rev. 438, 213888 (2021).

[99] J. M. Escoffre, et al., *New insights on the role of ROS in the mechanisms of sonoporation-mediated gene delivery*, Ultrason. Sonochem. 64, 104998 (2020).

5 Amino Acid, Peptide, and Protein Modifications by Plasma-Driven Processes

Katharina Stapelmann, María J. Herrera Quesada,
Sander Bekeschus, Kristian Wende

5.1 INTRODUCTION

Biomolecules, or biological molecules, are defined as any molecule produced by living organisms. They have a wide range of sizes and structures, perform a vast array of functions, and are usually subdivided into four main groups: lipids, carbohydrates, nucleic acids, and proteins. Biomolecules represent the majority of matter in cells, tissues, and the whole organism—besides water. In the context of redox biology, polar lipids (see Chapter 4) and proteins are of major interest. This chapter will focus on the plasma-driven modifications observed in proteins, peptides, and their building blocks, the amino acids. Proteins consist of carbon (C), hydrogen (H), oxygen (O), nitrogen (N), and sulfur (S) in a ratio of $C_1H_{1.58}N_{0.28}O_{0.30}S_{0.01}$ [1]. With few exceptions, twenty canonical amino acids and two common peculiarities (selenocysteine, pyrrolysine) are the conserved monomers all species on earth use. Given the potentially large numbers of amino acids in a protein—for example, the human muscle protein titin consists of 34,500—their free combination and further modification during cellular protein synthesis and reprocessing allows the conservative assumption that a human cell contains up to 1,000,000 different proteins (proteoforms), some present in only a single copy. Given the variety of chemical groups present in the amino acids and hence the proteins, these molecules represent a substantial target for enzymatic and chemical reactions, yielding post-translational modifications. Extensive research has been carried out on such modifications [2–7]. In the following, the concept of plasma-induced modifications of amino acids, peptides, and proteins, with a special focus on but not limited to thiol group oxidation [8], is presented and discussed.

5.2 COLD PLASMAS AS SOURCE FOR CHEMICALLY ACTIVE COMPONENTS

Cold non-equilibrium plasmas produce a variety of active components such as light, electromagnetic fields, free electrons, and neutral or charged chemical species [9–11]. Of particular interest in terms of protein modifications are the emitted reactive oxygen and nitrogen species (RONS) and light, especially in the form of ultraviolet radiation [12]. A number of processes contribute to the formation of RONS: primary processes in the active plasma zone, secondary processes in the transition zone between the active plasma and the ambient conditions, and processes at the interface between the gaseous phase and a target (liquid or tissue). Since the life times of the species may vary from a few μs to hours and days, downstream reactions between the species occur, and species-specific transport mechanisms modulate the actual pattern. Hence, a spatially and temporally dynamic multi-RONS pattern evolves. These patterns are distinct, depending on the plasma source used, the gas in which the plasma is ignited, and the path from the location of RONS generation to the target. The quantity and variety of RONS can be influenced by these factors and controlled by modulating plasma device settings. Once the ROS mixture reaches the target, chemical reactions with the biomolecules occur. Notably, the broadest RONS spectrum is present at the discharge-air interface, and the variety declines with increasing spatial or temporal distances. After the plasma device treating tissue or liquid is switched off, the short-lived species and all other plasma effects, such as UV radiation, are lost [13, 14] (see also Chapter 2).

Plasmas produce RONS similar to those involved in redox biology processes, including hydrogen peroxide (H_2O_2), superoxide anion radicals ($\bullet O_2^-$), singlet oxygen ($^1O_2$), hydroxyl radicals ($\bullet OH$), nitric oxide ($NO\bullet$), nitrogen dioxide ($\bullet NO_2$), peroxynitrite/peroxynitrate ions ($ONOO^-/O_2NOO-$), nitrite/nitrate ions (NO_2^-/NO_3^-), and protons (H^+/H_3O^+) [12]. In addition, plasma can produce ozone (O_3), atomic oxygen (O), and some higher oxidation products of nitrogen (e.g., N_2O_5), the latter of which are not produced in physiological or pathological processes in the human body [15]. In liquids or liquid-like systems such as hydrogels or tissues, the different types of RONS are not equal in their life times, chemical reactivity, and target selectivity, allowing a wide variety of chemical reactions. Short-lived reactive species such as atomic oxygen, $^1O_2$, or $\bullet OH$ are quenched quickly by relaxation or reaction with a partner and are active only in close proximity to their site of generation, such as the interface between the gas phase and a liquid target. In contrast, under physiological conditions, long-lived reactive species such as H_2O_2, NO_2^-/NO_3^- and, to some extent, $ONOO-$ can diffuse longer distances into the plasma-treated target or persist in a treated liquid, thereby extending the range of activity in time and space. Yet these species have limited chemical reactivity. In contrast, hypochlorite ions (OCl^-), which are strong oxidants formed by the reaction between $Cl-$ (if present) and O, mainly at the gas-liquid interface [16, 17], have an intermediate life time and reactivity and can migrate into the bulk of a liquid or a tissue. In addition, plasma-generated UV radiation contributes to the formation of reactive species via secondary processes, e.g., the photolytic cleavage of water or H_2O_2 yielding $\bullet OH$ (photo-Fenton). Furthermore, UV light can trigger photochemical reactions by exciting electrons of chemical bonds yielding to replacement and cleavage reactions [18] (see also chapters 2 and 3).

5.3   ROS-DRIVEN MODIFICATIONS OF PROTEINS

Whether generated by plasma or physiological processes, RONS affect peptides and proteins, causing either reversible (transient) or irreversible (permanent) changes in the molecular composition [19–21]. These changes alter the activity of a protein or its recognition and can be detected by downstream signal transducers. The sensitivity of the different amino acids varies widely, and electron-rich residues or atoms are the main targets for RONS-driven modifications. In particular, proteins containing sulfur in the form of the thiol(-SH)-containing amino acid cysteine or the thioether (R-S-R) methionine are prone to oxidation or nitrosylation [22, 23]. All major biological processes are partially influenced or controlled by the redox state of protein-bound cysteine residues. The tripeptide glutathione (GSH), for example, is the major cellular scavenger molecule relevant to maintaining functional cellular redox signaling [24]. Another example is the extracellular signaling protein high mobility group box 1 (HMGB1), which contains three redox-sensitive cysteine residues [25]. Depending on the dynamic redox state of each cysteine residue, the protein fulfills different functions such as, e.g., inflammatory signaling. The concept of regulating protein function by chemical modifications of amino acids is not limited to cysteine; other amino acids are also targeted by oxidation to modify biological processes. For instance, the hydroxylation of the protein hypoxia-inducible factor-1 alpha (HIF-1α) at the proline residues Pro402 and Pro564 controls downstream signaling that in normoxic conditions leads to the degradation of the protein, whereas in hypoxic conditions, a regulator protein binds to HIF-1α and helps the cell adapt to reduced oxygen levels [26]. Hence, introducing new chemical groups into a protein can change its function and recognition by, e.g., cellular or soluble receptors.

5.4   IMPACT OF PLASMA ON AMINO ACIDS WITH A FOCUS ON CYSTEINE

Initially, the aim of plasma bio-sciences was to understand plasma-driven liquid chemistry in water or aqueous buffers, and early work focused on investigating single amino acids. With growing understanding and awareness of the importance of plasma-driven biomolecule modifications, more complex models using small peptides and isolated proteins were investigated in experiments and model simulations.

Takai et al. performed an early in-depth study using high-resolution mass spectrometry to elucidate the chemical effects of plasma on the twenty canonical amino acids in aqueous solutions. Degradation was observed in the following order: Methionine (Met) > Cysteine (Cys) > Tryptophan (Trp) > Phenylalanine (Phe) > Tyrosine (Tyr) > Histidine (His) > others. These results showed that sulfur-containing (Cys and Met) and aromatic (Tyr, Phe, and Trp) amino acids are preferentially modified by plasma treatment [22]. The results were corroborated and expanded upon in further studies, showing that plasma-induced modifications mainly occur on the side chains of the amino acids and can be categorized into five general categories: oxidation, dimerization, hydroxylation, nitration, and dehydrogenation [27, 28]. Amino acid modifications following plasma treatment occur predominantly on sulfur-containing and aromatic amino acids, with basic, non-polar, and polar amino acids being modified to a lesser extent [29]. Due to their relevance in redox biology [20, 30, 31] and their propensity to plasma-induced modifications, cysteine modifications have been investigated in depth [15, 23, 28, 32, 33]. Most common is the oxidation of the thiol group to sulfinic acid ($R-SO_2H$) and sulfonic acid ($R-SO_3H$) and the break of the C-S bond yielding sulfite ions (SO_3^{2-}) or sulfate ions (SO_4^{2-}) and dehydroalanine. The initial oxidation product sulfenic acid (R-SOH), believed to play a key role in redox signaling, could not be detected after plasma treatment due to its instability (Figure 5.1) [28, 33]. If plasma processes that produce fewer strongly oxidizing species or plasma-treated liquids void of short-lived reactive species are used, the dimerization of cysteine forming the disulfide is dominant (cystine, RSSR). Cystine contains a disulfide bridge, a key structural element essential for protein structure and function [20]. Cystine, a sensitive sensor of the redox state of a cell, can be further oxidized or nitrosylated, leading to bioactive derivates that have anti-proliferative effects on cancer cells [28, 30, 34] and play a role in the NO• pathway, with implications for wound healing, cancer biology, and cardiac functioning [28, 35, 36]. A range of products, including those of non-stable and transient nature, can be found after plasma treatment [23, 28]. Several authors have observed the formation of cystine derivatives with a higher degree of oligomerization along with a considerable number of mixed S-oxidation products (S-oxides) [34, 37]. An interesting diagnostic role was determined for cysteine-S-sulfonate, a compound produced by reactions of •OH in the bulk of a plasma-treated liquid [38]. The bioactive reactive sulfur species (RSS), e.g., cyst(e)ine sulfoxides and cysteine-S-sulfonate, have the potential to influence cellular redox balance [30, 31, 34]. Figure 5.1 illustrates the above-discussed points by showing cysteine modifications observed when cysteine was treated with the same plasma device but different working gas conditions. Depending on the working gas conditions or plasma device, different modifications of the amino acid cysteine can be detected. Oxygen as a gas admixture leads preferably to higher oxidation states of cysteine and its derivates, while nitrogen (or weaker plasma sources) yields preferably cystine and its derivates. The use of air in the process gas results in the production of S-nitrosocysteine.

The precisely controllable cocktail of reactive species produced by plasmas offers the opportunity to generate reversible and irreversible modifications [31]. In conditions where atomic O that is not produced endogenously by biological systems is predominantly produced, the cysteine modifications tend towards irreversible oxidation states. By contrast, under plasma conditions that enrich pathways of the production of $•OH/H_2O_2$, species common to eukaryotic systems, reversible modifications are more prevalent, such as cystine, S-nitrosocysteine, or cysteine-S-sulfonate (Figure 5.1, compounds labeled in blue) [20, 39]. In experiments where chemically-produced H_2O_2 was used as a control substance to understand the contribution of that oxidant in plasma-induced cysteine modifications, only the dimer of cysteine, cystine, was detected. This indicates that (plasma-generated) short-lived species are necessary to form higher cysteine oxidations products. Due to its universal importance in biological systems and straightforward characterization, cysteine was suggested as a benchmark compound in plasma bio-science for testing the oxidant capacity of plasma discharges and weighing the impact of discharge condition adjustments [33, 40].

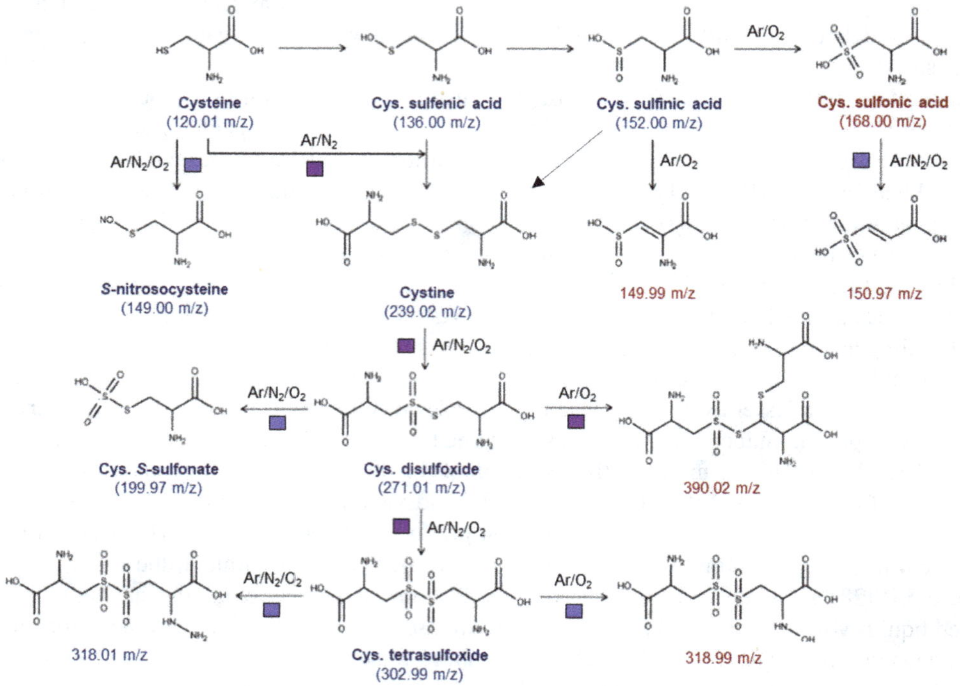

FIGURE 5.1 Oxidation products of cysteine formed by the reaction with an argon plasma jet using various working gas compositions (argon + 0.5% molecular gas admixture, blue square shielded effluent, purple square humidified working gas). Compounds labeled in red are stable compounds. m/z values are given if no common name was available. Adapted from [28].

When amino acids are plasma-treated in solutions containing chloride ions (Cl-), e.g., phosphate-buffered saline (PBS), monochloramines (R-NHCl$_2$) (observed for Phe and Tyr [41] as well as Leu [42]) are formed. The monochloramines result from the formation of OCl- in the liquid, followed by the chlorination of the amino acid (Figure 5.2). Oxidation occurs simultaneously, as described above, and oxidative chlorination occurs for certain amino acids. The modifications occur only when OCl- is formed, which requires O to react with Cl- [16, 17]. Thus, amino acid chlorination depends on the presence of Cl- in the solution and the production of O by the plasma. Interestingly, chlorination is one of the major modifications for otherwise not very reactive amino acids such as lysine, proline, and serine [29].

For plasma treatment of the aromatic amino acid tyrosine, chlorination is the primary modification under favorable conditions (presence of Cl- and O), while carbonylation combined with oxidation remains the primary modification when OCl- is not formed during plasma treatment of the liquid (e.g., in distilled water or phosphate-based buffers) [29]. It is worth mentioning that the chlorination is not stable under oxidative conditions. In prolonged or intensive plasma treatment conditions, oxidative de-chlorination occurs. Via the loss of HCl, a C=C double bond is introduced in the aliphatic chain of amino acids (e.g., dehydroalanine). In aromatic systems, oxygen replaces chlorine, yielding a phenol structure often oxidized to a carbonyl (and later a quinone). The thermodynamic driver of these reactions is the chloride ion's high symmetry, making it an ideal leaving group. The chloramines are comparably more stable and tend to decay less rapidly than the other chlorination products. The compounds exert an antibacterial effect [41, 42].

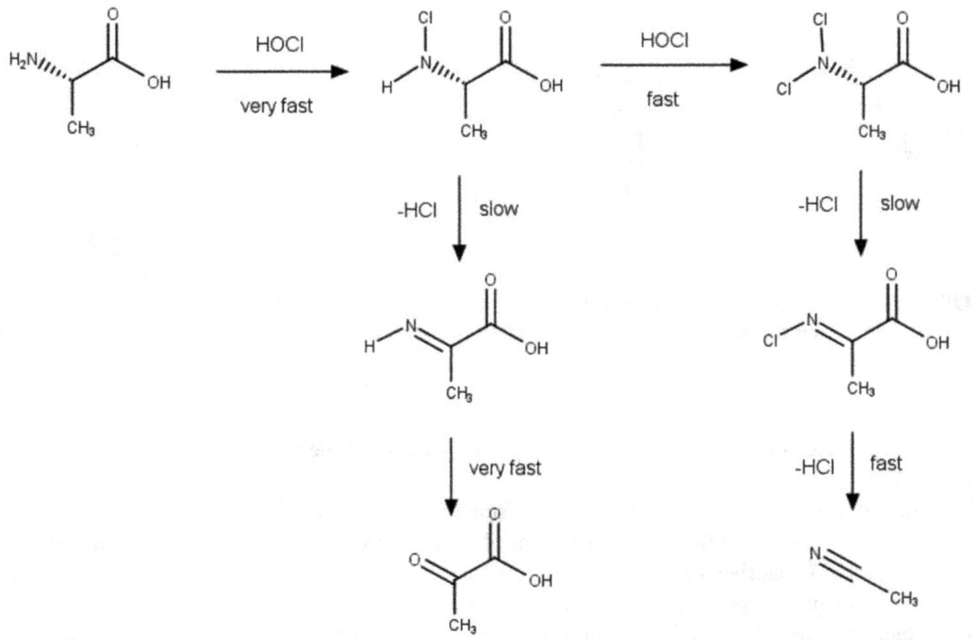

FIGURE 5.2 Chlorination of amino acids (example: alanine). HOCl/active chlorine attacks the α-amino group. Chloramines, dichloroamines, α-oxo acids, and nitriles occur as reaction products. Adapted from [43].

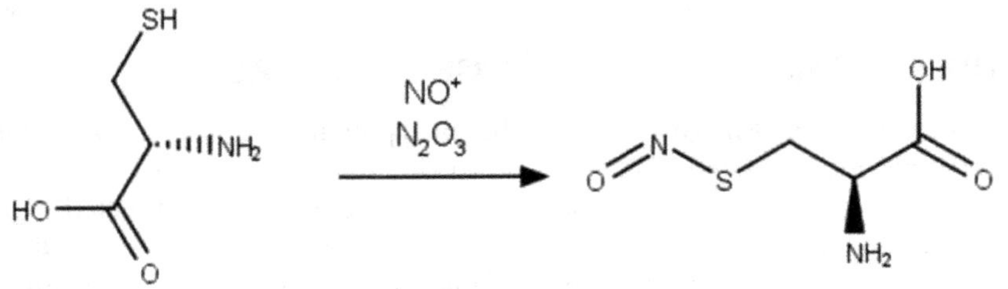

FIGURE 5.3 Nitrosylation of cysteine or glutathione (GSH). The nitronium ion and distickstofftrioxide (NO^+/N_2O_3) are the attacking species.

In biological systems, storage of the hormone-like vasodilating NO• can be realized via cysteine thiols forming nitrosocysteine (R-SNO). This compound was detected via mass spectrometry after treatment using two plasma sources and various conditions, but only in small amounts due to its instability in oxidizing conditions (Figure 5.3) [23].

The nitration of tyrosine is a common feature of oxidative stress in biological systems, and several antibodies are available to trace this as a marker compound in oxidized proteins. Plasma-generated RONS can nitrosate (R-NO$_2$) or nitrosylate (R-NO) tyrosine, and the according products have been detected by mass spectrometry [23]. However, comparable to the chlorine-containing compounds, the stability of these modifications is low, and oxidative de-nitrosation or de-nitrosylation occurs. In both situations, the NO$_2$/NO group is replaced by -OH or =O. Very rarely, nitrosylation of amino groups was observed (N-nitrosamines) [44]. Both the nitration and the chlorination of tyrosine changes the pKa of the tyrosine´s hydroxyl group and, with that, the reaction probability for the protein activation by enzymatic phosphorylation. So far, these modifications are not thought to be involved in signaling

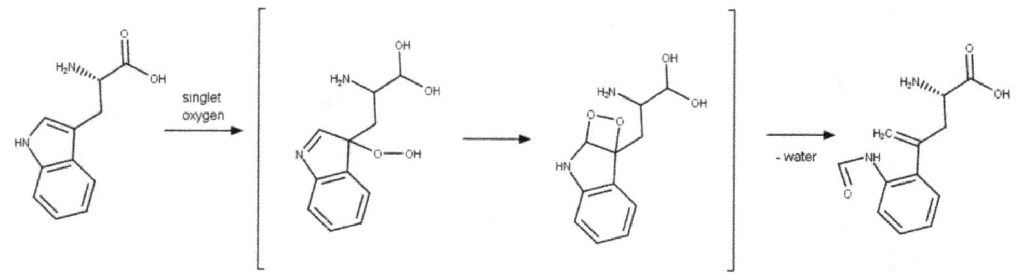

FIGURE 5.4 Oxidation of tryptophan via singlet oxygen $^1O_2$. After an initial hydroperoxide formation, a highly strained 4-ring marks the transition towards a ring-open form (N-formyl-kynurenin) that can further react to kynurenin [45].

processes, and there is no known "switching step" comparable to the reversible oxidative modifications of sulfur [30].

One amino acid that showed a remarkable sensitivity towards plasma-generated species is the bulky, apolar tryptophan. The electron-rich indol-type heterocyclic sidechain can react easily with all plasma-derived reactive species. Due to the lipophilicity of the sidechain, an enrichment and orientation towards the gas-liquid interface facilitates the reaction with gas-phase-derived short-lived species such as $^1O_2$ [45]. The tryptophan double oxidation (Trp+2O) is a typical intermediate that decays into the N-formyl-kynurenine or kynurenine end products (Figure 5.4). These particular products indicate the presence of $^1O_2$ generated in the plasma. Oxidation via OCl- or ONOO- does not contribute to the ring-open forms but instead yields the corresponding chlorination or nitration of the aromatic ring system of the indole moiety.

5.5 PEPTIDES AS TARGETS OF PLASMA-GENERATED RONS

Peptides, oligomers of amino acids but much smaller than proteins, are suitable model for investigating the impact of plasma-generated RONS in a model with only medium complexity. Peptides offer two relevant benefits: easy sample preparation for structure elucidation-directed high-resolution mass spectrometry and a variety of competing chemical structures attackable by RONS that reflect the circumstances in a protein. Despite the increasing complexity, cysteine-containing peptides show similar modifications after plasma treatment as compared to plasma treatment of individual cysteine molecules in solution. The redox pair glutathione (GSH) and glutathione disulfide (GSSG), which is physiologically responsible for detoxification in organisms and the most abundant intracellular redox couple [46, 47], is oxidized up to the final product GSO_3H, but all intermediate oxidation states are produced during plasma treatment [40].

In contrast to cystine alone, the corresponding nitrosylation product GSNO, a potential NO donor in organisms with opportunities for therapeutic applicability [48], is continuously produced in a pH-independent manner. In general, a rapid loss of the free thiols and the formation of intermediate, reversible (GSO_2H) and irreversible (GSO_3H) products were observed following plasma treatment. Adding the redox-active metal iron, the most abundant transition metal in healthy humans, to GSH and GSSG in the form of iron(II) and iron(III) complexes yields selective modifications of GSH and GSSG exclusively to GSO_3H [49]. When iron complexes are present, the nitroso-glutathione pathway and its derivatives are suppressed. Instead of a variety of products, as discussed above, the transition metal ion favors the production of the final oxidation state GSO_3H, likely via the formation of •OH via the Fenton reaction [49]. Due to its relevance in redox biology and the observed modifications after plasma treatment (which can be correlated to possible cellular effects), extracellular plasma-induced GSH modification was found in early studies to correlate with biological outcomes, as seen, e.g., with human T-cell viability and production of mitochondrial •O_2- following

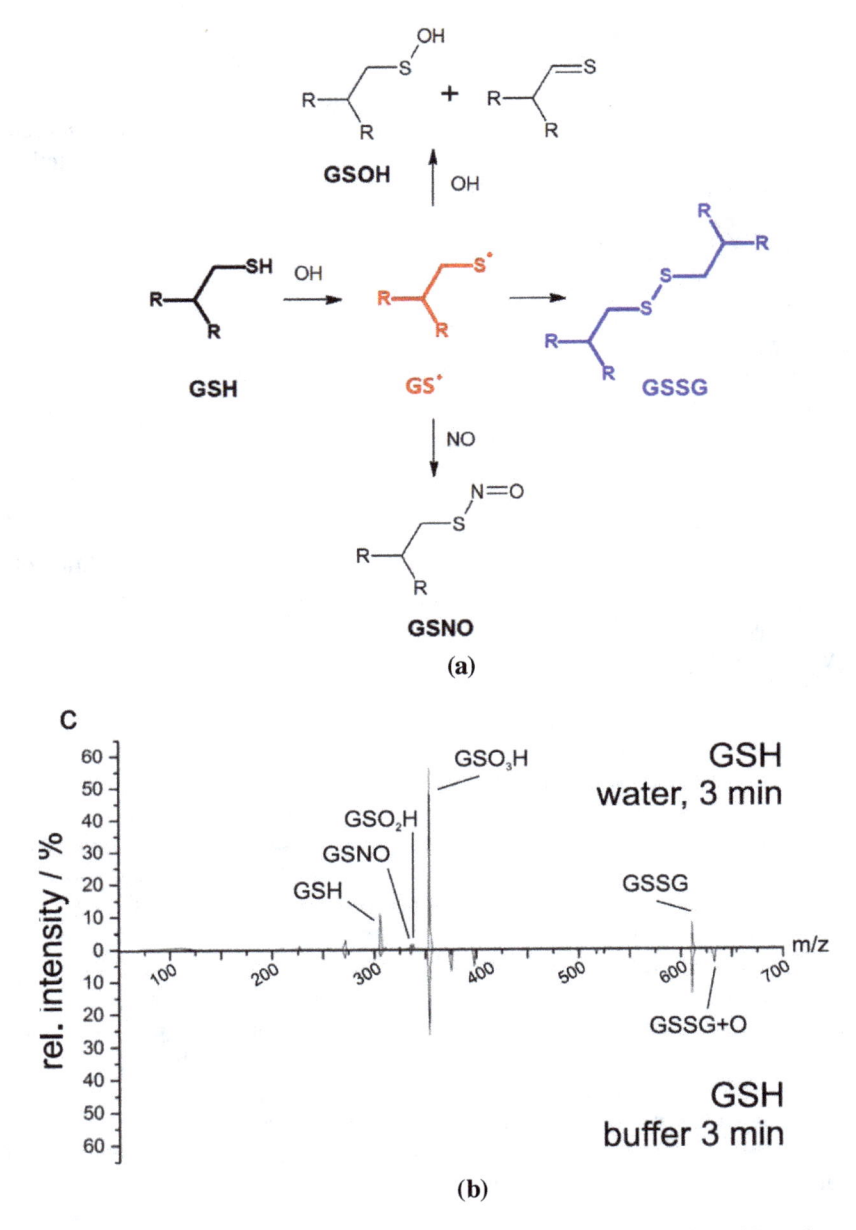

FIGURE 5.5 Glutathione oxidation by •OH (molecular dynamics simulation). Note the presence of the sulfenic acid derivative (G-S-OH) that is unstable and oxidized further to G-SO₂H and the end product G-SO₃H (left). The products are also displayed in the MS survey scan (right). Adapted from [40].

plasma treatment activity [50]. The potential correlation to cellular effects was then further refined using Raman spectroscopy-based tracking of GSH modification, which was proposed as a marker for plasma treatment outcomes, since they correlated with cell viability and mitochondrial •O₂₋ results in Jurkat T-lymphocytes exposed to different plasma sources [51].

In a series of experiments, ten different artificial decapeptides and two physiological peptides (fragments of angiotensin and bradykinin) were treated with two orthogonal plasma sources: the argon-based dielectric barrier jet kINPen [52] and a helium-based radiofrequency microjet (COST jet, [53]). In the first case, a central pin electrode was powered by a high voltage sinusoidal waveform (≈1 MHz) while the counter electrode was shielded by a dielectric. Charged species followed the vector of the pin electrode and the gas flow towards the target, interacting with ambient species. In the second case, the powered and the counter electrode faced each other—perpendicular to the gas flow. A radiofrequency high voltage waveform (13.56 MHz) ensured a homogeneous discharge. The effluent was small and did not contain charged species that are confined between the electrodes. Due to these significant differences in the plasma source design, the formation of primary and secondary species was disparate. When oxygen was admixed to the respective working gas, oxygen-based reactive species dominated in both cases, with atomic oxygen as a central hub in the COST jet and singlet oxygen/nitric oxide in the kINPen. The occurrence of oxidative modifications within the model peptides was recorded using high-resolution mass spectrometry. The results largely corroborated the knowledge of amino acid modification by plasma-derived species (Figure 5.6), but there were also differences.

The observed differences reflected the competition among the different amino acid side chains to react with RONS (Figure 5.6). As major targets, the sulfur-containing amino acids (Cys = Met) outcompeted the amino acids containing an aromatic system (Trp > Tyr > Phe). Basic amino acids (Lys > His > Arg) and the polar amino acids (Ser > Glu > Gln = Asp > Asp > Thr) were harder to attack and showed little impact of the plasma-derived reactive species. The non-polar, aliphatic amino acids were intermediately susceptible (Leu > Val > Ile = Ala = Pro). This reflects a radical-driven reaction mechanism, with an initial hydrogen abstraction and consecutive addition of •OH or chlorine atoms. Lysine was significantly susceptible to chlorination at the ε-amino group. Histidine contains a pyrrole ring structure that can be attacked by $^1O_2$, yielding ring cleavage products. The researchers produced an inclusion list indicating the major amino acids targeted and the observed oxidative modifications [29]. While the oxidation of several amino acids seems almost device-agnostic, some modifications can be achieved only under certain conditions. Nitration, for example, seems to occur only in conditions where ONOO- is present. When Cl- is present in the solution, chlorination occurs under conditions where atomic oxygen can form OCl- [17, 29]. $^1O_2$, which is not present in every plasma discharge in sufficient amounts, results in ring cleavage of, e.g., tryptophan [29]. The peptides bradykinin (containing Arg-Pro-Pro-Gly-Phe-Ser-Pro-Phe-Arg but no sulfur-containing amino acids), a small pro-inflammatory peptide in mammals, and angiotensin 1–7 (Asp-Arg-Val-Tyr-Ile-His-Pro), the amino-terminal fragment of angiotensin I/II involved in blood pressure regulation, showed nine different modifications [54]. Most frequently, oxidation, nitration, and ring cleavage was observed. In the absence of sulfur-containing amino acids, the main targets for plasma-induced modifications were tyrosine, tryptophan, histidine, phenylalanine, proline, and isoleucine [54]. Molecular dynamic (MD) simulations including only OH and ground state O_2 as reaction partners point towards the importance of other short-lived species, such as O, $•O_2^-$, $^1O_2$, and ONOO- [54, 55]. Newer MD simulations have included atomic oxygen and ozone and appear to agree better with experimental results, although not all aspects could be depicted in the simulations [56]. These results illustrate the unique reactive species composition of modification patterns generated in the plasma treatment of biomolecules such as peptides.

5.6 PROTEIN OXIDATION BY COLD PLASMA-DERIVED RONS

Experiments on plasma-treated single proteins or mixtures thereof were carried out, motivated by various aspects of protein biochemistry and application. These studies mainly looked into global effects on the protein, such as effects on the activity or structure of the protein. The abovementioned disulfide bonds in cystine as well as in GSSG are covalent links between pairs of cysteine residues. Most proteins in the extracellular environment contain disulfide bonds, a key structural element for stabilizing the protein structure [57]. Disulfide bonds hold the protein together but can also be responsible for controlling how the proteins work. This control is achieved by cleavage of the disulfide bonds, which in many cases is reversible, since the formation of disulfide bonds is typically associated with a lower entropy [57]. RNase A, for example, is a tough-to-inactivate protein that contains four intramolecular disulfide bonds crisscrossing the molecule, which are responsible

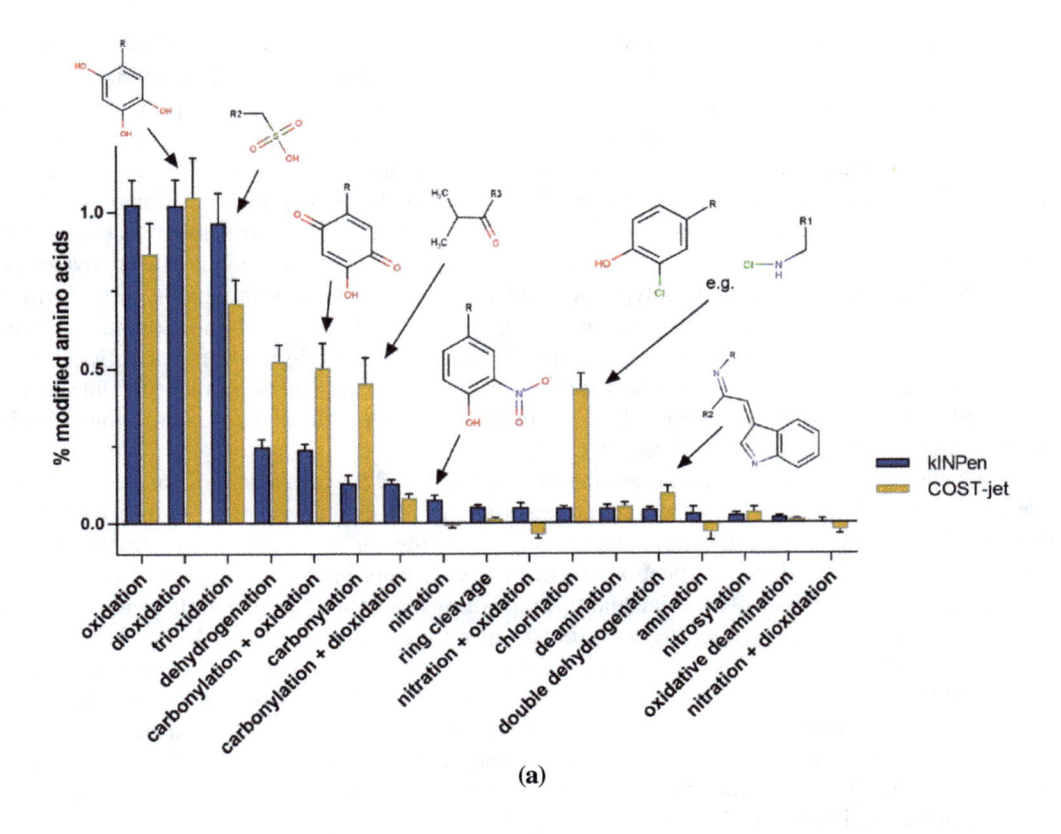

(a)

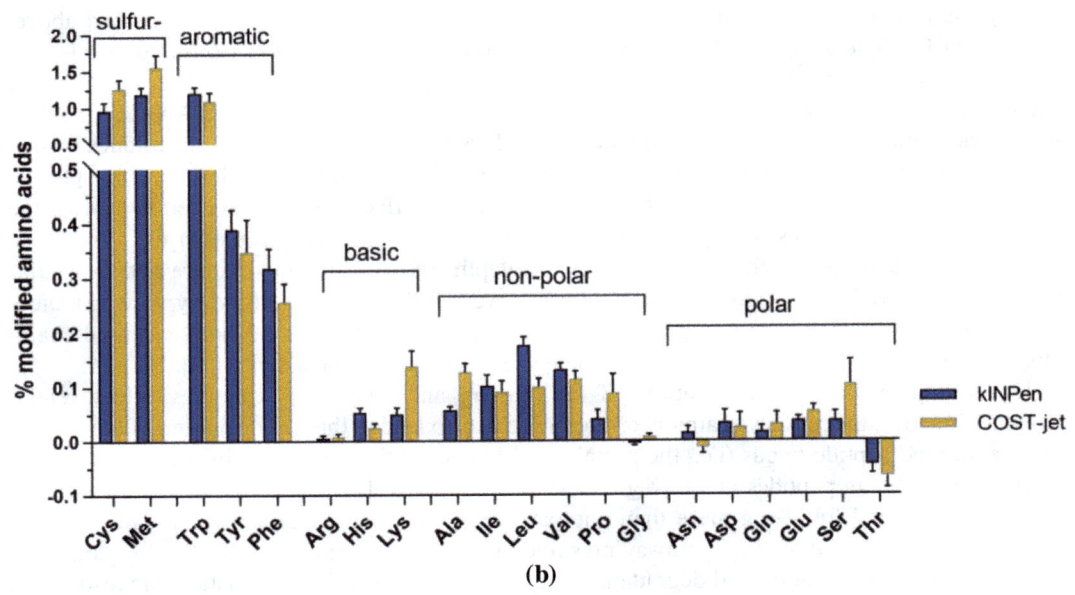

(b)

FIGURE 5.6 Overview of observed amino acid oxidation when part of a decapeptide. Main targets are electron-rich amino acids (top); major modification is the introduction of one to three oxygens (bottom). The argon plasma jet kINPen and the helium jet COST jet were compared. Note the differences in chlorination and nitration products. Adapted from [29].

for its extraordinary stability. Heat exposure above 90 °C is required to inactivate RNase A; exposure to 80 °C only leads to reversible denaturation [58]. While RNase A is difficult to inactivate by traditional methods, and high temperatures are required for terminal inactivation, plasma treatment of RNase A can result in complete inactivation of RNase A at low temperatures. Depending on the plasma device and treatment conditions, total inactivation and unfolding of RNase A can be observed. In addition to the over-oxidation of the disulfide bonds, methionine oxidation occurs [59]. The cleavage of disulfide bonds appears to be a common feature of plasma treatment, as it has been observed not only for RNase A but also for antibodies consisting of four polypeptides covalently linked by disulfide bonds and in polypeptides during nanoelectrospray ionization [60]. Similar observations were made in the plasma-treated skin component keratin [61] and, computationally as well as experimentally, in the signaling molecule human epidermal growth factor (hEGF) [62]. Structural changes in the protein due to disulfide bond cleavage may result in more flexible structures, and, as was shown in the case of hEGF, an increased oxidation level may lead to a lower binding free energy of the factor with its receptor [62, 63].

While cysteine-containing proteins are readily oxidized, and the cleavage or formation of disulfide bonds alters the protein structure and protein function, modifications of proteins are not limited to sulfur oxidation or thiol nitrosylation. By employing the inclusion list generated from a peptide library [29], further modifications in full proteins were observed by mass spectrometry. Among these, modifications of tyrosine and tryptophan were identified as major alternative targets. Oxidation (addition of one to three oxygen atoms) was most commonly observed.

Cysteine-containing proteins can be not only inactivated but, less comonly, also activated, as in the case of the redox-regulated chaperone heat shock protein Hsp33 of *Escherichia coli* [64]. Hsp33 binds to unfolded proteins and prevents protein aggregation [65]. The chaperone is usually activated by oxidation at elevated temperatures [66, 67], initiated by the formation of disulfide bonds. Plasma treatment of Hsp33 leads to the oxidation of its cysteine residues with a subsequent formation of disulfide bonds, activating the Hsp33 chaperone [64].

In general, plasma treatment of enzymes (e.g., lysozyme [68], RNase A (as discussed above) [59], GapDH [69], and superoxide dismutases (Sod) Sod A and Sod B [70]) is followed by a loss of structure and enzyme activity due to either direct modification of the amino acid(s) in the active site or changes in secondary structure that make the active site less accessible. Another example is the inactivation of the enzyme phospholipase A. This protein cleaves fatty acid residues from a phospholipid bilayer and is part of the inflammation-controlling machinery of the cell. Upon plasma treatment, the ability to attack a lipid bilayer was lost. This effect could be blocked by adding the amino acid histidine as a scavenger of $^1O_2$. Common reference oxidants, such as H_2O_2, $ONOO^-$, or OCl^-, did not result in enzyme inactivation. An in-depth study of protein structure and molecular composition by high-resolution mass spectrometry revealed the oxidation of the tryptophan moiety 128 (Figure 5.7). The resulting ring opening rendered the conformation of the C-terminal chain, blocking the docking to the lipid bilayer and hence disrupting enzyme activity [71].

In addition to specific modifications of the primary and secondary structures, which render enzymes inactive, the plasma treatment of proteins can also lead to the fragmentation of the protein by cleaving the peptide bonds (i.e., the protein's backbone) [72]. The plasma-initiated degradation of proteins results in peptides of varying length, pointing towards a non-selectivity of the position of the cleavage and following more than one pathway: a well-established diamide pathway [73], dependent on •OH, and another pathway presumed to be independent of •OH [72]. In addition, a combination of modification and degradation can occur [74], and this combination can also affect protein immunogenicity (see chapters 10 and 11).

Under certain conditions, enzymes inactivated by plasma treatment have shown a slight recovery in activity over time, as observed for plasma-treated lactate dehydrogenase (LDH) solution [75]. It was proposed that the enzyme's active site loses recognition and catalytic functions after the plasma-induced amino acid modifications and subsequent altering of the protein's secondary structure. Peptide chains polymerize, and the aggregation of molecules decreases the enzyme activity

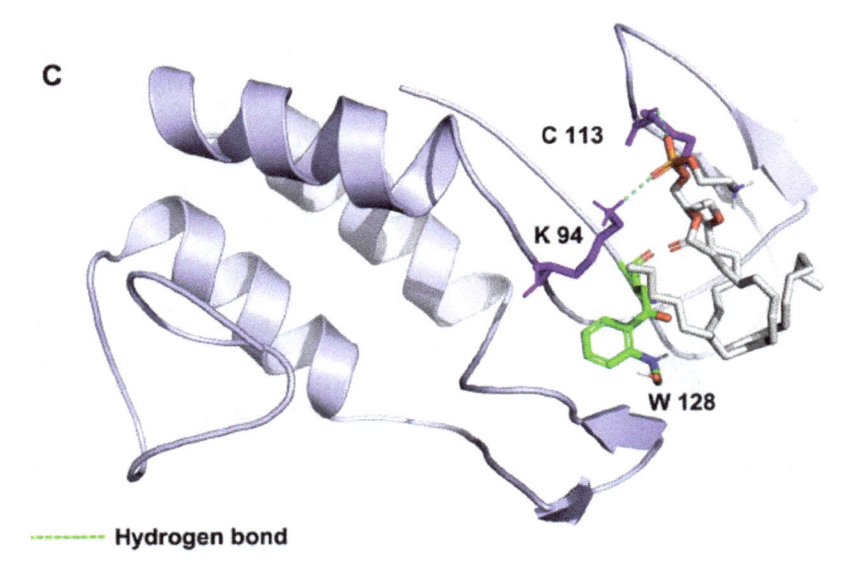

FIGURE 5.7 Phospholipase A_2 (*Apis mellifera*) is inactivated by cold plasma generated $^1O_2$. The oxidation yielded a ring opening of Trp128 and a subsequent change in conformation, blocking access to the catalytic center of the enzyme. Adapted from [71].

while protecting the active site from being modified. The so-called supramolecules can then spontaneously disintegrate over time and expose the active site, leading to the observed recovery of enzyme activity [75].

5.7 SUMMARY AND CONCLUSIONS

In summary, plasma-generated RONS can modify amino acids, peptides, and proteins. The sulfur-containing amino acids methionine and cysteine are most readily modified by the plasma-generated RONS, followed by tyrosine, tryptophan, and phenylalanine. Oxidation of the amino acid side chains is a common feature of plasma-produced RONS. More specific modifications, e.g., nitration or chlorination, require specific conditions or plasma sources. As some modifications are specific to certain conditions, plasma can be used as a controllable source to induce protein modifications. Many of the modifications will affect the secondary protein structure and the physicochemical properties of the amino acid residues, impacting the function and performance of the protein. More research is needed to identify the impact of plasma-modified proteins and amino acids on eukaryotic systems in general, and specifically, to investigate the hypothesis that biomolecule oxidation forms the mechanistic basis of biomedical plasma effects *in vivo*.

REFERENCES

1. Torabizadeh, H., *All Proteins Have a Basic Molecular Formula.* World Academy of Science, Engineering and Technology, 2011. **78**: p. 961–965.
2. Ramazi, S., and J. Zahiri, *Post-Translational Modifications in Proteins: Resources, Tools and Prediction Methods.* Database, 2021. **2021**.
3. Muller, M.M., *Post-Translational Modifications of Protein Backbones: Unique Functions, Mechanisms, and Challenges.* Biochemistry, 2018. **57**(2): p. 177–185.
4. Nussinov, R., et al., *Allosteric Post-Translational Modification Codes.* Trends in Biochemical Sciences, 2012. **37**(10): p. 447–455.

5. Khoury, G.A., R.C. Baliban, and C.A. Floudas, *Proteome-Wide Post-Translational Modification Statistics: Frequency Analysis and Curation of the Swiss-Prot Database.* Scientific Reports, 2011. **1**(1).

6. Witze, E.S., et al., *Mapping Protein Post-Translational Modifications with Mass Spectrometry.* Nature Methods, 2007. **4**(10): p. 798–806.

7. Gianazza, E., J. Crawford, and I. Miller, *Detecting Oxidative Post-Translational Modifications in Proteins.* Amino Acids, 2007. **33**(1): p. 51–56.

8. Heusler, T., et al., *Can the Effect of Cold Physical Plasma-Derived Oxidants be Transported Via Thiol Group Oxidation?* Clinical Plasma Medicine, 2019. **14**: p. 100086.

9. Reuter, S., T. von Woedtke, and K.D. Weltmann, *The kINPen-A Review on Physics and Chemistry of the Atmospheric Pressure Plasma Jet and Its Applications.* Journal of Physics D-Applied Physics, 2018. **51**(23): p. 233001.

10. Golda, J., et al., *Concepts and Characteristics of the 'COST Reference Microplasma Jet'.* Journal of Physics D: Applied Physics, 2016. **49**(8): p. 084003.

11. Murakami, T., et al., *Afterglow Chemistry of Atmospheric-Pressure Helium-Oxygen Plasmas with Humid Air Impurity.* Plasma Sources Science & Technology, 2014. **23**(2): p. 025005.

12. Graves, D.B., *The Emerging Role of Reactive Oxygen and Nitrogen Species in Redox Biology and Some Implications for Plasma Applications to Medicine and Biology.* Journal of Physics D-Applied Physics, 2012. **45**(26): p. 263001.

13. Wende, K., et al., *Chemistry and Biochemistry of Cold Physical Plasma Derived Reactive Species in Liquids.* Biological Chemistry, 2018. **400**(1): p. 19–38.

14. Freund, E., and S. Bekeschus, *Gas Plasma-Oxidized Liquids for Cancer Treatment: Preclinical Relevance, Immuno-Oncology, and Clinical Obstacles.* IEEE Transactions on Radiation and Plasma Medical Sciences, 2021. **5**(6): p. 761–774.

15. Stapelmann, K., et al., *Following O and OH in He/O2 and He/H2O Gas Mixtures—From the Gas Phase Through the Liquid Phase to Modifications on a Biological Sample.* Journal of Physics D: Applied Physics, 2021. **54**(43): p. 434003.

16. Wende, K., et al., *Identification of the Biologically Active Liquid Chemistry Induced by a Nonthermal Atmospheric Pressure Plasma Jet.* Biointerphases, 2015. **10**(2): p. 029518.

17. Jirásek, V., and P. Lukeš, *Formation of Reactive Chlorine Species in Saline Solution Treated by Non-Equilibrium Atmospheric Pressure He/O2 Plasma Jet.* Plasma Sources Science and Technology, 2019. **28**(3): p. 035015.

18. Bottecchia, C., and T. Noel, *Photocatalytic Modification of Amino Acids, Peptides, and Proteins.* Chemistry, 2019. **25**(1): p. 26–42.

19. von Woedtke, T., et al., *Plasma Medicine: A Field of Applied Redox Biology. In Vivo*, 2019. **33**(4): p. 1011–1026.

20. Poole, L.B., *The Basics of Thiols and Cysteines in Redox Biology and Chemistry.* Free Radical Biology and Medicine, 2015. **80**: p. 148–157.

21. Jones, D.P. and H. Sies, *The Redox Code.* Antioxid Redox Signal, 2015. **23**(9): p. 734–746.

22. Takai, E., et al., *Chemical Modification of Amino Acids by Atmospheric-Pressure Cold Plasma in Aqueous Solution.* Journal of Physics D-Applied Physics, 2014. **47**(28): p. 285403.

23. Lackmann, J.W., et al., *Nitrosylation vs. Oxidation—How to Modulate Cold Physical Plasmas for Biological Applications.* PLoS One, 2019. **14**(5): p. e0216606.

24. Sies, H., *Glutathione and Its Role in Cellular Functions.* Free Radical Biology and Medicine, 1999. **27**(9–10): p. 916–921.

25. Yang, H., et al., *Redox Modifications of Cysteine Residues Regulate the Cytokine Activity of HMGB1.* Molecular Medicine, 2021. **27**(1): p. 58.

26. Semenza, G.L., *Hydroxylation of HIF-1: Oxygen Sensing at the Molecular Level.* Physiology, 2004. **19**(4): p. 176–182.

27. Zhou, R., et al., *Interaction of Atmospheric-Pressure Air Microplasmas with Amino Acids as Fundamental Processes in Aqueous Solution.* PLoS One, 2016. **11**(5): p. e0155584.

28. Bruno, G., et al., *Cold Physical Plasma-Induced Oxidation of Cysteine Yields Reactive Sulfur Species (RSS).* Clinical Plasma Medicine, 2019. **14**: p. 100083.

29. Wenske, S., et al., *Reactive Species Driven Oxidative Modifications of Peptides-Tracing Physical Plasma Liquid Chemistry.* Journal of Applied Physics, 2021. **129**(19): p. 193305.

30. Giles, G.I., et al., *The Reactive Sulfur Species Concept: 15 Years On.* Antioxidants (Basel), 2017. **6**(2): p. 38.

31. Paulsen, C.E., and K.S. Carroll, *Cysteine-Mediated Redox Signaling: Chemistry, Biology, and Tools for Discovery.* Chemical Reviews, 2013. **113**(7): p. 4633–4679.

32. Kogelheide, F., et al., *FTIR Spectroscopy of Cysteine as a Ready-to-Use Method for the Investigation of Plasma-Induced Chemical Modifications of Macromolecules.* Journal of Physics D-Applied Physics, 2016. **49**(8): p. 084004.

33. Lackmann, J.W., et al., *Chemical Fingerprints of Cold Physical Plasmas—An Experimental and Computational Study Using Cysteine as Tracer Compound.* Scientific Reports, 2018. **8**(1): p. 7736.

34. Giles, G.I., K.M. Tasker, and C. Jacob, *Oxidation of Biological Thiols by Highly Reactive Disulfide-S-Oxides.* General Physiology and Biophysics, 2002. **21**(1): p. 65–72.

35. Ali, A.A., et al., *The Contribution of N2O3 to the Cytotoxicity of the Nitric Oxide Donor DETA/NO: An Emerging Role for S-Nitrosylation.* Bioscience Reports, 2013. **33**(2): p. e00031.

36. Belge, C., et al., *Nitric Oxide and the Heart: Update on New Paradigms.* Annals of the New York Academy of Sciences, 2005. **1047**(1): p. 173–182.

37. Farhadi, M., and F. Sohbatzadeh, *Influence of a Transient Spark Plasma Discharge on Producing High Molecular Masses of Chemical Products from l-Cysteine.* Scientific Reports, 2023. **13**(1): p. 2059.

38. Wende, K., et al., *On a Heavy Path—Determining Cold Plasma-Derived Short-Lived Species Chemistry Using Isotopic Labelling.* RSC Advances, 2020. **10**(20): p. 11598–11607.

39. Alvarez, B., et al., *Redox Chemistry and Biology of Thiols.* 2022: Academic Press.

40. Klinkhammer, C., et al., *Elucidation of Plasma-Induced Chemical Modifications on Glutathione and Glutathione Disulphide.* Scientific Reports, 2017. **7**(1): p. 13828.

41. Jirásek, V., B. Tarabová, and P. Lukeš, *Treatment of Phenylalanine and Tyrosine in Phosphate-Buffered Saline by Plasma-Supplied Oxygen Atoms: Chemical Characterization and Bactericidal Effects.* Plasma Processes and Polymers, 2022. **19**(11): p. 2200079.

42. Jirásek, V., et al., *Leucine Modifications by He/O2 Plasma Treatment in Phosphate-Buffered Saline: Bactericidal Effects and Chemical Characterization.* Journal of Physics D: Applied Physics, 2021. **54**(50): p. 505206.

43. How, Z.T., et al., *Chlorination of Amino Acids: Reaction Pathways and Reaction Rates.* Environmental Science & Technology, 2017. **51**(9): p. 4870–4876.

44. Bruno, G., et al., *On the Liquid Chemistry of the Reactive Nitrogen Species Peroxynitrite and Nitrogen Dioxide Generated by Physical Plasmas.* Biomolecules, 2020. **10**(12): p. 1687.

45. Ronsein, G.E., et al., *Tryptophan Oxidation by Singlet Molecular Oxygen [O2 (1Δg)]: Mechanistic Studies Using 18O-Labeled Hydroperoxides, Mass Spectrometry, and Light Emission Measurements.* Chemical Research in Toxicology, 2008. **21**(6): p. 1271–1283.

46. Giustarini, D., et al., *Assessment of Glutathione/Glutathione Disulphide Ratio and S-Glutathionylated Proteins in Human Blood, Solid Tissues, and Cultured Cells.* Free Radical Biology and Medicine, 2017. **112**: p. 360–375.

47. Monostori, P., et al., *Determination of Glutathione and Glutathione Disulfide in Biological Samples: An In-Depth Review.* Journal of Chromatography B, 2009. **877**(28): p. 3331–3346.

48. Broniowska, K.A., A.R. Diers, and N. Hogg, *S-Nitrosoglutathione.* Biochimica et Biophysica Acta (BBA)—General Subjects, 2013. **1830**(5): p. 3173–3181.

49. Śmiłowicz, D., et al., *Study on Chemical Modifications of Glutathione by Cold Atmospheric Pressure Plasma (Cap) Operated in Air in the Presence of fe (ii) and fe (iii) Complexes.* Scientific Reports, 2019. **9**(1): p. 1–13.

50. Bekeschus, S., et al., *Cold Physical Plasma Treatment Alters Redox Balance in Human Immune Cells.* Plasma Medicine, 2013. **3**(4): p. 267–278.

51. Ranieri, P., et al., *GSH Modification as a Marker for Plasma Source and Biological Response Comparison to Plasma Treatment.* Applied Sciences, 2020. **10**(6): p. 2025.

52. Weltmann, K.D., et al., *Atmospheric Pressure Plasma Jet for Medical Therapy: Plasma Parameters and Risk Estimation.* Contributions to Plasma Physics, 2009. **49**(9): p. 631–640.

53. Ellerweg, D., et al., *Characterization of the Effluent of a He/O2microscale Atmospheric Pressure Plasma jet by Quantitative Molecular Beam Mass Spectrometry.* New Journal of Physics, 2010. **12**(1): p. 013021.

54. Wenske, S., et al., *Nonenzymatic Post-Translational Modifications in Peptides by Cold Plasma-Derived Reactive Oxygen and Nitrogen Species.* Biointerphases, 2020. **15**(6): p. 061008.

55. Verlackt, C.C.W., et al., *Mechanisms of Peptide Oxidation by Hydroxyl Radicals: Insight at the Molecular Scale.* Journal of Physical Chemistry C, 2017. **121**(10): p. 5787–5799.

56. Ding, Y.-H., et al., *Numerical Study on Interactions of Atmospheric Plasmas and Peptides by Reactive Molecular Dynamic Simulations.* Plasma Processes and Polymers, 2023. **20**(3): p. 2200148.

57. Hogg, P.J., *Disulfide Bonds as Switches for Protein Function.* Trends in Biochemical Sciences, 2003. **28**(4): p. 210–214.

58. Hermans Jr, J., and H.A. Scheraga, *Structural Studies of Ribonuclease. v. Reversible Change of Configuration1–3*. Journal of the American Chemical Society, 1961. **83**(15): p. 3283–3292.

59. Lackmann, J.W., et al., *A Dielectric Barrier Discharge Terminally Inactivates RNase a by Oxidizing Sulfur-Containing Amino Acids and Breaking Structural Disulfide Bonds*. Journal of Physics D-Applied Physics, 2015. **48**(49): p. 494003.

60. Xia, Y., and R.G. Cooks, *Plasma Induced Oxidative Cleavage of Disulfide Bonds in Polypeptides during Nanoelectrospray Ionization*. Analytical Chemistry, 2010. **82**(7): p. 2856–2864.

61. Kartaschew, K., et al., *Unraveling the Interactions between Cold Atmospheric Plasma and Skin-Components with Vibrational Microspectroscopy*. Biointerphases, 2015. **10**(2): p. 029516.

62. Yusupov, M., et al., *Impact of Plasma Oxidation on Structural Features of Human Epidermal Growth Factor*. Plasma Processes and Polymers, 2018. **15**(8): p. 1800022.

63. Stapelmann, K., et al., *Computational and Experimental Study Of The Impact Of Plasma On The Human Epidermal Growth Factor and Its Implications For Wound Healing and Cancer Treatment*. Clinical Plasma Medicine, 2018. **9**: p. 12.

64. Krewing, M., et al., *The Molecular Chaperone Hsp33 is Activated by Atmospheric-Pressure Plasma Protecting Proteins from Aggregation*. Journal of The Royal Society Interface, 2019. **16**(155): p. 20180966.

65. Winter, J., et al., *Severe Oxidative Stress Causes Inactivation of DnaK and Activation of the Redox-Regulated Chaperone Hsp33*. Molecular Cell, 2005. **17**(3): p. 381–392.

66. Graf, P.C., et al., *Activation of the Redox-Regulated Chaperone Hsp33 by Domain Unfolding*. Journal of Biological Chemistry, 2004. **279**(19): p. 20529–20538.

67. Winter, J., et al., *Bleach Activates a Redox-Regulated Chaperone by Oxidative Protein Unfolding*. Cell, 2008. **135**(4): p. 691–701.

68. Choi, S., et al., *Structural and Functional Analysis of Lysozyme After Treatment with Dielectric Barrier Discharge Plasma and Atmospheric Pressure Plasma Jet*. Scientific Reports, 2017. **7**(1): p. 1027.

69. Lackmann, J.W., et al., *Photons and Particles Emitted from Cold Atmospheric-Pressure Plasma Inactivate Bacteria and Biomolecules Independently and Synergistically*. Journal of the Royal Society Interface, 2013. **10**(89): p. 20130591.

70. Krewing, M., et al., *Dielectric Barrier Discharge Plasma Treatment Affects Stability, Metal Ion Coordination, and Enzyme Activity of Bacterial Superoxide Dismutases*. Plasma Processes and Polymers, 2020. **17**(10): p. 2000019.

71. Nasri, Z., et al., *Singlet-Oxygen-Induced Phospholipase A2 Inhibition: A Major Role for Interfacial Tryptophan Dioxidation*. Chemistry—A European Journal, 2021. **27**(59): p. 14702–14710.

72. Krewing, M., B. Schubert, and J.E. Bandow, *A Dielectric Barrier Discharge Plasma Degrades Proteins to Peptides by Cleaving the Peptide Bond*. Plasma Chemistry and Plasma Processing, 2020. **40**(3): p. 685–696.

73. Stringfellow, H.M., et al., *Selectivity in ROS-Induced Peptide Backbone Bond Cleavage*. The Journal of Physical Chemistry A, 2014. **118**(48): p. 11399–11404.

74. Clemen, R., et al., *Gas Plasma Protein Oxidation Increases Immunogenicity and Human Antigen-Presenting Cell Maturation and Activation*. Vaccines (Basel), 2022. **10**(11): p. 1814.

75. Zhang, H., et al., *Effects and Mechanism of Atmospheric-Pressure Dielectric Barrier Discharge Cold Plasmaon Lactate Dehydrogenase (LDH) Enzyme*. Scientific Reports, 2015. **5**(1): p. 10031.

6 Hydrogels in Plasma Medicine

*Francesco Tampieri, Albert Espona-Noguera,
Inès Hamouda, Cristina Canal*

6.1 INTRODUCTION—PLASMA FOR SURFACE MODIFICATION OF POLYMERS

Polymers constitute the largest and most versatile class of biomaterials [1]. With appropriate chemical and physical properties, they have been extensively employed in a variety of biomedical applications [2–4]. Cold plasmas [5,6] have traditionally played an important role in the field of polymer treatment due to several advantages, including their versatility, which makes it possible to tailor physicochemical surface properties of the first nanometers of the surface without altering the bulk of the solid polymers [7]. A recurrent drawback of polymers for biomedical applications is their hydrophobicity [8], which is related to their low surface energy and which makes it difficult for the cells to attach, spread and proliferate in these materials [9]. In order to enhance their hydrophilicity or surface energy or to introduce new functional groups, the surface modification of polymers has been widely investigated [10], and cold plasmas are an important player among the different methods that have been explored to modify the chemistry on the surface of polymers [11–13]. Moving beyond the simple modification of wettability, cold plasmas have been widely employed to improve the biocompatibility of polymers designed to be in contact with biological tissues [14,15]. Research on plasma modification of polymers for biomedical applications has already received excellent reviews [6,16,25,17–24].

6.2 MOVING TOWARDS APPLICATIONS IN PLASMA MEDICINE

Beyond its use in modifying surfaces, the development of plasma sources at atmospheric pressure, such as atmospheric pressure plasma jets (APPJs) or dielectric barrier discharges (DBDs), has paved the way for new applications in medicine, creating a new branch of medical technology called plasma medicine [26]. This multidisciplinary field is based on the interactions among the different species of plasmas and the complex biochemistry involved in treatment of biological tissues or samples [27]. Today, plasma medicine raises great interest and includes three main fields: sterilization, wound healing and cancer therapy [28–30] (see Chapter 1).

In this chapter, we will describe the new role that hydrated polymers are playing in medical applications in the field of plasma medicine. The first application is relative to the direct plasma treatment on biological tissues, i.e. the skin, that is already in the clinics [31,32] (see Chapter 3). To improve understanding of the interactions between plasma and biological tissues, depth of penetration of plasma, etc., new pathways employing hydrogels have been explored [33].

In parallel, since reactive oxygen and nitrogen species (RONS) generated from plasma sources are thought to be primarily responsible for several biological responses [34], plasma treatment of liquids has been widely investigated as a transporter of RONS, as these liquids can be injected locally in a minimally invasive approach [35–39] (see chapters 3 and 13). In that, research has recently been initiated on the ability of plasma-treated polymer solutions to form hydrogels. These two new areas of investigation will be discussed in detail in this chapter.

6.3 HYDROGELS: DEFINITION AND MAIN USES

Hydrogels are three-dimensional (3D) networks based on crosslinked hydrophilic polymer/s (i.e. polysaccharides, proteins or synthetic polymers) that contain large amounts of water (> 90 %).

Hydrogels are employed in a wide range of medical and pharmaceutical applications due to their: i) high biocompatibility, ii) non-toxicity, iii) tunable mechanical properties that can be close to human tissues, and iv) ability to encapsulate and deliver biological agents (i.e. small molecules, nucleic acids, proteins and therapeutic cells) and water-soluble drugs [40]. Altogether, these features make hydrogels great candidates for the controlled delivery of biological agents and/or drugs in therapeutic approaches. Typically, hydrogels are formed from a polymer solution (liquid state) that turns to a solid-like structure when the polymeric chains crosslink between them, generating an intricate 3D network. There are different crosslinking mechanisms to form hydrogels and, for this reason, when developing a hydrogel-based approach, it is essential to select a crosslinking mechanism compatible with the biological and/or therapeutic agents to be embedded. Based on the crosslinking strategy, hydrogels can be obtained through physical or chemical mechanisms (Figure 6.1).

Physical crosslinking takes place through weak interactions between polymer networks (i.e. entangled chains, hydrogen bonding, charge/ionic interactions, and crystallization) that result in reversible hydrogels that can go from the liquid to the solid state and vice versa as many times as required [41]. In recent years, physically crosslinked hydrogels have attracted a lot of interest in the medical and pharmaceutical fields due to the avoidance of toxic crosslinking agents. This crosslinking mechanism is usually triggered through the addition of charged compounds or by pH or temperature changes. For instance, alginate, one of the most widely used polymers in the biomedical field, can form hydrogels by ionic interactions through the linkage of carboxylic groups in the mannuronic and glucuronic acid residues in the presence of calcium ions [42].

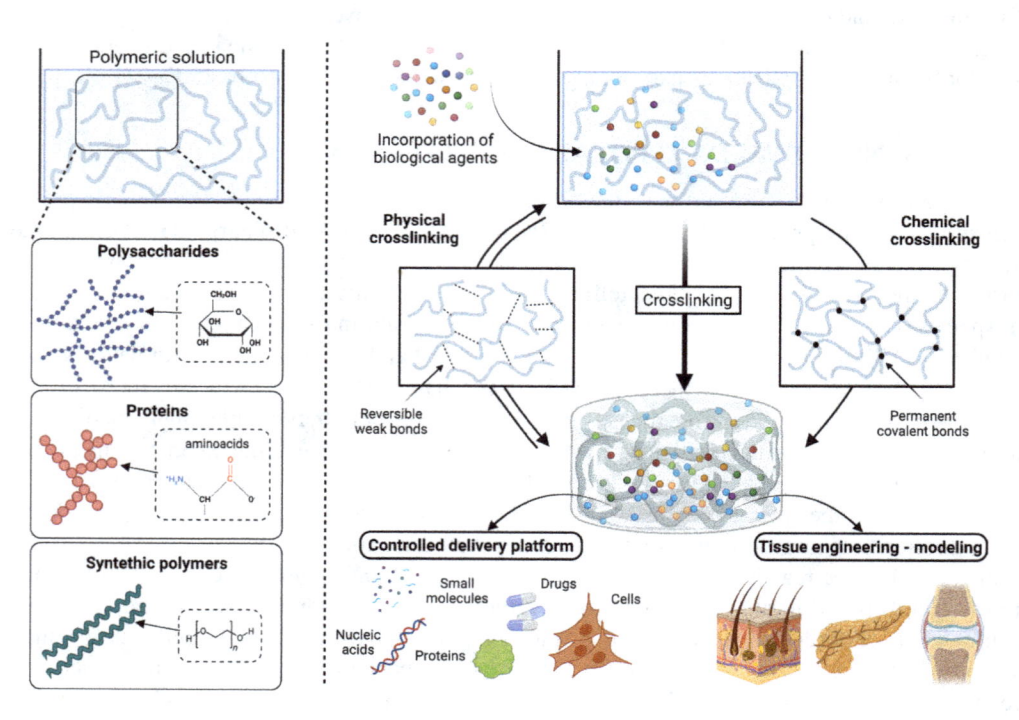

FIGURE 6.1 The main hydrophilic polymers used to form hydrogels in biomedical applications are polysaccharides, proteins and synthetic polymers, and they can form hydrogels through physical or chemical crosslinking. Biological agents can be incorporated into the polymer solution and remain entrapped within the hydrogel after the crosslinking process. Hydrogels can then be used as controlled delivery platforms of therapeutic agents, or to generate tissue-engineered constructs for tissue and/or diseases modeling. Figure created with Biorender®.

In this case, the crosslinking process can be carried out under mild conditions at room temperature and physiological pH, which is compatible with the incorporation of biological agents within the alginate matrix. For this reason, alginate hydrogels, as well as other hydrogels with mild crosslinking (i.e. gelatin, poloxamer, polyvinyl alcohol, chitosan, etc.), have been used in many applications such as the fabrication of scaffolds for the encapsulation and delivery of living cells (e.g. β-cells for insulin delivery in Type I diabetes mellitus [43] or cardiosphere-derived cells for cardiac tissue regeneration [44]), the development of biocompatible carriers for the delivery of therapeutic proteins and/or drugs (i.e. VEGF protein to promote wound healing [45] and chemotherapy agents for cancer treatment [46]), or the manufacture of protective patches to accelerate wound healing processes [47]. Another interesting example of a physically crosslinked hydrogel can be found in methylcellulose, a polymer with the ability to form thermosensitive hydrogels. In aqueous solution, methylcellulose is solubilized at cold temperature (4°C), while on heating up to body temperature (37 °C), the methoxy groups from methylcellulose chains crosslink and form a solid hydrogel [48]. Room temperature injectability, followed by hydrogel formation during subsequent heating to physiological temperature (in situ crosslinking hydrogels), allow the development of minimally invasive strategies to administer hydrogels to deliver therapeutic agents locally on any organ or tissue. Examples of injectable hydrogels in therapeutic applications include filler composites (containing hydroxyapatite particles) to enhance bone regeneration in bone defects [49], corrector dermal agents of depressed cutaneous scars [47], or even painkiller formulations for the treatment of painful knee osteoarthritis [50]. Therefore, physical hydrogels represent a fascinating tool for a large variety of medical and pharmaceutical applications.

However, despite the wide range of applications of physical hydrogels, it is sometimes difficult to tune their physicochemical properties as desired. In this regard, chemical crosslinking can generate hydrogels formed by covalent bonds between polymer chains usually catalyzed by small molecules (i.e. genipin), enzymatic reactions (i.e. horseradish peroxidase (HRP)/H_2O_2 system) or photocuring agents (acrylate or thiol groups/UV irradiation system) [51]. This crosslinking method generates irreversible hydrogels and enables precise control of their key properties, such as gelation time, mesh pore size, stiffness, potential for chemical functionalization, and degradation rate. Photo-crosslinking is one of the most frequently used strategies to generate chemically crosslinked hydrogels. Briefly, a polymer (e.g. gelatin, polyethylene glycol, hyaluronic acid, etc.) is functionalized with photosensitive functional groups (i.e. acrylate or methacrylate groups). Next, in the presence of a photoinitiator, free radicals are generated upon visible light irradiation and react with the photosensitive functional groups, leading to chemical bonds between the polymer chains, and subsequently, to the formation of a solid hydrogel. By modifying the proportion of polymer, functional groups, photoinitiator and light exposure time, the properties of the resulting hydrogels can be tuned as desired. In the same vein, the other chemical crosslinking methods mentioned here also permit tuning of the physicochemical properties of the hydrogels by modulating the parameters involved in the reaction, thus enabling design of hydrogels for specific applications. For example, chemical hydrogels have been explored for developing degradable platforms for a controlled and prolonged delivery of siRNA to enhance the osteogenic differentiation in bone regeneration [52], and also for the fabrication of healthy and pathologic tissue-specific *in vitro* models with similar mechanical properties to soft physiological tissues (i.e. cartilage, vascular tissue, skin, etc.) as drug-screening platforms [53]. The application of hydrogels remains a highly active area of research with a growing number of applications in the medical field.

With the expansion of plasma medicine, the great versatility and resourcefulness of hydrogels has recently caught the attention of the scientific community. Up to now, hydrogels have been explored for two specific objectives of interest in the plasma medicine field (Figure 6.2): i) hydrogels as model systems for living tissues, meaning that gelled or crosslinked hydrogels are treated with plasma; and ii) hydrogels to store therapeutic RONS and deliver them to the living tissues.

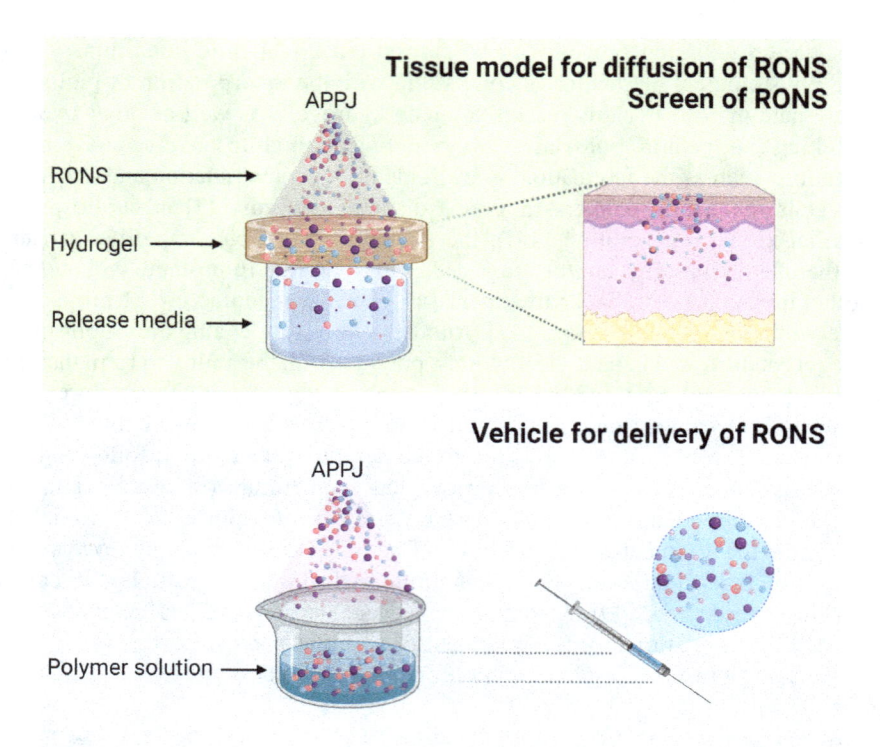

FIGURE 6.2 Current strategies related to treatment of hydrogels for applications in plasma medicine. (Top) Hydrogels are used as surrogates of tissues to investigate the penetration depth of plasmas, or also as screens for certain RONS during plasma treatment. (Bottom) Biopolymers in liquid solution with ability to crosslink are employed to generate and deliver RONS locally to the diseased site. Figure reproduced from [54].

6.4   HYDROGELS AS TISSUE MODELS AND AS SHIELDS FOR DIFFUSION OF RONS

Within the first area of study of hydrogels in plasma medicine, as model systems for living tissues, the main focus has been on understanding the radius of action, surface distribution [55–64] and depth of penetration of the plasma species within the tissues [55,60–66]. The main aim has been to study the transport of RONS though tissues by measuring their concentration in fluids below [58,62,65–72] and employing theoretical modelling to explain the diffusion of RONS in tissues [73,74]. Agarose [55,56,75,76,57–59,65,67–70] and gelatin [61–64,66,71,72,77] were the main hydrogels used to understand potential mechanisms involved during plasma treatment of skin and of soft tissues. Some details about the works published in this field are collected in Table 6.1.

The distribution of plasma-generated RONS on the surface of solid hydrogels was studied using colorimetric and fluorescent probes that were incorporated within the hydrogels before the treatment. The main methods used were KI-starch for generic RONS [58,63,64], Griess reagent for nitrite ions [59], Ti(IV) oxysulphate and ortho-phenylenediamine or Amplex Red for hydrogen peroxide [55,56,59,62], indigo for ozone [59], terephthalate for OH radicals [59] and fluorescein-based probes for pH and generic RONS [55,57,61]. In most cases, the treatment of the hydrogel surface using plasma jets generated a radial distribution of reactive species, with intensity concentration usually decreasing moving away from the center [55,57,58,60,63,64]. However star-shaped patterns were also reported by treating a gelatin layer containing a fluorescent dye with a helium plasma jet [61,62]. The distribution of RONS on the surface of hydrogels was also observed to depend on the feed gas, argon being more efficient in the transport of RONS than helium, and on the orientation

TABLE 6.1

Hydrogels employed as tissue surrogates to investigate the effects of atmospheric pressure plasmas, with indication of the kind of polymer, concentration (in %$_{w/v}$), thickness (mm) of the hydrogel evaluated, type of plasma, type of RONS measured, aim of the work and main results.

Polymer	Concentration and thickness	Plasma type (feed gas)	RONS measured	Action of the hydrogel with respect to RONS	Objective	Main results	Ref.
Agarose	0.6%, 1.5%, 3% 1 mm	DBD (air)	H_2O_2 pH	Distribution Penetration	Simulation of an open wound. Analysis of H_2O_2 and pH distribution and permeation.	Penetration depth of H_2O_2 in the tissue for different polymer concentrations. Validation of the model.	[55]
	1.5% 5-10 mm	DBD APPJ (He)	H_2O_2	Distribution	Propagation of plasma bullets inside gel tubes mimicking tissue.	Plasma bullet propagation in conductive tubes with different shapes and generation of H_2O_2 and ·OH.	[56]
	2% 4 mm	DBD APPJ (He)	H_2O_2 NO_2^- NO_3^-	Transport	Monitor RONS transport through a tissue surrogate into deionized water.	Identification and quantification of RONS passing through the hydrogel layer via UV absorption spectroscopy. Reactivity of RONS within the hydrogel.	[67]
	0.7%, 1%, 2% 1, 2.2, 3.2 mm	DBD APPJ (He)	H_2O_2 NO_2^- NO_3^-	Transport	Monitor RONS transport through a tissue surrogate into deionized water.	RONS transport through and generation in hydrogel layer. Effects of hydrogel concentration and thickness. Effects of plasma treatment time and distance.	[68]
	2% 3mm	APPJ (He, Ar)	H_2O_2 NO_2^- NO_3^-	Transport	Monitor RONS transport through a tissue surrogate into deionized water.	RONS and O_2 transport through and generation in hydrogel layer. Effect of different feed gases.	[69]
	2% 3.2 mm	APPJ (He)	Gas-phase species Ions	Transport	Monitor transport of gaseous neutral and ionic species through a tissue surrogate with MS.	Transport of neutral gaseous species and plasma generated ions through hydrogel layer. Accumulation effect slows the release.	[70]
	2% 2 mm	APPJ (He)	pH	Distribution	Investigation of acidification in the surface of a tissue model.	Visualization and quantification of plasma-induced acidification of the hydrogel surface.	[57]
	2% 2 mm	APPJ (Ar)	RONS	Penetration Transport	Comparison of UV photolysis *vs* plasma jet in the generation of RONS in a tissue surrogate.	Ability of plasma jets to generate reactive species deep within model tissues.	[65]

(Continued)

TABLE 6.1 *(Continued)*

Hydrogels employed as tissue surrogates to investigate the effects of atmospheric pressure plasmas, with indication of the kind of polymer, concentration (in $\%_{w/v}$), thickness (mm) of the hydrogel evaluated, type of plasma, type of RONS measured, aim of the work and main results.

Polymer	Concentration and thickness	Plasma type (feed gas)	RONS measured	Action of the hydrogel with respect to RONS	Objective	Main results	Ref.
	1.5% 2 mm	APPJ (He)	H_2O_2 RONS	Distribution Transport	Monitor RONS transportation though and distribution in phantom tissue.	Quantification of RONS distributed on the surface and penetrated within the hydrogel layer.	[58]
	2% n.d.	DBD (air)	—	—	*In vitro* model for onychomycosis treatment.	Plasma is an effective, non-invasive therapy for onychomycosis. Validation of the model, need for clinical studies.	[75]
	1% 1 mm	sDBD (He, air)	$\cdot OH$, O_3, H_2O_2, NO_2^-, $OONO^-$	Distribution	Study of the distribution of RONS in the surface of a tissue model for inactivation of yeast.	Different distributions of different RONS according to the feed gas. Effective yeast inactivation.	[59]
	1.5% 3 mm	DBD (air)	—	—	Decomposition of tattoo inks dispersed in a skin model.	Visible degradation of inks within the hydrogel via oxidation of the pigments.	[76]
Collagen	2.5% 2-3 mm	DBD (air)	H_2O_2 NO_2^- NO_3^-	Distribution Penetration	Treatment of lung cancer cells in a 3D collagen matrix. Analysis of RONS penetration.	Better performance by 3D model than 2D models. Reduction of lung cancer viability. High RONS penetration.	[60]
Gelatin	5% 2mm	APPJ (He)	RONS	Distribution Penetration	Model for studying the interaction of plasma with biological systems using fluorescent dye.	Effect of treatment time, distance and angle. Heterogeneous effect on the gelatin surface. Penetration within the hydrogel.	[61]
	10% 1.5 mm	APPJ (He)	H_2O_2 ROS	Distribution Penetration Transport	Study of the delivery of ROS through a tissue model and into the surface	Heterogeneous effect on the surface. Penetration in the hydrogel. Molecularly specific reactions of ROS with matrix.	[62]
	5% 1 mm	APPJ (He)	$\cdot OH$ H_2O_2 NO_2^-	Transport	Study of the influence of protein and O_2 on the generation and transport RONS in tissue.	Modification of the amount of RONS due to reactions with proteins and O_2, passing through organic peroxides generation.	[71]

1.5-30% 1, 12 mm	sDBD (air)	H_2O_2, O_3 NO_2^- NO_3^-	Penetration Transport	Investigation of the barrier effects of tissue models on the plasma-generated RONS.	Measurement of RONS permeation in and transport through the hydrogel layer. Small electric fields reduce the barrier effect.	[66]
10% 2 mm	APPJ (He)	H_2O_2	Transport	Establishing a methodology to assess plasma-induced DNA-strand breaks in a tissue model.	Plasma induces DNA-strand breaks in the tissue model. Effect reduced after the application of hydrogel dressing. Transportation of H_2O_2 through the dressing.	[72]
10%, 30% 30 mm	APPJ (He+O_2)	ROS	Distribution Penetration	Studying the distribution and permeation of ROS in a model tissue using He+O_2 plasma.	Different surface distributions by changing the treatment parameters. ROS diffusion and penetration in the hydrogel.	[63]
10%, 15%, 30% 30 mm	APPJ (Ar+O_2)	ROS	Distribution Penetration	Studying the distribution and permeation of ROS in a model tissue using Ar+O_2 plasma.	Different surface distributions by changing the treatment parameters. Higher ROS transport area using Ar as feed gas.	[64]
15% 2 mm	APPJ (He, Ar)	—	—	Decomposition of tattoo inks dispersed in a skin model.	Gelatin scavenges RONS. No degradation of pigment. Thermal effects on gelatin.	[77]

DBD: dielectric barrier discharge; sDBD: surface dielectric barrier discharge; APPJ: atmospheric pressure plasma jet and distance of the plasma source relative to the hydrogel surface [63,64]. Moreover, different distributions were measured by employing specific probes for different RONS [59].

The same chemical probes that were used to study RONS distribution on the hydrogel surfaces were also used to study their penetration depth. By treating agarose with a DBD air plasma, several millimoles of H_2O_2 are produced in the hydrogels, diffusing at depths of 1.5–3.5 mm. With gelatin and collagen, smaller penetration depth was measured, depending on the type of plasma used. In 10 % gelatin, depths of 0.250 mm were reached using an Ar jet for 4 min [64], 0.400 mm using a He jet [63] and 2 mm using a surface DBD in air [66]. The penetration depth of RONS has been reported to increase with the plasma treatment time [55,62–64,66]. A clear inverse dependence of the penetration depth on the biopolymer concentration was also observed in the case of gelatin hydrogels [62–64,66], while for agarose this dependence was not clear [55]. The application of small external electric fields (voltage 5–20 V, positive or negative) was proposed to increase the permeation of RONS through tissues, showing that the penetration depth of hydrogen peroxide and nitrite ions in gelatin increased by 20–30%, depending on the polymer concentration [66].

One of the main uses of solid hydrogels in plasma medicine is to simulate the transport of reactive species through living tissues such as skin. To that aim, hydrogel thin layers are put on top of a vessel containing a water solution, and the plasma treatment is applied directly to the hydrogel surface. The RONS that are able to enter the liquid after passing through the hydrogel layer can be detected directly by UV absorption spectroscopy [65,67–69] or indirectly using the colorimetric and fluorescent probes discussed earlier [58,62,65,66,71,72]. Some studies showed how the type and concentrations of RONS detected in water solution after being transported through a hydrogel thin layer depend on the hydrogel itself (composition, concentration, thickness) and on the plasma treatment parameters (feed gas, treatment time, distance). In [68] it was reported that the total amount of RONS passing through an alginate layer—estimated using the total UV absorbance between 190 and 340 nm, where the main plasma-generated RONS absorb—increases by decreasing the hydrogel thickness and does not depend significantly on its concentration, similarly to what was observed in the permeation experiments using alginate described earlier [55]. In the case of gelatin, [66] reports a clear decrease in the RONS concentration as the polymer concentration increases. In general, the amount of RONS was increased by increasing treatment time and decreasing the gas flow and distance between the plasma nozzle and the hydrogel [67,68]. Some researchers compared the type and concentration of RONS transferred in the liquid phase by a helium APPJ with and without an agarose layer in between [67–69] and found that, in absence of the hydrogel layer, the concentration of RONS increased linearly with the treatment time and remained stable once the plasma was turned off, while in the experiments with the agarose layer, there was a lag period of about 20 min before the amount of RONS in solution started to increase, and the rate of increase was slower than in the condition without agarose. Then, when the jet was switched off, the increase of RONS continued with a steeper slope. This difference may be due to slow transport of species through the polymer layer and/or to specific reactions happening within the polymer phase between primary RONS and the polymer itself that generate secondary RONS and release them in the liquid.

The transport of RONS through a thin agarose layer was also studied in the gas phase by mass spectrometry [70]. A 3.2 mm-thick agarose target was placed between a helium plasma jet and an ambient mass spectrometer. This experiment enabled measurement of the transport of neutral gas species (N_2, O_2, H_2O) and of plasma-generated cations (H^+, H_2O^+, O_2^+, O^+, N_2^+, NO^+, HN_2^+) and anions (H^-, O^-, O_2^-, O_3^-, OH^-, NO_2^-, NO_3^-). Similar to what was observed in liquid experiments, the authors observed a lag time before detecting the species through the hydrogel layer, indicating an accumulation and subsequent release.

Besides the studies listed above, which focused on the reactive species generated by plasma, some articles also reported other applications of hydrogels as tissue models within the field of plasma medicine. Park et al. used agarose hydrogels loaded with a fluorescent dye to study the propagation of plasma bullets within small channels to explore the possibility of treating internal diseases [56]. Szili and co-workers studied the interaction of plasma with DNA and its ability to break DNA strands within a synthetic gelatin tissue, allowing the design of simpler, faster and more cost-effective experiments than by using real tissues [72]. Karki et al. published the treatment of lung cancer cells (A549) within a collagen matrix, showing better performance and reliability than the usual 2D cancer models [60]. Regarding antimicrobial applications, Bulson et al. used a toe/nail-plate model made of cadaver nails and agarose media inoculated with *Candida albicans* to study the potential of plasma in the treatment of onychomycosis [75], and Du and co-workers studied the inactivation of *Saccharomyces cerevisiae* in an agarose substrate [59]. Finally, two recent works used agarose- and gelatin-based hydrogel loaded with tattoo inks to explore the feasibility of plasma-assisted tattoo removal [76,77]. While some measurable discoloration effect was reported in the case of alginate, this was not the case with the gelatin models because the amino-acid-based polymer acted as a scavenger of plasma-generated RONS, possibly due to its higher chemical complexity compared with a polysaccharide.

6.5 HYDROGELS AS VEHICLE FOR DELIVERY OF PLASMA-GENERATED RONS

Interest in the use of hydrogels for storage and delivery of plasma-generated RONS derives from an important intrinsic limitation of plasma-treated liquids: their physical state (see also Chapter 13). The therapeutic effect of plasma-treated liquids (PTLs) depends, among other factors, on the time they remain in contact with the target. If this contact time can easily be adjusted in *in vitro* experiments, this is not always true in real applications. When PTLs are injected in the intended patient site, i.e. in a tumor environment, their permanence is limited because they are ultimately washed away by body fluids. Therefore, the therapeutic effect of the RONS that are contained in these liquids may substantially decrease. Using hydrogels for storage and delivery of plasma-generated RONS can help overcome this limitation since gels can remain in the application site for a longer time and eventually release RONS more slowly than liquids can for prolonged therapies.

At present, only a few studies deal with plasma-treated biopolymer solutions and hydrogels with the aim of developing biomaterials acting as vehicles of RONS (Table 6.2).

Biopolymers' solutions and hydrogels can be loaded with plasma-generated RONS in two ways (Figure 6.3): i) by first treating liquids (water, saline solution, cell media, etc.) to generate PTLs, and then mixing them with biopolymers (powder or solution) [78], or ii) by directly treating biopolymers solutions [79–82] or gels [59,83] with plasma.

In the first case, the diffusion of plasma-generated RONS from the gas to the liquid phase is more efficient because of the lower viscosity of liquids compared with biopolymer-containing solutions or hydrogels.

Another difference between the two loading modalities is that, if biopolymers are added to solutions already treated by plasma, they are exposed only to plasma-generated long-lived species (H_2O_2, HNO_2/NO_2^-), while in the second case they interact with all the components that are generated during a direct treatment (short- and long-lived species, electromagnetic radiation). This is a more reactive environment that can partially modify the biopolymer's structure and functionality. Two recent papers demonstrated how non-thermal plasma treatment is able to break methylcellulose and alginate chains to generate fragments with very wide molecular weight distribution [81,84]. This plasma-induced fragmentation has been related to a decrease in the viscosity of the treated solution [84]. In the case of alginate, plasma has also been reported to modify the carboxylic groups of the polymer that are responsible for the gel formation by interaction with divalent ions, thus decreasing the cross-linking ability of the polymer, and to generate organic peroxides along the chains [84]. The capability of non-thermal plasma to oxidize some chemical groups and to break chains has been reported in the literature also for other biocompatible oligomers and polymers [85–89]. The interaction of short-lived RONS with biopolymers has been observed also during the treatment of solid agarose and gelatin hydrogels used as tissue models, as described in the previous section [62,65,67–69,71].

The direct treatment of biopolymers solutions and therefore their interaction with short-lived RONS can have an effect also on the amount of long-lived species that are generated in solution.

TABLE 6.2

Summary of polymer solutions evaluated in plasma treatment, their crosslinking method and conditions to obtain the hydrogel, together with the aim of the work and main results obtained.

Polymer	Crosslinking	Concentration and volume	Plasma type (feed gas)	Objective	Main results	Ref.
Alginate	Physical (ionic interactions with Ca^{2+})	0.5 %$_{w/v}$ 200 µL	APPJ (He, Ar)	Vehicle for RONS storage and delivery; Biocompatibility and cytotoxicity	• Enhanced RONS generation • No change in physico-chemical properties • Low RONS stability during crosslinking • Fast RONS release • Poor cytotoxicity with healthy cells	[79]
		Solid state	DBD	Antimicrobial wound dressing	• Strong, long-term biocidal effect • No cytotoxicity • No surface damage by plasma	[83]
		0.5 %$_{w/v}$ 1 mL	APPJ (He)	Study the interaction between alginate and plasma	• Enhanced RONS generation • Fragmentation of alginate chains • Modification of carboxylic groups • Production of organic peroxides	[84]
Agarose	Physical (Thermal)	1 %$_{w/v}$ 1.05 mL	sDBD (He, air)	Vehicle for RONS storage and delivery for antimicrobial applications	• Different distribution of RONS in the gel • Yeast inactivation	[59]
Gelatin/alginate composite	Physical (Ionic interaction with Ca^{2+})	1/0.25 %$_{w/v}$ 200 µL	APPJ (Ar)	Vehicle for RONS storage and delivery for cancer therapy and bone regeneration	• Selective anticancer activity • Local RONS delivery • Safety in *in vitro* and *in vivo* experiments	[82]
Gelatin	Physical (Thermal)	1 %$_{w/v}$ 200 µL	APPJ (He, Ar)	Vehicle for RONS storage and delivery for cancer therapy	• Enhanced RONS generation • Partial RONS stability • Fast RONS release • Selective anticancer activity	[80]
Hydroxyethyl cellulose	Physical (Thermal)	3 %$_{w/v}$	sDBD	Treatment of vitiligo	• Improvement of lesions • No side effects	[78]
Matrigel	Physical (Thermal)	— 1 mL	DBD (air)	Effect of plasma treatment on *in vivo* bone formation	• Modification of the extracellular matrix • Influence on cellular behavior accelerating or inhibiting chondrogenesis and endochondral ossification	[91]
Methylcellulose	Physical (Thermal)	1 %$_{w/w}$ 1 mL	APPJ (Ar)	Vehicle for RONS storage and delivery; Anticancer activity	• Tunable gel point • Fragmentation of polymer chains • Enhanced RONS generation • Good stability of RONS in solution • Fast RONS release • Cytotoxicity (cancer cells)	[81]

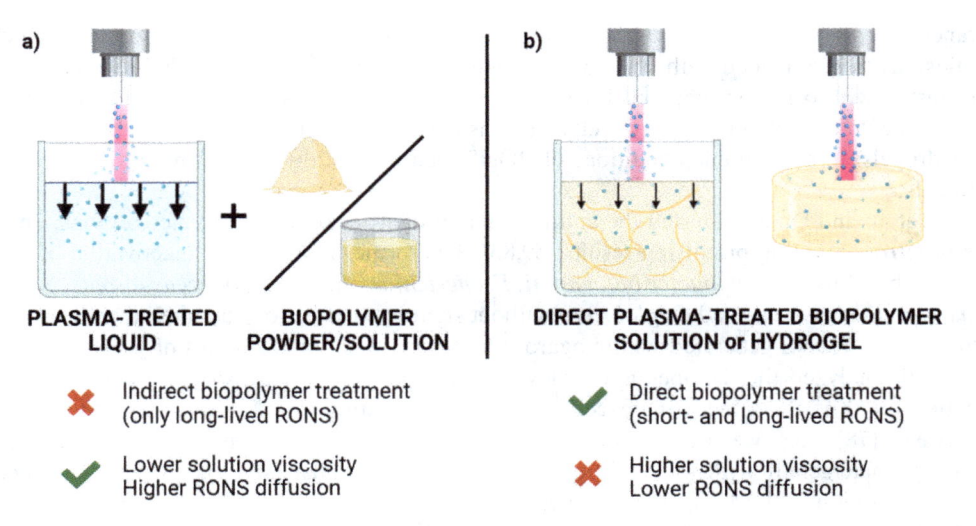

FIGURE 6.3 Methods for preparation of RONS-loaded biopolymer solutions and hydrogels with associated advantages or drawbacks. Figure created with Biorender®.

Alginate, gelatin and methylcellulose have been reported to enhance the generation of long-lived RONS when exposed to direct plasma treatment using helium or argon jets, if compared with polymer-free solutions [79–81,84]. Two recent papers reported the quantification of hydrogen peroxide and nitrite ions generated in $0.5\%_{w/w}$ alginate or $2\%_{w/w}$ gelatin solutions and Ringer's saline or ultrapure water using two atmospheric plasma jets working with helium and argon and explored different gas-flow rate and treatment distances [79,80]. The concentration of both species was more than one order of magnitude higher in the presence of the polymer. This difference was mainly attributed to the buffering effect of the polymers that prevented the pH from dropping, because of the plasma treatment, to a region where hydrogen peroxide and nitrite ions are not stable. In similar experiments from the same groups, $0.5\%_{w/w}$ alginate and $1\%_{w/w}$ methylcellulose were dissolved in phosphate buffer at neutral pH instead of non-buffered media [81,84]. In this case, the presence of polymers enhanced the amount of RONS in solution, albeit to a lesser extent, if compared to the previous cases that employed non-buffered solvents, and the difference was attributable not to pH differences but to the very nature of the polymers. At the same time, the polymers acted as scavengers for short-lived RONS (OH radicals and O atoms) [81,84]. These apparently conflicting effects—on the one hand the scavenging of short-lived RONS, and on the other the enhancement of long-lived ones—are due to the activation of new chemical pathways in solution that lead to the generation of secondary RONS. This effect was more marked in the case of gelatin than in alginate, likely due to its higher chemical complexity and reactivity. The mechanisms behind these effects are still largely unknown. A hypothesis was recently proposed in the case of alginate, involving the generation of organic peroxides [84], and with some specific amino acids, i.e. cysteine and tyrosine, involving the reactivity of the thiol group and the aromatic ring, respectively [90].

The stability and release kinetics of long-lived reactive species have also been investigated in alginate, methylcellulose and gelatin solutions and hydrogels [79–81]. Both hydrogen peroxide and nitrite ions are stable in polysaccharide solutions (alginate and methylcellulose) for at least 72 hours at physiological conditions (37 °C and neutral pH), but, critically, more than half of the amount was lost during the crosslinking process. In gelatin hydrogels, hydrogen peroxide was stable for up to 72 hours, and nitrite ions were reduced by 20–30%. The release of RONS from the hydrogels was studied at physiological conditions, and in all cases was it reached the maximum value within 20 min.

The release of RONS from hydrogels can be of interest in several biomedical applications, such as tissue engineering, cancer therapy, antimicrobial protection and regenerative medicine. For instance, *in vitro* tests have shown the selective anticancer activity of alginate, gelatin and methyl-cellulose hydrogels loaded with plasma-generated RONS [79–82]. These RONS-loaded hydrogels were tested with osteosarcoma cell lines (SaOs-2, MG-63) and human mesenchymal stem cells, showing a cytotoxic effect on cancer cells that was dependent on the amount of RONS delivered by the hydrogels. In comparable conditions of RONS loading, healthy cells were not affected by the hydrogels.

Poor et al. and Du et al. tested the antimicrobial properties of plasma-treated alginate and aga-rose *in vitro*, obtaining promising results [59,83]. The former reported the inactivation of various pathogen biofilms (*Acinetobacter baumannii*, *Escherichia coli*, *Staphylococcus aureus*, *S. epider-midis*, *Candida albicans* and *C. glabrata*) without significant effect on endothelial cells. The latter correlated the RONS generated within agarose hydrogels with the outcomes of yeast inactivation. Zhai et al. analyzed the therapeutic efficacy of a plasma-treated hydroxyethyl cellulose hydrogel in patients with vitiligo, reporting excellent efficacy in treating this skin disorder and no relevant side effects [78]. Solé-Martí et al. showed, in an *in vivo* study using rabbits, the ability of plasma-treated alginate/gelatin hydrogel, infiltrated in a calcium phosphate scaffold, to not affect healthy bone regeneration [82]. Matrigel, a commercially available extra-cellular matrix, was employed in [91] to study whether plasma treatment could potentially favor cell growth in views of applications in bone and cartilage tissue engineering. In this work, matrigel hydrogels treated with different plasma sources (micro-second plasma and nano-second plasma) were compared in terms of the cell response in a murine ossification model. It was reported that micro-second plasma treatment significantly increased bone formation while nano-second plasma treatment decreased it. These different biological responses are related to the varying amounts of RONS generated from the dif-ferent plasma types.

6.6 CONCLUSIONS

Although plasma treatment of polymers has been used for several decades, the advent of more simple and compact plasma devices that work at atmospheric pressure has opened new avenues in the field. It has moved from treating solid polymers, with the main aim of introducing new chemical moieties on the surface to tune cell behavior (adhesion, proliferation), to delivering and entrapping the reactive species from plasmas in hydrated polymers and also employing them to mimic biologi-cal tissues.

This shift is reflected in the inclusion of this area within plasma medicine, wherein treatment of hydrogels with atmospheric plasmas is emerging as useful in a variety of areas, including for inves-tigating the behavior of plasmas in the treatment of real tissues and their use as delivery systems for RONS. In summary, these novel areas of research are paving the way for exciting discoveries.

REFERENCES

[1] J.B.L.R.S. Park, Biomaterials, 3rd Edition, Springer, New York, NY, 2007. https://doi.org/10.1007/978-0-387-37880-0.
[2] L.G. Griffith, Polymeric biomaterials, Acta Mater. 48 (2000) 263–277. https://doi.org/10.1016/S1359-6454(99)00299-2.
[3] M.S. Shoichet, Polymer scaffolds for biomaterials applications, Macromolecules. 43 (2010) 581–591. https://doi.org/10.1021/ma901530r.
[4] D.L. Elbert, J.A. Hubbell, Surface treatments of polymers for biocompatibility, Annu. Rev. Mater. Sci. 26 (1996) 365–394. https://doi.org/10.1146/annurev.ms.26.080196.002053.
[5] J.M. Goddard, J.H. Hotchkiss, Polymer surface modification for the attachment of bioactive compounds, Prog. Polym. Sci. 32 (2007) 698–725. https://doi.org/10.1016/j.progpolymsci.2007.04.002.
[6] T. Desmet, R. Morent, N. De Geyter, C. Leys, E. Schacht, P. Dubruel, Nonthermal plasma technology as a versatile strategy for polymeric biomaterials surface modification: A review, Biomacromolecules. 10 (2009) 2351–2378. https://doi.org/10.1021/bm900186s.

[7] E. Stoffels, A.J. Flikweert, W.W. Stoffels, G.M.W. Kroesen, Plasma needle: A non-destructive atmospheric plasma source for fine surface treatment of (bio)materials, Plasma Sources Sci. Technol. 11 (2002) 383–388. https://doi.org/10.1088/0963-0252/11/4/304.

[8] A. Hezi-Yamit, C. Sullivan, J. Wong, L. David, M. Chen, P. Cheng, D. Shumaker, J.N. Wilcox, K. Udipi, Impact of polymer hydrophilicity on biocompatibility: Implication for DES polymer design, J. Biomed. Mater. Res.—Part A. 90 (2009) 133–141. https://doi.org/10.1002/jbm.a.32057.

[9] R.M. Thurston, J.D. Clay, M.D. Schulte, Effect of atmospheric plasma treatment on polymer surface energy and adhesion, J. Plast. Film Sheet. 23 (2007) 63–78. https://doi.org/10.1177/87560879 07078698.

[10] H. Schonhorn, R.H. Hansen, Surface treatment of polymers for adhesive bonding, J. Appl. Polym. Sci. 11 (1967) 1461–1474. https://doi.org/10.1002/app.1967.070110809.

[11] O. Neděla, P. Slepička, V. Švorčík, Surface modification of polymer substrates for biomedical applications, Materials (Basel). 10 (2017) 1115. https://doi.org/10.3390/ma10101115.

[12] Y. Bu, J. Ma, J. Bei, S. Wang, Surface modification of aliphatic polyester to enhance biocompatibility, Front. Bioeng. Biotechnol. 7 (2019) 98. https://doi.org/10.3389/fbioe.2019.00098.

[13] D. Katti, R. Vasita, K. Shanmugam, Improved biomaterials for tissue engineering applications: Surface modification of polymers, Curr. Top. Med. Chem. 8 (2008) 341–353. https://doi.org/10.2174/156802608783790893.

[14] H. Yasuda, M. Gazicki, Biomedical applications of plasma polymerization and plasma treatment of polymer surfaces, Biomaterials. 3 (1982) 68–77. https://doi.org/10.1016/0142-9612(82)90036-9.

[15] D.B. Graves, Low temperature plasma biomedicine: A tutorial review, Phys. Plasmas. 21 (2014) 080901. https://doi.org/10.1063/1.4892534.

[16] P. Favia, E. Sardella, R. Gristina, R. D'Agostino, Novel plasma processes for biomaterials: Microscale patterning of biomedical polymers, Surf. Coat. Technol. 169–170 (2003) 707–711. https://doi.org/10.1016/S0257-8972(03)00174-9.

[17] R. D'Agostino, P. Favia, C. Oehr, M.R. Wertheimer, Low-temperature plasma processing of materials: Past, present, and future, Plasma Process. Polym. 2 (2005) 7–15. https://doi.org/10.1002/ppap.200400074.

[18] E.M.M. Liston, L. Martinu, M.R.R. Wertheimer, Plasma surface modification of polymers for improved adhesion: A critical review, J. Adhes. Sci. Technol. 7 (1993) 1091–1127. https://doi.org/10.1163/156856193X00600.

[19] J.H. Loh, Plasma surface modification in biomedical applications., Med. Device Technol. 10 (1999) 24–30. www.ncbi.nlm.nih.gov/pubmed/10344871.

[20] P.K. Chu, Plasma-treated biomaterials, IEEE Trans. Plasma Sci. 35 (2007) 181–187. https://doi.org/10.1109/TPS.2006.888587.

[21] R. Morent, N. De Geyter, T. Desmet, P. Dubruel, C. Leys, Plasma surface modification of biodegradable polymers: A review, Plasma Process. Polym. 8 (2011) 171–190. https://doi.org/10.1002/ppap.201000153.

[22] N. Gomathi, A. Sureshkumar, S. Neogi, RF plasma-treated polymers for biomedical applications, Curr. Sci. 94 (2008) 1478–1486. www.jstor.org/stable/24100504.

[23] F. Poncin-Epaillard, G. Legeay, Surface engineering of biomaterials with plasma techniques, J. Biomater. Sci. Polym. Ed. 14 (2003) 1005–1028. https://doi.org/10.1163/156856203769231538.

[24] S. Cheruthazhekatt, M. Černák, P. Slavíček, J. Havel, Gas plasmas and plasma modified materials in medicine, J. Appl. Biomed. 8 (2010) 55–66. https://doi.org/10.2478/v10136-009-0013-9.

[25] P. Cools, R. Ghobeira, S. Van Vrekhem, N. De Geyterand, R. Morent, Non-Thermal Plasma Technology for the Improvement of Scaffolds for Tissue Engineering and Regenerative Medicine—A Review, in: T. Mieno (Ed.), Plasma Sci. Technol.—Prog. Phys. States Chem. React., InTech, 2016. https://doi.org/10.5772/62007.

[26] S. Kalghatgi, D. Dobrynin, G. Fridman, M. Cooper, G. Nagaraj, L. Peddinghaus, M. Balasubramanian, K. Barbee, A. Brooks, V. Vasilets, A. Gutsol, A. Fridman, G. Friedman, Applications of Non Thermal Atmospheric Pressure Plasma in Medicine, in: NATO Secur. through Sci. Ser. A Chem. Biol., Springer, Dordrecht, 2008: pp. 173–181. https://doi.org/10.1007/978-1-4020-8439-3_15.

[27] M. Laroussi, Plasma medicine: A brief introduction, Plasma. 1 (2018) 47–60. https://doi.org/10.3390/plasma1010005.

[28] G. Fridman, G. Friedman, A. Gutsol, A.B. Shekhter, V.N. Vasilets, A. Fridman, Applied plasma medicine, Plasma Process. Polym. 5 (2008) 503–533. https://doi.org/10.1002/ppap.200700154.

[29] M. Keidar, Plasma for cancer treatment, Plasma Sources Sci. Technol. 24 (2015) 033001. https://doi.org/10.1088/0963-0252/24/3/033001.

[30] M. Laroussi, Low-temperature plasma jet for biomedical applications: A review, IEEE Trans. Plasma Sci. 43 (2015) 703–712. https://doi.org/10.1109/TPS.2015.2403307.

[31] G. Isbary, T. Shimizu, Y.F. Li, W. Stolz, H.M. Thomas, G.E. Morfill, J.L. Zimmermann, Cold atmospheric plasma devices for medical issues, Expert Rev. Med. Devices. 10 (2013) 367–377. https://doi.org/10.1586/erd.13.4.

[32] H.-R. Metelmann, C. Seebauer, V. Miller, A. Fridman, G. Bauer, D.B. Graves, J.-M. Pouvesle, R. Rutkowski, M. Schuster, S. Bekeschus, K. Wende, K. Masur, S. Hasse, T. Gerling, M. Hori, H. Tanaka, E. Ha Choi, K.-D. Weltmann, P.H. Metelmann, D.D. Von Hoff, T. von Woedtke, Clinical experience with cold plasma in the treatment of locally advanced head and neck cancer, Clin. Plasma Med. 9 (2018) 6–13. https://doi.org/10.1016/j.cpme.2017.09.001.

[33] E.J.J. Szili, S.-H.H.H.S.H.H. Hong, J.S.S.J.-S. Oh, N. Gaur, R.D.D. Short, Tracking the penetration of plasma reactive species in tissue models, Trends Biotechnol. 36 (2018) 594–602. https://doi.org/10.1016/j.tibtech.2017.07.012.

[34] D.B. Graves, The emerging role of reactive oxygen and nitrogen species in redox biology and some implications for plasma applications to medicine and biology, J. Phys. D. Appl. Phys. 45 (2012) 263001. https://doi.org/10.1088/0022-3727/45/26/263001.

[35] Y. Sato, S. Yamada, S. Takeda, N. Hattori, K. Nakamura, H. Tanaka, M. Mizuno, M. Hori, Y. Kodera, Effect of plasma-activated lactated ringer's solution on pancreatic cancer cells *in vitro* and *in vivo*, Ann. Surg. Oncol. 25 (2018) 299–307. https://doi.org/10.1245/s10434-017-6239-y.

[36] D. Yan, J.H.H. Sherman, M. Keidar, The application of the cold atmospheric plasma-activated solutions in cancer treatment, Anticancer. Agents Med. Chem. 18 (2017) 769–775. https://doi.org/10.2174/1871520617666170731115233.

[37] D. Yan, H. Cui, W. Zhu, N. Nourmohammadi, J. Milberg, L.G. Zhang, J.H. Sherman, M. Keidar, The specific vulnerabilities of cancer cells to the cold atmospheric plasma-stimulated solutions, Sci. Rep. 7 (2017) 4479. https://doi.org/10.1038/s41598-017-04770-x.

[38] S. Takeda, S. Yamada, N. Hattori, K. Nakamura, H. Tanaka, H. Kajiyama, M. Kanda, D. Kobayashi, C. Tanaka, T. Fujii, M. Fujiwara, M. Mizuno, M. Hori, Y. Kodera, Intraperitoneal administration of plasma-activated medium: Proposal of a novel treatment option for peritoneal metastasis from gastric cancer, Ann. Surg. Oncol. 24 (2017) 1188–1194. https://doi.org/10.1245/s10434-016-5759-1.

[39] W. Van Boxem, J. Van der Paal, Y. Gorbanev, S. Vanuytsel, E. Smits, S. Dewilde, A. Bogaerts, Anticancer capacity of plasma-treated PBS: Effect of chemical composition on cancer cell cytotoxicity, Sci. Rep. 7 (2017) 16478. https://doi.org/10.1038/s41598-017-16758-8.

[40] S. Correa, A.K. Grosskopf, H. Lopez Hernandez, D. Chan, A.C. Yu, L.M. Stapleton, E.A. Appel, Translational applications of hydrogels, Chem. Rev. 121 (2021) 11385–11457. https://doi.org/10.1021/acs.chemrev.0c01177.

[41] F. Rizzo, N.S. Kehr, Recent advances in injectable hydrogels for controlled and local drug delivery, Adv. Healthc. Mater. 10 (2021) 2001341. https://doi.org/10.1002/adhm.202001341.

[42] A. Espona-Noguera, J. Ciriza, A. Cañibano-Hernández, L. Fernandez, I. Ochoa, L. Saenz del Burgo, J.L. Pedraz, Tunable injectable alginate-based hydrogel for cell therapy in Type 1 Diabetes Mellitus, Int. J. Biol. Macromol. 107 (2018) 1261–1269. https://doi.org/10.1016/j.ijbiomac.2017.09.103.

[43] A. Espona-Noguera, J. Etxebarria-Elezgarai, L. Saenz del Burgo, A. Cañibano-Hernández, H. Gurruchaga, F.J. Blanco, G. Orive, R.M. Hernández, F. Benito-Lopez, J. Ciriza, L. Basabe-Desmonts, J.L. Pedraz, Type 1 Diabetes Mellitus reversal via implantation of magnetically purified microencapsulated pseudoislets, Int. J. Pharm. 560 (2019) 65–77. https://doi.org/10.1016/j.ijpharm.2019.01.058.

[44] C. Báez-Díaz, V. Blanco-Blázquez, F.M. Sánchez-Margallo, E. López, H. Martín, A. Espona-Noguera, J.G. Casado, J. Ciriza, J.L. Pedraz, V. Crisóstomo, Intrapericardial delivery of APA-microcapsules as promising stem cell therapy carriers in an experimental acute myocardial infarction model, Pharmaceutics. 13 (2021) 1824. https://doi.org/10.3390/pharmaceutics13111824.

[45] L. Wei, J. Tan, L. Li, H. Wang, S. Liu, J. Chen, Y. Weng, T. Liu, Chitosan/alginate hydrogel dressing loaded FGF/VE-cadherin to accelerate full-thickness skin regeneration and more normal skin repairs, Int. J. Mol. Sci. 23 (2022) 1249. https://doi.org/10.3390/ijms23031249.

[46] B. Reig-Vano, B. Tylkowski, X. Montané, M. Giamberini, Alginate-based hydrogels for cancer therapy and research, Int. J. Biol. Macromol. 170 (2021) 424–436. https://doi.org/10.1016/j.ijbiomac.2020.12.161.

[47] A. Mandal, J.R. Clegg, A.C. Anselmo, S. Mitragotri, Hydrogels in the clinic, Bioeng. Transl. Med. 5 (2020) e10158. https://doi.org/10.1002/btm2.10158.

[48] B. Niemczyk-Soczynska, A. Gradys, D. Kolbuk, A. Krzton-Maziopa, P. Sajkiewicz, Crosslinking kinetics of methylcellulose aqueous solution and its potential as a scaffold for tissue engineering, Polymers (Basel). 11 (2019) 1772. https://doi.org/10.3390/polym11111772.

[49] L. Deng, Y. Liu, L. Yang, J.-Z. Yi, F. Deng, L.-M. Zhang, Injectable and bioactive methylcellulose hydrogel carrying bone mesenchymal stem cells as a filler for critical-size defects with enhanced bone regeneration, Colloids Surf. B Biointerfaces. 194 (2020) 111159. https://doi.org/10.1016/j.colsurfb.2020.111159.

[50] R.D. Altman, J.E. Rosen, D.A. Bloch, H.T. Hatoum, P. Korner, A double-blind, randomized, saline-controlled study of the efficacy and safety of EUFLEXXA® for treatment of painful osteoarthritis of the knee, with an open-label safety extension (the FLEXX trial), Semin. Arthritis Rheum. 39 (2009) 1–9. https://doi.org/10.1016/j.semarthrit.2009.04.001.

[51] R. parhi, Cross-linked hydrogel for pharmaceutical applications: A review, Adv. Pharm. Bull. 7 (2017) 515–530. https://doi.org/10.15171/apb.2017.064.

[52] M.K. Nguyen, A. McMillan, C.T. Huynh, D.S. Schapira, E. Alsberg, Photocrosslinkable, biodegradable hydrogels with controlled cell adhesivity for prolonged siRNA delivery to hMSCs to enhance their osteogenic differentiation, J. Mater. Chem. B. 5 (2017) 485–495. https://doi.org/10.1039/C6TB01739H.

[53] J.R. Choi, K.W. Yong, J.Y. Choi, A.C. Cowie, Recent advances in photo-crosslinkable hydrogels for biomedical applications, Biotechniques. 66 (2019) 40–53. https://doi.org/10.2144/btn-2018-0083.

[54] I. Adamovich, S. Agarwal, E. Ahedo, L.L. Alves, S. Baalrud, N. Babaeva, A. Bogaerts, A. Bourdon, P.J. Bruggeman, C. Canal, E.H. Choi, S. Coulombe, Z. Donkó, D.B. Graves, S. Hamaguchi, D. Hegemann, M. Hori, H.-H. Kim, G.M.W. Kroesen, M.J. Kushner, A. Laricchiuta, X. Li, T.E. Magin, S. Mededovic Thagard, V. Miller, A.B. Murphy, G.S. Oehrlein, N. Puac, R.M. Sankaran, S. Samukawa, M. Shiratani, M. Šimek, N. Tarasenko, K. Terashima, E. Thomas Jr, J. Trieschmann, S. Tsikata, M.M. Turner, I.J. van der Walt, M.C.M. van de Sanden, T. von Woedtke, The 2022 plasma roadmap: Low temperature plasma science and technology, J. Phys. D. Appl. Phys. 55 (2022) 373001. https://doi.org/10.1088/1361-6463/ac5e1c.

[55] D. Dobrynin, G. Fridman, G. Friedman, A. Fridman, Deep penetration into tissues of reactive oxygen species generated in Floating-Electrode Dielectric Barrier Discharge (FE-DBD): An *in vitro* agarose gel model mimicking an open wound, Plasma Med. 2 (2012) 71–84. https://doi.org/10.1615/PlasmaMed.2013006218.

[56] D. Park, G. Fridman, A. Fridman, D. Dobrynin, Plasma bullets propagation inside of agarose tissue model, IEEE Trans. Plasma Sci. 41 (2013) 1725–1730. https://doi.org/10.1109/TPS.2013.2265373.

[57] G. Busco, A.V. Omran, L. Ridou, J.M. Pouvesle, E. Robert, C. Grillon, Cold atmospheric plasma-induced acidification of tissue surface: Visualization and quantification using agarose gel models, J. Phys. D. Appl. Phys. 52 (2019) 24LT01. https://doi.org/10.1088/1361-6463/ab1119.

[58] A. Omran, G. Busco, S. Dozias, C. Grillon, J.-M. Pouvesle, E. Robert, A. Valinataj Omran, G. Busco, S. Dozias, J. Pouvesle, E. Robert, Distribution and Penetration of Reactive Oxygen and Nitrogen Species through a Tissue Phantom After Plasma Gun Treatment, in: 24th Int. Symp. Plasma Chem., Napoli, 2019. https://hal.archives-ouvertes.fr/hal-02263879.

[59] M. Du, H. Xu, Y. Zhu, R. Ma, Z. Jiao, A comparative study of the major antimicrobial agents against the yeast cells on the tissue model by helium and air surface micro-discharge plasma, AIP Adv. 10 (2020) 025036. https://doi.org/10.1063/1.5110972.

[60] S.B. Karki, T.T. Gupta, E. Yildirim-Ayan, K.M. Eisenmann, H. Ayan, Investigation of non-thermal plasma effects on lung cancer cells within 3D collagen matrices, J. Phys. D. Appl. Phys. 50 (2017) 315401. https://doi.org/10.1088/1361-6463/aa7b10.

[61] S.E. Marshall, A.T.A. Jenkins, S.A. Al-Bataineh, R.D. Short, S.-H. Hong, N.T. Thet, J.-S. Oh, J.W. Bradley, E.J. Szili, Studying the cytolytic activity of gas plasma with self-signalling phospholipid vesicles dispersed within a gelatin matrix, J. Phys. D. Appl. Phys. 46 (2013) 185401. https://doi.org/10.1088/0022-3727/46/18/185401.

[62] E.J. Szili, J.W. Bradley, R.D. Short, A "tissue model" to study the plasma delivery of reactive oxygen species, J. Phys. D. Appl. Phys. 47 (2014) 152002. https://doi.org/10.1088/0022-3727/47/15/152002.

[63] D. Liu, T. He, Z. Liu, S. Wang, Z. Liu, M. Rong, M.G. Kong, Spatial-temporal distributions of ROS in model tissues treated by a He+O2 plasma jet, Plasma Process. Polym. 15 (2018) 1800057. https://doi.org/10.1002/ppap.201800057.

[64] T. He, D. Liu, Z. Liu, S. Wang, Z. Liu, M. Rong, M.G. Kong, Transportation of ROS in model tissues treated by an Ar + O 2 plasma jet, J. Phys. D. Appl. Phys. 52 (2019) 045204. https://doi.org/10.1088/1361-6463/aaed6f.

[65] B. Ghimire, E.J. Szili, P. Lamichhane, R.D. Short, J.S. Lim, P. Attri, K. Masur, K.D. Weltmann, S.H. Hong, E.H. Choi, The role of UV photolysis and molecular transport in the generation of reactive species in a tissue model with a cold atmospheric pressure plasma jet, Appl. Phys. Lett. 114 (2019) 093701. https://doi.org/10.1063/1.5086522.

[66] T. He, D. Liu, H. Xu, Z. Liu, D. Xu, D. Li, Q. Li, M. Rong, M.G. Kong, A "tissue model" to study the barrier effects of living tissues on the reactive species generated by surface air discharge, J. Phys. D. Appl. Phys. 49 (2016) 205204. https://doi.org/10.1088/0022-3727/49/20/205204.

[67] J.S. Oh, E.J.J. Szili, N. Gaur, S.H. Hong, H. Furuta, R.D. Short, A. Hatta, In-situ UV absorption spectroscopy for monitoring transport of plasma reactive species through agarose as surrogate for tissue, J. Photopolym. Sci. Technol. 28 (2015) 439–444. https://doi.org/10.2494/photopolymer.28.439.

[68] E.J.J. Szili, J.-S.S. Oh, S.-H.H. Hong, A. Hatta, R.D.D. Short, Probing the transport of plasma-generated RONS in an agarose target as surrogate for real tissue: Dependency on time, distance and material composition, J. Phys. D. Appl. Phys. 48 (2015) 202001. https://doi.org/10.1088/0022-3727/48/20/202001.

[69] J.-S. Oh, E.J. Szili, N. Gaur, S.-H. Hong, H. Furuta, H. Kurita, A. Mizuno, A. Hatta, R.D. Short, How to assess the plasma delivery of RONS into tissue fluid and tissue, J. Phys. D. Appl. Phys. 49 (2016) 304005. https://doi.org/10.1088/0022-3727/49/30/304005.

[70] J.S. Oh, E.J. Szili, S.H. Hong, N. Gaur, T. Ohta, M. Hiramatsu, A. Hatta, R.D. Short, M. Ito, Mass spectrometry analysis of the real-time transport of plasma-generated ionic species through an agarose tissue model target, J. Photopolym. Sci. Technol. 30 (2017) 317–323. https://doi.org/10.2494/photopolymer.30.317.

[71] N. Gaur, E.J. Szili, J.S. Oh, S.H. Hong, A. Michelmore, D.B. Graves, A. Hatta, R.D. Short, Combined effect of protein and oxygen on reactive oxygen and nitrogen species in the plasma treatment of tissue, Appl. Phys. Lett. 107 (2015) 103703. https://doi.org/10.1063/1.4930874.

[72] E.J. Szili, N. Gaur, S.H. Hong, H. Kurita, J.S. Oh, M. Ito, A. Mizuno, A. Hatta, A.J. Cowin, D.B. Graves, R.D. Short, The assessment of cold atmospheric plasma treatment of DNA in synthetic models of tissue fluid, tissue and cells, J. Phys. D. Appl. Phys. 50 (2017) 274001. https://doi.org/10.1088/1361-6463/aa7501.

[73] C. Chen, D.X. Liu, Z.C. Liu, A.J. Yang, H.L. Chen, G. Shama, M.G. Kong, A model of plasma-biofilm and plasma-tissue interactions at ambient pressure, Plasma Chem. Plasma Process. 34 (2014) 403–441. https://doi.org/10.1007/s11090-014-9545-1.

[74] W. Tian, M.J. Kushner, Atmospheric pressure dielectric barrier discharges interacting with liquid covered tissue, J. Phys. D. Appl. Phys. 47 (2014) 165201. https://doi.org/10.1088/0022-3727/47/16/165201.

[75] J.M.M. Bulson, D. Liveris, I. Derkatch, G. Friedman, J. Geliebter, S. Park, S. Singh, M. Zemel, R.K.K. Tiwari, Non-thermal atmospheric plasma treatment of onychomycosis in an *in vitro* human nail model, Mycoses. 63 (2020) 225–232. https://doi.org/10.1111/myc.13030.

[76] E. Çukur, U.K. Ercan, Degradation of tattoo inks by cold atmospheric plasma treatment: A proof-of-concept study, Plasma Med. 12 (2022) 1–21. https://doi.org/10.1615/PlasmaMed.2022046239.

[77] F. Tampieri, A.G. Araguz, C. Canal, Can we remove tattoos with non-thermal atmospheric plasma?, Plasma Process. Polym. 19 (2022) 2100188. https://doi.org/10.1002/ppap.202100188.

[78] S. Zhai, M. Xu, Q. Li, K. Guo, H. Chen, M.G. Kong, Y. Xia, Successful treatment of vitiligo with cold atmospheric plasma–activated hydrogel, J. Invest. Dermatol. 141 (2021) 2710–2719. https://doi.org/10.1016/j.jid.2021.04.019.

[79] C. Labay, I. Hamouda, F. Tampieri, M.-P. Ginebra, C. Canal, Production of reactive species in alginate hydrogels for cold atmospheric plasma-based therapiesle, Sci. Rep. 9 (2019) 16160. https://doi.org/10.1038/s41598-019-52673-w.

[80] C. Labay, M. Roldán, F. Tampieri, A. Stancampiano, P.E. Bocanegra, M.P. Ginebra, C. Canal, Enhanced generation of reactive species by cold plasma in gelatin solutions for selective cancer cell death, ACS Appl. Mater. Interfaces. 12 (2020) 47256–47269. https://doi.org/10.1021/acsami.0c12930.

[81] X. Solé-Martí, T. Vilella, C. Labay, F. Tampieri, M.-P. Ginebra, C. Canal, Thermosensitive hydrogels to deliver reactive species generated by cold atmospheric plasma: A case study with methylcellulose, Biomater. Sci. 10 (2022) 3845–3855. https://doi.org/10.1039/D2BM00308B.

[82] X. Solé-Martí, C. Labay, Y. Raymond, J. Franch, R. Benitez, M. Ginebra, C. Canal, Ceramic-hydrogel composite as carrier for cold-plasma reactive-species: Safety and osteogenic capacity *in vivo*, Plasma Process. Polym. 20 (2023) 2200155. https://doi.org/10.1002/ppap.202200155.

[83] A.E. Poor, U.K. Ercan, A. Yost, A.D. Brooks, S.G. Joshi, Control of multi-drug-resistant pathogens with non-thermal-plasma-treated alginate wound dressing, Surg. Infect. (Larchmt). 15 (2014) 233–243. https://doi.org/10.1089/sur.2013.050.

[84] F. Tampieri, A. Espona-Noguera, C. Labay, M. Ginebra, M. Yusupov, A. Bogaerts, C. Canal, Does non-thermal plasma modify biopolymers in solution? A chemical and mechanistic study for alginate, Biomater. Sci. 11 (2023) 4845–4858. https://doi.org/10.1039/D3BM00212H.

[85] M. Krewing, B. Schubert, J.E. Bandow, A dielectric barrier discharge plasma degrades proteins to peptides by cleaving the peptide bond, Plasma Chem. Plasma Process. 40 (2020) 685–696. https://doi.org/10.1007/s11090-019-10053-2.

[86] I. Hamouda, C. Labay, M.P. Ginebra, E. Nicol, C. Canal, Investigating the atmospheric pressure plasma jet modification of a photo-crosslinkable hydrogel, Polymer (Guildf). 192 (2020) 122308. https://doi.org/10.1016/j.polymer.2020.122308.

[87] M. Yusupov, J.-W. Lackmann, J. Razzokov, S. Kumar, K. Stapelmann, A. Bogaerts, Impact of plasma oxidation on structural features of human epidermal growth factor, Plasma Process. Polym. 15 (2018) 1800022. https://doi.org/10.1002/ppap.201800022.

[88] M. Yusupov, A. Privat-Maldonado, R.M. Cordeiro, H. Verswyvel, P. Shaw, J. Razzokov, E. Smits, A. Bogaerts, Oxidative damage to hyaluronan—CD44 interactions as an underlying mechanism of action of oxidative stress-inducing cancer therapy, Redox Biol. 43 (2021) 101968. https://doi.org/10.1016/j.redox.2021.101968.

[89] M. Yusupov, D. Dewaele, P. Attri, U. Khalilov, F. Sobott, A. Bogaerts, Molecular understanding of the possible mechanisms of oligosaccharide oxidation by cold plasma, Plasma Process. Polym. 20 (2023) 2200137. https://doi.org/10.1002/ppap.202200137.

[90] V. Veronico, P. Favia, F. Fracassi, R. Gristina, E. Sardella, The active role of organic molecules in the formation of long-lived reactive oxygen and nitrogen species in plasma-treated water solutions, Plasma Process. Polym. 19 (2022) e2100158. https://doi.org/10.1002/ppap.202100158.

[91] P. Eisenhauer, N. Chernets, Y. Song, D. Dobrynin, N. Pleshko, M.J. Steinbeck, T.A. Freeman, Chemical modification of extracellular matrix by cold atmospheric plasma-generated reactive species affects chondrogenesis and bone formation, J. Tissue Eng. Regen. Med. 10 (2016) 772–782. https://doi.org/10.1002/term.2224.

7 Cold Atmospheric Pressure Plasma

A Novel Approach for Virus Treatment Through Reactive Oxygen and Nitrogen Species

Pasquale Isabelli, Matteo Gherardi, Romolo Laurita

7.1 INTRODUCTION

The rapid and extensive global spread of the SARS-CoV-2 virus in 2019 had profound consequences, causing significant disruption for over two years. This event resulted in substantial social, human, and economic costs, impacting the healthcare system and the global economy. The magnitude of this crisis has revitalized global interest in innovative and alternative strategies for mitigating the transmission of viruses. The first article on the use of cold atmospheric plasmas (CAPs) against viruses dates to 2007, but there has been a notable surge in research on using CAPs for viral inactivation since the outbreak of the COVID-19 pandemic. CAPs have been recognized for their antimicrobial properties, with numerous studies investigating their interaction with viruses even before the emergence of SARS-CoV-2. In this chapter, we will provide an overview of viral infections and their impacts on human health, examining current approaches to controlling the spread of viruses. Furthermore, we will evaluate CAP as an alternative approach for inactivating viruses and treating infections, as schematically reported in Figure 7.1. The chapter will delve into the fundamental properties of CAP, including its generation methods, and discuss its current state-of-the-art use in contrasting viral infections. Additionally, we will explore the primary mechanisms underlying the interaction between reactive oxygen and nitrogen species (RONS) generated by CAP and viruses.

7.2 VIRAL INFECTION AND TRANSMISSION ROUTES

Viruses are the most prevalent entities in the biosphere and are ubiquitously distributed, and each living organism encounters millions of viral particles daily. Although viruses have been recognized as distinct biological entities for only slightly more than a century, indications of viral infections can be traced back to the earliest documented records of human activities [1]. Furthermore, strategies to combat viral diseases were implemented well before the initial identification of any virus. Efforts to comprehend and manage these significant pathogens began in the 20th century [1]. A virus is a tiny infectious particle that exhibits an obligatory dependency on host cells for replication. Its structure comprises genetic material, DNA or RNA, enclosed within a proteinaceous capsid. Viruses sometimes possess an additional lipid envelope [1,2]. History offers numerous illustrations of viral diseases that have inflicted widespread devastation, claiming the lives of millions and leaving an indelible legacy in the annals of human civilization. For example, influenza viruses caused several pandemics, such as the Spanish flu (1918–1919) [3], the Asian flu (1957–1958) [4], and the H1N1 influenza pandemic (2009–2010) [5]. The human immunodeficiency virus (HIV) and acquired

DOI:10.1201/9781003328056-8

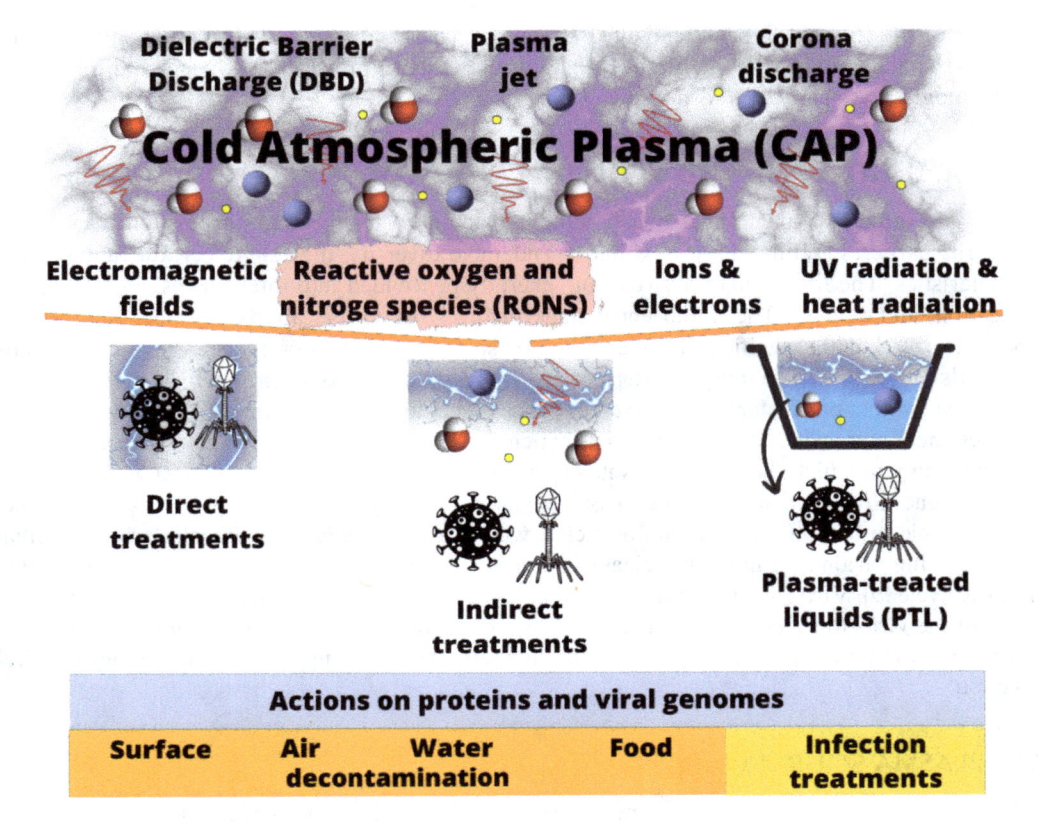

FIGURE 7.1 Cold atmospheric plasma interaction with viruses.

immunodeficiency syndrome (AIDS) have profoundly impacted global health since their identification in the 1980s [6]. Another example is the Ebola Virus Disease, which caused severe and often fatal illness during the West African outbreak in 2014–2016 [7]. Most recently, the spread of the SARS-CoV-2 virus has impacted the world, causing millions of deaths and significant disruptions in societies, healthcare systems, and the global economy.

Viruses can be transmitted through various modes. Many viruses, such as influenza or SARS-CoV-2, are spread through respiratory droplets when an infected person coughs, sneezes, talks, or breathes. Others can inhale these droplets, leading to infections (respiratory transmission). Other viruses (e.g., herpes simplex virus or human papillomavirus) are transmitted through direct physical contact with infected individuals [8]. These are the two principal modalities of virus transmission; nevertheless, numerous alternative routes exist, including fecal-oral or bloodborne transmission. It is crucial to underscore that viruses exhibit diverse transmission modes, with certain viruses capable of engaging in multiple modes concurrently. Prevention and treatment of viral infections are essential to reduce the spread of viruses. Viral infections are treated using antiviral drugs or medications to relieve symptoms while the immune response fights the virus. However, the emergence of antiviral resistance has attenuated the efficacy of antiviral drugs, augmenting the significance of viral infection prevention. Antiviral resistance presents formidable challenges in the treatment of viral infections. Prevention strategies include vaccination, immune-system care, and good hygiene. Regardless of the circumstances, individuals are constantly exposed to viruses through various means of transmission, as highlighted previously. Consequently, preventive measures must be implemented to inhibit the transmission and subsequent occurrence of viral infections. Conventional approaches involving chemical agents, such as disinfectants, and physical techniques, such as UV radiation or filtration, have been employed to contrast virus transmission. The rapid spread of the COVID-19 pandemic in 2019 highlighted that current systems are still insufficient and innovative methodologies to thwart the transmission of viruses must be explored.

7.3   VIRAL LIFE CYCLE AND VIRUS INACTIVATION

Viruses are infectious agents that hijack cellular machinery for the sole purpose of propagating new virus particles. While viruses exhibit considerable diversity, they share certain fundamental characteristics. The life cycle of a virus can generally be divided into three stages: entry, genome replication, and exit [9]. The initial entry stage involves the virus's ability to infect specific cell types selectively. This specificity is determined by specific receptors on the surface of target cells (host cells). Each cell has a unique pattern of protein receptors on its surface, which viruses exploit to target and infect particular cells. The entry process encompasses attachment, during which virus particles encounter and bind to the cell surface; penetration, in which the virus enters the cytoplasm; and uncoating, in which the virus sheds its capsid. After uncoating, the naked viral genome is utilized for gene expression and replication. Subsequently, as viral proteins and genomes accumulate, they assemble to form progeny virion particles, which are then released extracellularly. This final step, involving virion assembly and release from the infected cell, constitutes the exit phase of the viral life cycle [9]. Virus disinfection involves the disruption of viral infectivity, leading to the incapacity of the virus to successfully infect host cells. To accomplish this objective, it is imperative to intervene in the viral replication cycle, thereby impeding or inhibiting key steps necessary for viral infection.

7.4   PLASMA SOURCES

Cold atmospheric plasma (CAP) is a specific state of plasma that is typically generated at or near room temperature and at atmospheric pressure. [10] This unique characteristic renders CAP particularly suitable for biomedical applications in which high temperatures can potentially damage biological tissue. Moreover, the ability to generate CAP at atmospheric pressure confers practical advantages, as it simplifies the production process and reduces associated costs compared to other plasma methods that require low-pressure conditions [11]. CAP comprises a reactive mix of free electrons, atoms, and molecules in neutral and excited states, including reactive oxygen species (ROS) and reactive nitrogen species (RNS), UV radiation, and electric fields, which have several effects on the biological matter [10,12]. All these components, and especially ROS and RNS, have antimicrobial properties, and they can effectively destroy bacteria, viruses, fungi, and spores. CAP can be generated using corona discharge, dielectric barrier discharge (DBD), plasma jets, etc. DBD plasma sources are among the most widely used for producing CAP. They use an alternating or pulsed electric field, with a dielectric layer covering at least one of the electrodes [13]. The discharge can be generated in the gap separating the two electrodes or, in the case of surface DBD, at the surface of one of the two electrodes. In corona plasma sources, a pin or thin wire is required as a high-voltage electrode, and the electric field is highest near the edge of the electrode but decreases rapidly with increasing distance from the tip [10]. Plasma jets are a distinctive category of CAP sources, distinguished by their capacity to generate a rapid flow of chemically reactive species. These plasma sources have gained significant prominence owing to their ability to produce plasmas that are not spatially constrained or limited by electrode confinement. Among the primary configurations of plasma jets, the DBD jet and the corona jet are two prominent examples [12]. Along with their diverse generation strategies, CAPs can be produced using a variety of gases and controlled by modulating factors such as voltage, frequency, power, and treatment time. The control of these parameters directly influences the CAP's composition and reactivity, which can be optimized for specific applications (see also Chapter 2). Overall, the properties and adaptability of CAPS suggest their significant potential for reducing the spread of viruses across various contexts.

7.5 COLD ATMOSPHERIC PRESSURE PLASMA FOR VIRUS INACTIVATION

The versatility of CAP in interacting with several matrices, whether biological or non-biological, stems from its ability to be generated at atmospheric pressure and near-ambient temperatures. It is crucial to acknowledge that each study conducted within this domain can be considered unique in employing distinct plasma sources (e.g., dielectric barrier discharge, plasma jets, corona, etc.), varying operational conditions (e.g., power, treatment times, etc.), environmental variables (such as temperature and relative humidity), the inherent properties of the matrix (e.g., surfaces, liquids, bioaerosols, etc.), and the specific characteristics of the viruses under scrutiny. Each of these factors exerts a significant influence on the efficacy of plasma treatment. Furthermore, it is feasible to differentiate between various treatment modalities (see Figure 7.1). With direct treatment, the contaminated matrix of microorganisms is directly exposed to the plasma discharge. With indirect treatment, in contrast, there is no direct contact between the plasma discharge and the matrix of microorganisms, which is exposed to an atmosphere abundant in (primarily long-lived) reactive species. An alternative approach uses plasma treated liquids (PTLs) as an intermediary medium for virus inactivation.

Here we present some examples of potential applications of cold atmospheric pressure plasma (CAP) in viral decontamination. Numerous studies have evaluated the efficacy of CAP in decontaminating both biologically and non-biologically based surfaces that are contaminated by viruses [14–22]. Velebit et al. assessed the effectiveness of CAP in inactivating murine norovirus (MNV) and hepatitis A virus (HAV) on aerosol-inoculated raspberries [23]. MNV, a member of the *Caliciviridae* family, is a non-enveloped virus with a single-stranded positive-RNA genome. It is commonly used as a surrogate for the highly contagious Human Norovirus (HuNoV) in research studies. Similarly, HAV belongs to the *Picornaviridae* family and is also a non-enveloped virus with a single-stranded RNA genome. Both HuNoV and HAV have been implicated in numerous outbreaks of foodborne illnesses [23]. The study employed a corona discharge system supplied with synthetic air to generate the CAP. The results demonstrated that CAP treatment led to a significant reduction of 4 $\log_{10}$ in infectivity within a treatment time of fewer than five minutes for MNV and approximately ten minutes for HAV [23]. Moldgy et al. compared four different CAP sources operated in air to decontaminate wet and dry surfaces (stainless steel discs) from feline calicivirus (FCV) [24]. FCV is a highly contagious, single-stranded, positive RNA virus that primarily affects cats; however, it is frequently used as a surrogate for Human Norovirus (HuNoV) in research studies and holds a prominent position as one of the most extensively investigated viruses in the field of CAP-virus interactions. The plasma sources investigated are a DBD source in direct contact with the substrate and three indirect plasma sources (a micro-hollow DBD (2D DBD), a volumetric DBD, and a gliding arc discharge). The authors indicate that surface humidity is crucial in increasing the efficacy of all four plasma-based decontamination methods, whereas only direct plasma treatment yields positive results for dry samples [24]. The state-of-the-art of CAP technology contains documented evidence regarding the efficacy of CAP in activating bioaerosol-containing viruses [17], [25–33]. A study by Bisag et al. demonstrated the effectiveness of a direct DBD source operating at an average discharge power of less than 5 W against SARS-CoV-2 virus bioaerosol [25], another single-stranded positive RNA virus that was responsible for the global COVID-19 pandemic. Nayak et al. used a wind tunnel to investigate the efficacy of a packed-bed DBD plasma source against the airborne, single-stranded, positive RNA porcine reproductive and respiratory virus (PRRV), one of the most significant pathogens in the pork industry. The results indicated a substantial reduction of 3.5 Log_{10} in viral load within a few milliseconds [27]. Xia et al. presented another packed-bed DBD, successfully constructed and implemented to deactivate MS2 bacteriophages present in aerosolized form. The experimental results demonstrated a significant reduction of 2.3 Log_{10} in viral load [28]. CAPs can potentially be used as an alternative to waterborne virus inactivation [34–36]. For example, Filipić et al. employed a corona plasma jet to decontaminate water samples contaminated with pepper mild mottle virus (PMMoV), a highly robust, waterborne Tobamovirus capable of withstanding passage through the human digestive system. They showed that CAPP can completely inactivate PMMoV in water samples after five minutes [34]. Additionally, in previous work the authors used the same plasma source to treat samples contaminated with Potato virus Y (PVY), which is a common drinking-water contaminant, accompanied by the presence of organic plant material, representing a more organically polluted sample [36]. The results revealed complete inactivation of PVY within a five-minute treatment period. In tap-water samples lacking organic residues, a reduced treatment time of one minute was adequate for the complete inactivation of Potato virus Y (PVY)[36]. An alternative strategy uses CAPs to produce liquids enriched with RONS [37,38], which can be employed as an intermediary medium for the inactivation of viruses on, for example, surfaces. Cortázar et al. evaluated the sensitivity of SARS-CoV-2 and PR8 H1N1 virus (a specific strain of influenza A that has been widely employed in scientific investigations as a valuable tool for comprehensively studying the influenza virus) to PTL produced by a DBD jet [39]. In this experimental setup, 200 ml of deionized or buffered deionized water was subjected to direct exposure to a CAP discharge for 30 minutes. The coaxial electrodes, separated by a quartz tube, were placed in contact with the liquid surface and used an airflow of 5 l/min during the exposure process. The experiments revealed reduced viral infectivity based on vitro assessments using both isolated virus and infected cells. Moreover, the authors observed that the antiviral effect was minimally influenced by changes in pH induced by the plasma treatment of the liquid. Additionally, it was observed that the PTL exhibited a non-inflammatory impact on the treated cells [39].

Plasma jets have gained significant popularity in fundamental research as they offer several advantages, such as their compact size and ability to transport reactive species away from the electrode zone through gas flow. These characteristics make plasma jets one of the most commonly used plasma sources for studying fundamental aspects of CAP-virus interactions. Such studies investigate the basic mechanisms underlying plasma-mediated virus inactivation [40–45].

7.6 MECHANISMS OF VIRUS INACTIVATION BY REACTIVE OXYGEN AND NITROGEN SPECIES (RONS)

This section will delve into the principal RONS involved in viral inactivation, based on the current state of knowledge. The critical components of viruses that merit examination include the capsid, envelope (if present), proteins, and genetic material (DNA or RNA). CAP exposure or PTL treatments can lead to modification and/or degradation of proteins, nucleic acids, and lipids of viral envelopes [46]. It is widely accepted that RONS are prominently involved in mechanisms that lead to the inactivation of viruses, as evidenced in the scientific literature [46,47,48]. As previously emphasized, a diverse range of RONS are generated in CAPs, and their abundance is influenced by various factors such as the plasma source, operational conditions, and gas composition. Evaluating the individual contributions of each RONS is an exceedingly intricate task. This complexity arises from the simultaneous production of multiple RONS, many of which have short lifespans, making their analysis challenging. Nevertheless, it is established that RONS have the capability to target proteins and viral genomes, thereby exerting their antiviral effects. Among the RONS generated by CAPs, various RONS exhibit potential antiviral properties. Notable examples of these include hydroxyl radical ($OH^{\bullet}$), superoxide radical ($O_2^{-\bullet}$), single oxygen ($^1O_2$), nitric oxide radicals ($NO^{\bullet}$), nitrogen dioxide radicals ($NO_2^{\bullet}$), and excited nitrogen (N^*). These reactive species arise from the interaction between free electrons and gaseous molecules (see also Chapter 2). Although their concentration can reach significant levels, these species are characterized as short-lived; they are highly reactive, posing challenges in their precise quantification and assessment. Beyond the short-lived reactive species, there is the potential for forming long-lived species, which are characterized by their ability to survive for extended durations before engaging in chemical reactions or undergoing transformative processes. Notable among these long-lived species are ozone (O_3), nitrogen dioxide (NO_2), nitrous oxide (N_2O), and dinitrogen pentoxide (N_2O_5). Also, PTLs harbor several RONS (see also Chapter 3), among the most extensively investigated of which are hydrogen peroxides (H_2O_2), nitrites (NO_2^-), and peroxynitrite ($ONOO^-$).

Several illustrative examples from the existing literature are provided below. CAP treatment with abundant singlet oxygen and hydroxyl radical effectively hindered viral entry mechanisms of SARS-CoV-2. Notably, the conformation of the viral spike protein was altered, impairing its capacity to bind to cellular receptors. As a consequence of CAP exposure, the infectivity of SARS-CoV-2 was also reduced, and the disruption of cell membranes occurred, leading to oxidative damage of viral RNA within the cells [49]. A significant reduction in the ability of the spike protein to bind to the human ACE2 protein was observed after five minutes of exposure to the CAP plasma [42]. Additionally, Khanikar et al. demonstrated that the CAP plasma has the ability to degrade the RNA of the SARS-CoV-2 virus. These effects were associated with a high concentration of hydrogen peroxide [42]. Shi et al. investigated the interaction between CAP and Hepatitis B virus (HBV), a viral pathogen known to cause chronic liver disease through transmission via bodily fluids. The study revealed that key antigens of HBV exhibited susceptibility to damage induced by RONS subsequent to exposure to CAP [50]. Sutter et al. comprehensively documented and summarized the potential applications of CAPs as a promising approach for *Herpes simplex* virus type 1 (HSV-1) infections [51]. CAPs can exhibit direct antiviral activity through the generation of RONS, which effectively target and inhibit HSV-1. Xia et al. investigated the efficacy of various PTLs in inactivating NDV. Their investigation revealed that short-lived reactive species $\cdot$OH and $\cdot$NO, along with long-lived H_2O_2, induced morphological changes in the virus, disrupted its RNA structure, and degraded its protein components [52]. Sakudo et al. propose H_2O_2 as the main actor and oxidation as the primary mechanism underlying the degradation and inactivation of influenza A and B viruses that cause human infection annually during the epidemic season. Their study demonstrates that treatment with N_2 gas plasma leads to the degradation of various viral proteins, including nucleoprotein, hemagglutinin, and neuraminidase. N_2 gas plasma treatment also induces damage to the viral RNA genome [53]. Aboubakr et al. examined the effect of CAP on the FCV. Longer exposure to CAP (two minutes) led to the disintegration of viral capsids due to strong oxidation of the capsid protein. Shorter exposure (15 seconds) did not damage the capsid structure but oxidized specific amino acids in crucial regions of the capsid involved in virus attachment and entry into host cells, preventing the host cells' infection. The study also found that CAP-induced oxidation damaged the viral RNA once the capsid was destroyed, but that this had a lesser impact on virus inactivation compared to capsid damage [54]. A previous study explored the mechanisms of FCV inactivation mediated by CAP, demonstrating that singlet oxygen plays a pivotal role in inducing oxidative modifications of capsid proteins [55]. A study by Yamashiro et al. further supports the involvement of singlet oxygen in the inactivation of FCV. Their research not only confirmed the role of singlet oxygen but also emphasized the contribution of peroxynitrite (ONOO−) in disrupting the viral RNA [56]. In addition to singlet oxygen, Nayak et al. emphasize that the inactivation of FCV in the gas phase is attributed to the synergistic effects of ozone and RNS, potentially including nitrogen oxides (NOx). Conversely, the mechanism of FCV inactivation in the liquid phase is influenced by pH and primarily mediated by RNS, particularly acidified nitrite, while the impact of ozone is minimal [57]. Guo et al. investigated the effect of CAP exposure and PTL on different bacteriophages (T4, φ174, and MS2)[58]. Bacteriophages are a distinct category of viruses that possess the exclusive capability to infect and replicate within bacterial cells. These viruses have been used as surrogates in scientific investigations to assess the transmission dynamics of viruses in an open-air environment [59]. The analysis of DNA and proteins provided evidence that the reactive species generated by the CAP resulted in damage to both nucleic acids and proteins. This observation was supported by morphological examination, which revealed the aggregation of bacteriophages after CAP treatment. Furthermore, the effectiveness of singlet oxygen scavengers in reducing the inactivation of bacteriophages suggests that singlet oxygen plays a central role in the inactivation process [58]. In addition to viral decontamination, several studies have assessed the impact of CAPs and PTLs on already infected cells. For example, the antiviral potential of CAP was assessed in HIV-1-infected cells. CAP treatment reduced HIV-1 infectivity, hindered virus-cell fusion, and disrupted viral assembly. Moreover, it was hypothesized that CAP exposure triggered the activation of cellular factors in macrophages, creating an environment that inhibited further viral entry [60]. Also, reduction in SARS-CoV-2 viral titer was evidenced following the use of PTLs on infected cells [39]. CAPs can also induce immunomodulatory changes in infected cells, leading to the stimulation of anti-HSV-1 adaptive immune responses. These combined effects make CAPs a promising therapeutic option for combating HSV-1 infections [51].

7.7 CONCLUSIONS

Numerous studies conducted by various scientists have presented compelling evidence of the effectiveness of CAPs in deactivating viral particles. These experimental investigations have utilized diverse sources of CAPs, such as DBDs, corona discharges, and plasma jets, with distinct structural configurations. Among these sources, DBDs, including DBD jets, direct DBDs, and surface DBDs, have gained significant popularity. Current research has focused primarily on RNA viruses, as they pose a higher risk to public health due to their propensity for genetic mutations. Despite these advances, further investigation is still needed to better understand the precise mechanisms underlying viral inactivation by CAPs. It has been established that RONS primarily target viral proteins and nucleic acids, impeding viral attachment, host cell entry, and replication, but there is ongoing debate about the contribution and significance of individual reactive RONS in the viral inactivation process. According to the key points discussed in this chapter, CAPs have the potential to be a viable strategy for controlling virus spread and preventing viral infections and outbreaks of emerging diseases. CAPs possess promising capabilities for decontaminating inanimate surfaces through direct exposure to plasma discharges or RONS-rich atmospheres or via the utilization of PTLs, and potentially for fighting contact-based infections. Furthermore, CAPs could combat the transmission of foodborne diseases by treating the surfaces of food products. However, careful consideration must be given to the potential undesired effects of CAPs on food materials. CAPs have also demonstrated effectiveness in mitigating airborne transmission of viruses, as seen during the COVID-19 pandemic. Commercial CAP-based solutions designed to purify air have emerged in the market. It is crucial, though, to closely monitor ozone production by these devices. While ozone possesses potent antimicrobial properties, its presence in high concentrations can pose health risks to humans. Therefore, special attention should be paid to ensure ozone concentrations remain within safe and regulated limits. CAPs offer also opportunities for combating problematic viruses found in water sources, including enteric viruses and plant viruses affecting agricultural production. The ability of CAPs to inactivate these viruses offers a promising avenue for ensuring water safety and protecting both human consumers and farming crops. However, the potential adverse effects of CAP-activated water, including genotoxic impacts and cytotoxic effects on human cells and plants, must be thoroughly evaluated to determine the safety and viability of utilizing CAP for water decontamination. Comprehensive studies and risk assessments are crucial for establishing guidelines and protocols to maximize the benefits of CAPs while minimizing risks to human health and the environment. Despite encouraging initial findings, such as the promising results observed with HSV-1, the application of cold plasma in therapeutic interventions still needs extensive investigation and developmental progression before reaching clinical implementation. Overall, using CAPs in various applications shows promise in preventing viral infections, controlling the spread of viruses, and safeguarding public health. However, further research and rigorous evaluation are necessary to fully understand the potential benefits and risks of using CAPs in different contexts.

REFERENCES

[1] J. Flint, V. R. Racaniello, G. F. Rall, A. M. Skalka, and L. W. Enquist, *Principles of Virology*, Volume 1, 4th edition. ASM Press, 2015.

[2] E. V. Koonin, and P. Starokadomskyy, 'Are viruses alive? The replicator paradigm sheds decisive light on an old but misguided question', *Stud. Hist. Philos. Sci. Part C Stud. Hist. Philos. Biol. Biomed. Sci.*, vol. 59, pp. 125–134, 2016, doi: 10.1016/j.shpsc.2016.02.016.

[3] A. Trilla, G. Trilla, and C. Daer, 'The 1918 "Spanish Flu" in Spain', *Clin. Infect. Dis.*, vol. 47, no. 5, pp. 668–673, 2008, doi: 10.1086/590567.

[4] C. Jackson, 'History lessons: The Asian Flu pandemic', *Br. J. Gen. Pract.*, vol. 59, no. 565, pp. 622–623, 2009, doi: 10.3399/bjgp09X453882.

[5] N. Vousden, and M. Knight, 'Lessons learned from the A (H1N1) influenza pandemic', *Best Pract. Res. Clin. Obstet. Gynaecol.*, vol. 76, pp. 41–52, 2021, doi: 10.1016/j.bpobgyn.2020.08.006.

[6] L. Montagnier, 'A history of HIV discovery', *Science*, vol. 298, no. 5599, pp. 1727–1728, 2002, doi: 10.1126/science.1079027.

[7] S. T. Jacob *et al.*, 'Ebola virus disease', *Nat. Rev. Dis. Primers.*, vol. 6, no. 1, pp. 1–31, 2020, https://doi.org/10.1038/s41572-020-0147-3.

[8] J. Louten, 'Chapter 5 — Virus Transmission and Epidemiology', in *Essential Human Virology*, pp. 71–92, 2016, doi.org/10.1016/B978-0-12-800947-5.00005-3.

[9] W-S. Ryu, *Molecular Virology of Human Pathogenic Viruses*, 1st edition. Academic Press, 2017.

[10] A. Fridman, *Plasma Chemistry*. Cambridge University Press, 2008.

[11] M. Domonkos, P. Tichá, J. Trejbal, and P. Demo, 'Applications of cold atmospheric pressure plasma technology in medicine, agriculture and food industry', *Appl. Sci.*, vol. 11, no. 11, pp. 4809, 2021, doi: 10.3390/app11114809.

[12] M. Laroussi, 'Cold plasma in medicine and healthcare: The new frontier in low temperature plasma applications', *Front. Phys.*, vol. 8, no. March, pp. 1–7, 2020, doi: 10.3389/fphy.2020.00074.

[13] U. Kogelschatz, 'Dielectric-barrier discharges: Their history, discharge physics, and industrial applications', *Plasma Chem. Plasma Process.*, vol. 23, no. 1, pp. 1–46, 2003, doi: 10.1023/A:1022470901385.

[14] J. He *et al.*, 'Plasma-generated reactive water mist for disinfection of N95 respirators laden with MS2 and T4 bacteriophage viruses', *Sci. Rep.*, vol. 12, no. 1, pp. 1–8, 2022, doi: 10.1038/s41598-022-23660-5.

[15] P. Bhartiya *et al.*, 'Nonthermal plasma-generated ozone inhibits human coronavirus 229E infectivity on glass surface', *Plasma Process. Polym.*, vol. 19, no. 11, p. 2200054, 2022, doi: 10.1002/ppap.202200054.

[16] B. Kramer, D. Warschat, and P. Muranyi, 'Disinfection of an ambulance using a compact atmospheric plasma device', *J. Appl. Microbiol.*, vol. 133, no. 2, pp. 696–706, 2022, doi: 10.1111/jam.15599.

[17] M. Nur *et al.*, 'A successful elimination of indonesian SARS-CoV-2 variants and airborne transmission prevention by cold plasma in fighting COVID-19 pandemic: A preliminary study', *Karbala Int. J. Mod. Sci.*, vol. 8, no. 3, pp. 446–454, 2022, doi: 10.33640/2405-609X.3234.

[18] K. Obrová *et al.*, 'Decontamination of high-efficiency mask filters from respiratory pathogens including SARS-CoV-2 by non-thermal plasma', *Front. Bioeng. Biotechnol.*, vol. 10, pp. 1–13, 2022, doi: 10.3389/fbioe.2022.815393.

[19] F. Capelli *et al.*, 'Decontamination of food packages from SARS-CoV-2 RNA with a cold plasma-assisted system', *Appl. Sci.*, vol. 11, no. 9, p. 4177, 2021, doi: 10.3390/app11094177.

[20] E. S. Lee, Y. J. Jeon, and S. C. Min, 'Microbial inactivation and quality preservation of chicken breast salad using atmospheric dielectric barrier discharge cold plasma treatment', *Foods*, vol. 10, no. 6, p. 1214, 2021, doi: 10.3390/foods10061214.

[21] Y. M. Huang, W. C. Chang, and C. L. Hsu, 'Inactivation of norovirus by atmospheric pressure plasma jet on salmon sashimi', *Food Res. Int.*, vol. 141, no. January, p. 110108, 2021, doi: 10.1016/j.foodres.2021.110108.

[22] M. S. Choi *et al.*, 'Virucidal effects of dielectric barrier discharge plasma on human norovirus infectivity in fresh oysters (Crassostrea gigas)', *Foods*, vol. 9, no. 12, pp. 1–13, 2020, doi: 10.3390/foods9121731.

[23] B. Velebit *et al.*, 'Efficacy of cold atmospheric plasma for inactivation of viruses on raspberries', *Innov. Food Sci. Emerg. Technol.*, vol. 81, p. 103121, 2022, doi: 10.1016/j.ifset.2022.103121.

[24] A. Moldgy, G. Nayak, H. A. Aboubakr, S. M. Goyal, and P. J. Bruggeman, 'Inactivation of virus and bacteria using cold atmospheric pressure air plasmas and the role of reactive nitrogen species', *J. Phys. D. Appl. Phys.*, vol. 53, no. 43, p. 434004, 2020, doi: 10.1088/1361-6463/aba066.

[25] A. Bisag *et al.*, 'Cold atmospheric plasma decontamination of SARS-CoV-2 bioaerosols', *Plasma Process. Polym.*, vol. 19, no. 3, pp. 1–11, 2022, doi: 10.1002/ppap.202100133.

[26] A. Bisag *et al.*, 'Cold atmospheric plasma inactivation of aerosolized microdroplets containing bacteria and purified SARS-CoV-2 RNA to contrast airborne indoor transmission', *Plasma Process. Polym.*, vol. 17, no. 10, pp. 1–8, 2020, doi: 10.1002/ppap.202000154.

[27] G. Nayak *et al.*, 'Rapid inactivation of airborne porcine reproductive and respiratory syndrome virus using an atmospheric pressure air plasma', *Plasma Process. Polym.*, vol. 17, no. 10, p. 1900269, 2020, doi: 10.1002/ppap.201900269.

[28] T. Xia *et al.*, 'Inactivation of airborne Porcine Reproductive and Respiratory Syndrome Virus (PRRSv) by a packed bed dielectric barrier discharge non-thermal plasma', *J. Hazard. Mater.*, vol. 393, no. February, p. 122266, 2020, doi: 10.1016/j.jhazmat.2020.122266.

[29] C. Schiappacasse *et al.*, 'Inactivation of aerosolized newcastle disease virus with non-thermal plasma', *Appl. Eng. Agric.*, vol. 36, no. 1, pp. 55–60, 2020, doi: 10.13031/aea.13650.

[30] T. Xia, Z. Lin, E. M. Lee, K. Melotti, M. Rohde, and H. L. Clack, 'Field operations of a pilot scale packed-bed Non-Thermal Plasma (NTP) reactor installed at a pig barn on a michigan farm to inactivate airborne viruses', *2019 IEEE Ind. Appl. Soc. Annu. Meet.*, pp. 1–4, 2019, doi: 10.1109/IAS.2019.8912457.

[31] H. A. Aboubakr *et al.*, 'Virucidal effect of cold atmospheric gaseous plasma on feline calicivirus, a surrogate for human norovirus', *Appl. Environ. Microbiol.*, vol. 81, no. 11, pp. 3612–3622, 2015, doi: 10.1128/AEM.00054-15.

[32] O. Terrier *et al.*, 'Cold oxygen plasma technology efficiency against different airborne respiratory viruses', *J. Clin. Virol.*, vol. 45, no. 2, pp. 119–124, 2009, doi: 10.1016/j.jcv.2009.03.017.

[33] H. Nojima *et al.*, 'Novel atmospheric pressure plasma device releasing atomic hydrogen: Reduction of microbial-contaminants and OH radicals in the air', *J. Phys. D. Appl. Phys.*, vol. 40, no. 2, pp. 501–509, 2007, doi: 10.1088/0022-3727/40/2/026.

[34] A. Filipić *et al.*, 'Inactivation of pepper mild mottle virus in water by cold atmospheric plasma', *Front. Microbiol.*, vol. 12, no. January, pp. 1–12, 2021, doi: 10.3389/fmicb.2021.618209.

[35] A. S. El-Kalliny *et al.*, 'Efficacy of cold atmospheric plasma treatment on chemical and microbial pollutants in water', *ChemistrySelect*, vol. 6, no. 14, pp. 3409–3416, 2021, doi: 10.1002/slct.202004716.

[36] A. Filipić *et al.*, 'Cold atmospheric plasma as a novel method for inactivation of potato virus Y in water samples', *Food Environ. Virol.*, vol. 11, no. 3, pp. 220–228, 2019, doi: 10.1007/s12560-019-09388-y.

[37] P. J. Bruggeman *et al.*, 'Plasma-liquid interactions: A review and roadmap', *Plasma Sources Sci. Technol.*, vol. 25, no. 5, pp. 1–59, 2016, doi: 10.1088/0963-0252/25/5/053002.

[38] R. Laurita, D. Barbieri, M. Gherardi, V. Colombo, and P. Lukes, 'Chemical analysis of reactive species and antimicrobial activity of water treated by nanosecond pulsed DBD air plasma', *Clin. Plasma Med.*, vol. 3, no. 2, pp. 53–61, 2015, doi: 10.1016/j.cpme.2015.10.001.

[39] O. D. Cortázar, A. Megía-Macías, S. Moreno, A. Brun, and E. Gómez-Casado, 'Vulnerability of SARS-CoV-2 and PR8 H1N1 virus to cold atmospheric plasma activated media', *Sci. Rep.*, vol. 12, no. 1, pp. 1–12, 2022, doi: 10.1038/s41598-021-04360-y.

[40] D. M. Mrochen *et al.*, 'Toxicity and virucidal activity of a neon-driven micro plasma jet on eukaryotic cells and a coronavirus', *Free Radic. Biol. Med.*, vol. 191, pp. 105–118, 2022, doi: 10.1016/j.freeradbiomed.2022.08.026.

[41] S. Sasaki *et al.*, 'Human coronavirus inactivation by', *J. Phys. D. Appl. Phys.*, vol. 55, no. 29, pp. 295203, 2022, doi: 10.1088/1361-6463/ac6a8c.

[42] R. R. Khanikar *et al.*, 'Cold atmospheric pressure plasma for attenuation of SARS-CoV-2 spike protein binding to ACE2 protein and the RNA deactivation', *RSC Adv.*, vol. 12, no. 15, p. 9466, 2022, doi: 10.1039/d2ra00009a.

[43] E. J. Szili *et al.*, 'On-demand cold plasma activation of acetyl donors for bacteria and virus decontamination', *Appl. Phys. Lett.*, vol. 119, no. 5, pp. 1–6, 2021, doi: 10.1063/5.0062787.

[44] Y. Miao *et al.*, 'Cold atmospheric plasma increases IBRV titer in MDBK cells by orchestrating the host cell network', *Virulence*, vol. 12, no. 1, pp. 679–689, 2021, doi: 10.1080/21505594.2021.1883933.

[45] O. Bunz *et al.*, 'Cold atmospheric plasma as antiviral therapy—Effect on human herpes simplex virus type 1', *J. Gen. Virol.*, vol. 101, no. 2, pp. 208–215, 2020, doi: 10.1099/jgv.0.001382.

[46] A. Filipić, I. Gutierrez-Aguirre, G. Primc, M. Mozetič, and D. Dobnik, 'Cold plasma, a new hope in the field of virus inactivation', *Trends Biotechnol.*, vol. 38, no. 11, pp. 1278–1291, 2020, doi: 10.1016/j.tibtech.2020.04.003.

[47] N. Kaushik *et al.*, 'The inactivation and destruction of viruses by reactive oxygen species generated through physical and cold atmospheric plasma techniques: Current status and perspectives', *J. Adv. Res.*, vol. 43, pp. 59–71, 2023, doi: 10.1016/j.jare.2022.03.002.

[48] J. Fu, Y. Xu, E. J. Arts, Z. Bai, Z. Chen, and Y. Zheng, 'Viral disinfection using nonthermal plasma: A critical review and perspectives on the plasma-catalysis system', *Chemosphere*, vol. 309, p. 136655, 2022, doi: 10.1016/j.chemosphere.2022.136655.

[49] T. Jin, Y. Xu, C. Dai, X. Zhou, Q. Xu, and Z. Wu, 'Cold atmospheric plasma: A non-negligible strategy for viral RNA inactivation to prevent SARS-CoV-2 environmental transmission', *AIP Adv.*, vol. 11, no. 8, p. 085019, 2021, doi: 10.1063/5.0060530.

[50] X. M. Shi *et al.*, 'Effect of low-temperature plasma on deactivation of hepatitis B virus', *IEEE Trans. Plasma Sci.*, vol. 40, no. 10, pp. 2711–2716, 2012, doi: 10.1109/TPS.2012.2210567.

[51] J. Sutter, P. J. Bruggeman, B. Wigdahl, F. C. Krebs, and V. Miller, 'Manipulation of oxidative stress responses by non-thermal plasma to treat herpes simplex virus type 1 infection and disease', *Int. J. Mol. Sci.*, vol. 24, no. 5, 2023, p. 4673, doi: 10.3390/ijms24054673.

[52] X. Su *et al.*, 'Inactivation efficacy of nonthermal plasmaactivated solutions against Newcastle disease virus', *Appl. Environ. Microbiol.*, vol. 84, no. 9, pp. 1–12, 2018, doi: 10.1128/AEM.02836-17.

[53] Y. I. Akikazu Sakudo, T. Misawa, N. Shimizu, 'N2 gas plasma inactivates influenza virus mediated by oxidative stress', *Front. Biosci. 6*, vol. 6, pp. 69–79, 2014, doi: 10.2741/e692.

[54] H. A. Aboubakr *et al.*, 'Cold argon-oxygen plasma species oxidize and disintegrate capsid protein of feline calicivirus', *PLoS One*, vol. 13, no. 3, pp. 1–24, 2018, doi: 10.1371/journal.pone.0194618.

[55] H. A. Aboubakr, U. Gangal, M. M. Youssef, S. M. Goyal, and P. J. Bruggeman, 'Inactivation of virus in solution by cold atmospheric pressure plasma: Identification of chemical inactivation pathways', *J. Phys. D. Appl. Phys.*, vol. 49, no. 20, p. 204001, 2016, doi: 10.1088/0022-3727/49/20/204001.

[56] R. Yamashiro, T. Misawa, and A. Sakudo, 'Key role of singlet oxygen and peroxynitrite in viral RNA damage during virucidal effect of plasma torch on feline calicivirus', *Sci. Rep.*, vol. 8, no. 1, pp. 1–13, 2018, doi: 10.1038/s41598-018-36779-1.

[57] G. Nayak, H. A. Aboubakr, S. M. Goyal, and P. J. Bruggeman, 'Reactive species responsible for the inactivation of feline calicivirus by a two-dimensional array of integrated coaxial microhollow dielectric barrier discharges in air', *Plasma Process. Polym.*, vol. 15, no. 1, p. 1700119, 2018, doi: 10.1002/ppap.201700119.

[58] M. G. K. Li Guo, R. Xu, L. Gou, Z. Liu, Y. Zhao, D. Liu, L. Zhang, and H. Chen, 'Mechanism of virus inactivation by cold atmospheric-pressure plasma and plasma-activated water', *Appl. Environ. Microbiol.*, vol. 84, no. 17, pp. e00726–18, 2018, doi: 10.1128/AEM.00726-18.

[59] G. T. Machado, C. R. de C. Pinto, L. A. V. da Fonseca, T. C. dos S. Ramos, T. F. P. Paggi, and B. Spira, 'Bacteriophages as surrogates for the study of viral dispersion in open air', *Arch. Microbiol.*, vol. 203, no. 7, pp. 4041–4049, 2021, doi: 10.1007/s00203-021-02382-8.

[60] O. Volotskova, L. Dubrovsky, M. Keidar, and M. Bukrinsky, 'Cold atmospheric plasma inhibits HIV-1 replication in macrophages by targeting both the virus and the cells', *PLoS One*, vol. 11, no. 10, pp. 1–9, 2016, doi: 10.1371/journal.pone.0165322.

Part II

Plasma Cell Biology and Medical Therapy

8 The Role of Electric Fields in Plasma Treatment of Cells

Eric Robert

8.1 ELECTRIC FIELD OR ELECTRIC FIELDS

In most laboratory situations, plasma is generated by applying voltage across an electrode gap to produce an electric field. Defined as the voltage drop-to-gap ratio in units of V/m, the so-called Laplacian electric field is probably the most critical physical quantity to consider for generating plasma with gas discharge setups. Derived from this electric field amplitude, another fundamental quantity defined as the reduced electric field expresses the ratio of the electric field strength to the gas density, often reported in the unit of Townsend (Td with $1 \text{ Td} = 10^{-21} \text{ Vm}^2$). Most excitation or ionization cross sections, i.e., the probability of such phenomena occurring when collisions occur in gas or gas mixtures excited with electric discharge, are tabulated as a function of this reduced electric field. Not only the electric field amplitude, driving the kinetic energy gain during the time of application of the voltage, but its balance with the energy losses during collisions or interactions with the surrounding species will indeed eventually impose the energy transfer and thus the generation of excited levels of the gas or gas mixture constituents in a plasma. A first deviation in the analysis of plasma generation with its correlation with the applied Laplacian electric field occurs when plasma is in contact with physical walls as in a plasma reactor, or when interacting with any surface. Typically, in low-pressure discharges, the Debye electrostatic sheath consists of a layer with a positive charge excess, while the surface experiences a negative counterbalancing negative charge. This electrostatic interaction results in a local electric field near the surfaces where specific plasma excitation features will occur. The streamer discharge is generated for higher pressure discharges, such as dielectric barrier discharges (DBD) and plasma jets (PJ) developed for biomedical applications. Among others, a specificity of such streamer discharge is the generation of the so-called space charge electric field at the tip of the propagating streamer, resulting from ions and electron spatial splitting, the light electron drifting much faster than heavy ions. It is worth mentioning that there is no direct correlation between the Laplacian field amplitude and the space charge electric field. The latter is likely to have a much higher amplitude, typically from a few to a few tens of kV/cm, than the applied electric field when 1- to 5-kV voltage is applied across a 0.5 to 1 cm gap, resulting in a Laplacian field smaller than 10 kV/cm. When DBD or PJ are developed to generate plasma in contact with any surface, as the streamer tip—also known as the ionization front—approaches closer and closer to this surface, an increasing electric field amplitude will be generated. Following the ionization front, a plasma column or streamer channel resulting from the recombination of ionized particles is generated. A "local" electric field at the tip of the streamer is then defined as the ratio of this voltage transferred across the more or less resistive streamer channel to the surface gap. Typically, a few kV are applied in a gap, gradually decreasing from a few mm to a fraction of a mm when the ionization front contacts the surface, generating a transient but significant local electric field.

These transient high amplitude electric fields (illustrated in Figure 8.1) all play a fundamental role in the generation of reactive species and thus in redox chemistry, and they will likely interact

DOI: 10.1201/9781003328056-10

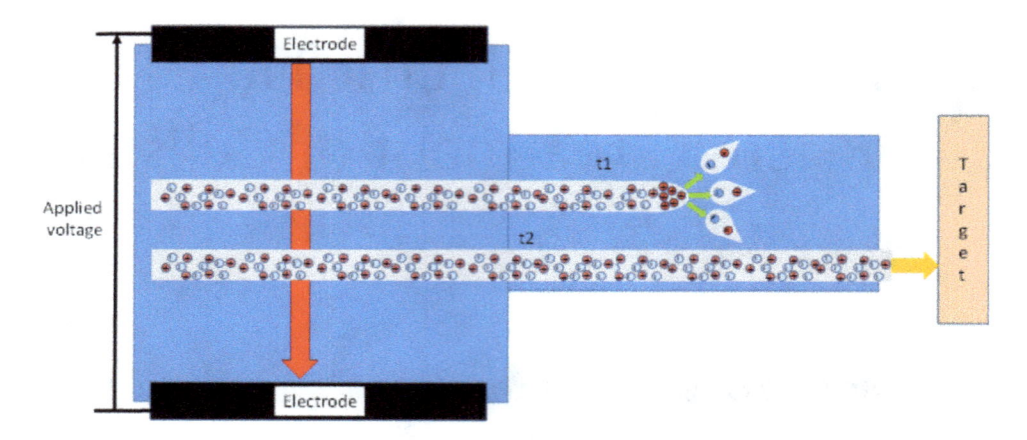

FIGURE 8.1   The applied voltage across the electrodes generates the Laplacian electric field (red arrow). At time t1, the streamer propagates towards the target (inside a capillary or at its outlet for a plasma jet device or in the air for dielectric barrier discharge setup). The space charge electric field is generated at the tip of the streamer (green arrow). Later, at time t2, the applied voltage is partly transferred along the streamer channel to its tip very close to the target and generates the local electric field (orange arrow). The target will sense all these fields. Please note that this is a schematic picture. The length or color of the three electric fields do not correlate with actual field strength.

with biological tissues. The following two paragraphs document some examples of these two modes of action in the context of biomedical applications.

8.2   THE ROLE OF ELECTRIC FIELDS IN RONS GENERATION

As previously mentioned, the reduced electric field is a crucial parameter for the population of ionic and excited atomic and molecular species by direct electron impact. The balance of this primary plasma excitation then determines the initial conditions for the successive chemical reactions during and immediately after the voltage pulse application. These chemical reactions, including volume collisions, surface interactions, and photon radiation, will govern the plasma composition in the glow and afterglow phases. The kinetics of this atmospheric pressure plasma includes many reactions related to collisional or radiative recombination, charge transfer reactions, molecular ions or dimer formation, and relaxation in the translational, rotational, and vibrational distribution of many reactive species. Non-thermal plasmas developed for biomedical applications include operation with rare gases (helium, argon, etc.) to produce species that, together with air environment (nitrogen, oxygen, water vapor) constituents, lead to so-called kinetics schemes including hundreds of reactions.

In the case of plasma jets or DBD discharges in contact with targets, at least two additional effects involving the electric field should be mentioned. In the context of plasma jet development in ambient air or impingement on various targets, it has been recently demonstrated that strong plasma discharge-gas flow interactions may occur during the voltage pulse application. Rare gas flow channeling, shift from turbulent to laminar regime and vice versa, and jet-to-jet interplay have been reported when multi-jet devices are operated [1–6]. The basic phenomena evidenced in these studies are the so-called electro-hydrodynamic (EHD) force and gas heating processes. The first one, studied for decades in the context of plasma actuators for different environment compositions, various voltage waveforms, polarity, duty cycles, and over different targets, includes the ionic wind. The ionic wind is generated as charged species drift in the electric field, acquiring energy that they then transfer to neutrals during collision. This ionic wind results in a gas flow induced by electrostatic forces. As expressed in the previous section, all kinds of electric fields may be involved in generating this ionic wind. Beyond the fundamental study of this EHD phenomenon for the plasma discharge description, the rare gas flow modulations (turbulence, unsteady regime, gas flow velocity increase) have a drastic impact on the "mixing" of the rare gas feeding gas flow with the ambient air environment when the plasma discharge is ignited. This results in dramatic changes in the plasma propagation and the production of reactive oxygen and nitrogen species (RONS). The feeding gas flow conditions and the interaction with targets of various natures (free flow, dielectric, or metallic substrates) play an essential role in developing the plasma jet and in the subsequent generation of reactive species. The same phenomena, but in a more confined space zone and with an initially static air situation in the gap between the DBD reactor and the target, are also relevant for plasma actuators.

The second class of electric field phenomena is based on charge deposition. When positive or negative streamers propagate inside dielectric capillaries or impinge on targets, they deposit ionic species. This charge deposition may drastically impact the pulse-to-pulse plasma propagation in high repetition rate regimes. These residual charges may take a few minutes to a few hours before recombining on some substrate. Besides the impact on streamer propagation, demonstrated inside capillary [7] and over target [2], such charge deposition will generate an electric field right over the top surface of the sample. In the case of biological tissues or plated cell cultures, this electric field will favor charged species transport, in analogy with microcurrent involved in iontophoresis [8], or may trigger electropermeabilization/electroporation processes. Charge deposition over targets has also resulted in strong electrostatic phenomena between the plasma jet or DBD streamers and the surface. Discharge patterning [9], plasma jet repulsion, and periodic oscillation have been reported [2] with the consecutive impact on pulse-to-pulse fluctuation, long-term discharge development over a target, and—again—in the delivery of RONS.

8.3 COMMON FEATURES AND DIFFERENCES BETWEEN PLASMA JETS/DBD AND PULSED ELECTRIC FIELDS DEVELOPED FOR ELECTROPORATION AND ELECTROPERMEABILIZATION

The more intuitive way to demonstrate the role of plasma electric fields is to expose biological tissues solely to these fields. The electric field is the only component of the plasma, among reactive species, photons, charges, or even heat, likely to be fully transmitted through a dielectric opaque layer; thus, many researchers have tried to expose tissues in such a protocol. Unfortunately, to the best of our knowledge, almost no effects from the electric field-only exposure were reported in the literature. Only seed germination improvement following long-term exposure was initially observed, but this finding has not yet been fully confirmed and published [10]. This is an astonishing situation, as electric field strength, duration, frequency, and polarizations should match those used and optimized for years in electroporation/electropermeabilization studies in many fields, including animal and plant cells, *in vivo*, and clinical applications. These studies are based on delivering pulsed electric fields (PEFs) in cuvette or through needles or flat electrodes implanted in tissues over an extensive range of characteristics to trigger reversible or irreversible phenomena in cell membranes. Broadly speaking, electric field strength has been reported as ranging from a few tens of V/cm to a few tens of kV/cm associated with a pulse duration ranging from sub-nanosecond to a few tens of milliseconds, at repetition rates ranging from a few Hz to a few kHz. In the range of electric fields measured with non-thermal plasma devices, i.e., for field strength ranging from a few tens of V/cm to a few tens of kV/cm and pulse duration ranging from a few nanoseconds to a few microseconds, and depending on the number of plasma pulses applied, either no effect or reversible or irreversible permeabilization should be expected (see Figure 8.2).

Thus, non-thermal plasma devices should offer potential alternatives to PEFs for developing irreversible electroporation, electrochemotherapy, or gene electrotransfer with dedicated plasma devices.

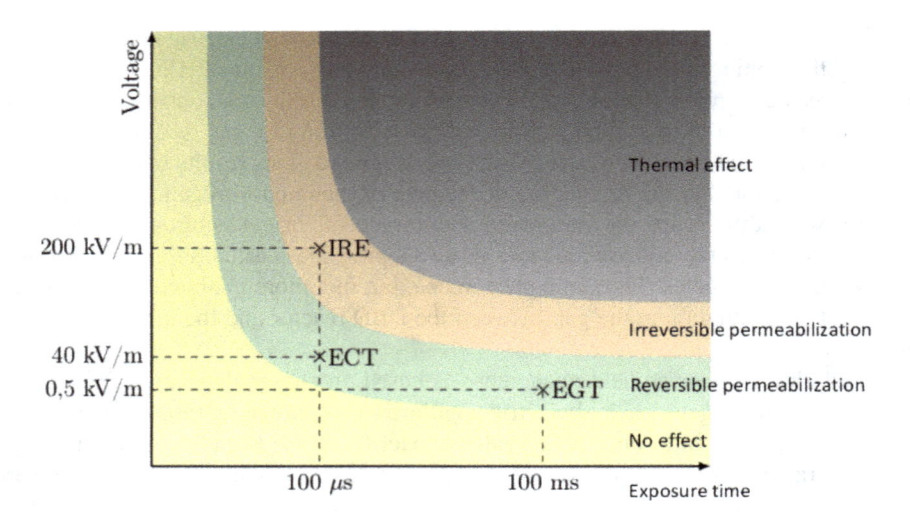

FIGURE 8.2 Mapping the effects induced by pulsed electric fields (PEFs); ECT, EGT, and IRE are electrochemotherapy, gene electrotransfer, and irreversible electroporation, respectively. Adapted from Michael LEGUÈBE, Ph.D. dissertation, Université de Bordeaux (2014), https://people.bordeaux.inria.fr/michael.leguebe/phd.html

Multiple research groups reported such demonstrations but with exposure to various plasmas, rather than solely to the electric field of these plasmas. Most of the time, the authors conclude that the electric field and reactive species are also involved in the measured action on different cells. As discussed in [11]:

> Jinno et al. [12] studied how physical, chemical, and biochemical factors played roles in plasma-induced cell membrane permeabilization. They concluded that chemical factors (such as RONS production), while essential to membrane permeabilization, are not sufficient for an effective permeabilization. Indeed, plasma electric field would also play a key role in that process suggesting a synergistic effect of both reactive species production and plasma electric field on transient cell membrane permeabilization. Synergistic action of plasma and pulsed electric fields [13] but also combinatory action of plasma generated reactive species, plasma associated transient electric fields, and charged particles [14] were shown to be likely to explain the interaction with biological membranes leading to an efficient antibacterial action during such protocols. It turns out that according to the literature the use of plasma can also allow gene transfer (high-weight molecular uptake). Ogawa et al. [15] initiated this technique in 2005, thus transfecting neuronal cells (known to be difficult to transfect using electroporation) with plasmid DNA encoding GFP by plasma treatment. The plasma used in that study was an air plasma from a device initially used for treating non-biological surfaces. Following this first study, Sakai et al. [16] also used a plasma generator for transfection and flow cytometry to measure transfection efficiency. They stated that the mechanisms involved in plasma-induced transfection could be similar to electroporation. The locally generated plasma electric field would allow transient cell membrane permeability. In 2009, Leduc et al. [17] studied the effect of an atmospheric pressure glow discharge-torch on biological targets. They observed HeLa cell membrane permeabilization of a small 10-kDa Dextran molecule [17]. They observed a local treatment of the cell layer surface. Kaneko et al. [18] have permeabilized fibroblasts with helium plasma using YOYO-1, a DNA intercalating fluorescent dye inside the nucleus. They observed a local cell detachment with high permeabilization on the border and the middle of the treated area.

With the PEF technology, critical roles in membrane permeabilization have reportedly been played by not only electric field strength and pulse duration but also electric field polarization and pulse repetition rate. Most experiments used a rectilinear polarization, generated by applying a constant voltage drop across planar electrodes, and the combination of successive positive and negative polarity, imposed by symmetric voltage pulses with a fixed delay in between them, was also shown to trigger a cancellation effect that may compromise the electropermeabilization process [19]. Pulse repetition rates around 1 Hz are the most appropriate and have been set as standard for clinical therapies where reversible electroporation is achieved in combination with other therapeutic (e.g., chemotherapeutic) agents. Such low repetition rate operation is not set when operating non-thermal plasma sources, which probably explains why plasma electric fields have not yet been adequately used. First, with non-thermal plasma, both longitudinal and radial electric fields are delivered with a very fast time shift between these two orthogonal components [20,21]. Second, if the voltage pulse is not specifically designed, as in most situations, sinusoidal but also Gaussian-like and square voltage pulses with successive rising and decay fronts will lead to the successive quick application of electric field with reverse polarization. The cancellation effect might then, unfortunately, be activated. Finally, most plasma jets or DBDs are operated at a very high repetition rate, from kHz to MHz regime, which may decrease if not compromise their efficiency. The last but not least difference between plasma delivery and PEF application is that the current amplitude during a plasma pulse ranges from a few to a few tens of mA, while it is of a few hundreds of mA with PEF applicators. Even if delivered at a high repetition rate, the low-duty cycle for kHz plasma jets or DBDs results in a mean current lower than a few mA. This is required to generate non-thermal plasma but may be the critical reason why membrane permeabilization is inefficient with plasma exposure only. Nevertheless, the advantage of plasma pulse delivery is that not only transient intense electric fields are delivered at the spot of plasma impact but also, at the same time and in the same place, a controlled dose of RONS.

As plasma delivery may present a unique opportunity for contactless electric field delivery on cell culture or tissues, a key question is whether this electric field may somehow penetrate. The simulation of the intracellular electric fields produced by dielectric barrier discharge treatment of skin models was reported in [22]. The delivery of DBD streamers generated in air and in direct contact with cell layers mimicking human skin structure, i.e., stacking cells with different geometry and dielectric permittivity, has been computed. The charging of the skin model surface and the intracellular production of electrical currents result in the penetration of electric fields. The authors then conclude that the penetrating electric field, originating from a DBD device powered with moderate voltage peak amplitude, may reach amplitude as high as 100 kV/cm, in excess of the reported threshold for electroporation.

8.4 RECENT DEMONSTRATIONS OF COMBINED TREATMENTS BASED ON PLASMA AND PEF EXPOSURE

In a comprehensive review paper, Wolff et al. [23] assess the opportunities and challenges of combining non-thermal plasma delivery with PEFs *in vitro* and, more excitingly, in clinical applications, in the context of cancer therapy. Same biological responses (apoptosis, necrosis, intracellular RONS increase, blood flow modulation, and local vascular topology modulation) have been demonstrated using either PEFs or plasma treatment. The authors claim that stable but highly oxidative short-living species generated by non-thermal plasma devices could be combined before or after PEF applications for either microsecond or nanosecond PEF pulses. The main likely additive or synergistic perspectives would be either to plasma-oxidize the cell membrane before PEF pulse application or, conversely, to benefit from transient pore or membrane permeabilization induced by ns/µs PEF delivery to critically enhance the plasma-derived RONS uptake.

In a pioneering study, Zhuang et al. [13] investigated the synergistic effects of treatments with low-temperature plasma jet and pulsed electric fields for antibacterial action, including the inactivation of *S. aureus* by a non-thermal argon plasma jet and by microsecond pulsed electric fields. The two modalities were used separately or in combination with each other. For their respective operating conditions, none of the single treatments (plasma or PEFs) led to a complete inactivation for *S. aureus*. The same 2-log reduction was measured for the longest plasma exposure time of 3 min or for the application of 300 consecutive electric field pulses with 100-μs duration and 15-kV/cm amplitude, which both represent quite strong treatment protocols. Conversely, the combined treatment with non-thermal plasma and PEFs induced significant synergistic antibacterial effects when samples were treated with plasma first. When samples were instead treated first with pulsed electric fields and subsequently with the argon plasma jet additive action, slight synergistic effects were measured. The main reason for potential synergistic action was associated with plasma treatment-induced bacteria suspension acidification. Such acidification, a universal impact of plasma delivery in solutions or on tissue, results from the delivery of RONS and their recombination to generate their conjugated acids. As expressed at the end of the previous paragraph, it is probably membrane oxidation triggered by plasma acidification that subsequently favors a much more efficient action of PEFs.

Chung et al. [24] studied the *in vitro* combination of plasma-treated solution with PEFs in the context of electrochemotherapy for cancer treatment. Microsecond-duration PEFs were used to increase the uptake of cytotoxic drugs that transiently permeabilize the cell membrane. Clinical applications with electrochemotherapy are effective and standardized palliative care, e.g., melanoma ablation [25, 26]. Nevertheless, this tumor treatment still suffers from some drawbacks due to the characteristics of the intense electric pulses used. It is worth investigating the potential capabilities of the combined application of indirect (plasma-treated solution) non-thermal plasma treatment and microsecond PEFs (μsPEFs) to outperform *in vitro* cell electropermeabilization as a first step towards clinical electrochemotherapy. Phosphate-buffered saline (PBS) was plasma-treated (pPBS) and used afterward to explore the effects of its combination with μsPEFs. Analysis of two different cell lines (DC-3F Chinese hamster lung fibroblasts and malignant B16-F10 murine melanoma cells) by flow cytometry revealed that this combination resulted in significant increases in the level of cell membrane electropermeabilization, even at very low electric field amplitude. The B16-F10 cells were more sensitive than DC-3F cells to the combined treatment. Notably, the percentage of permeabilized cells reached values similar to those of cells exposed to classical electroporation field amplitude (1100 V/cm) when the cells were treated with pPBS before and after being exposed only to very low PEF amplitude (600 V/cm). Although the level of permeabilization of the cells that are treated by the pPBS and the PEFs at 600 V/cm is lower than the level reached after the exposure to μsPEFs alone at 1100 V/cm, the combined treatment opens the possibility of reducing the amplitude of the PEF, potentially allowing for a novel ECT with reduced side-effects.

Mentheour and Machala [27] recently reported a further step in analyzing the coupled antibacterial effects of plasma-treated water and pulsed electric field for the *E. coli* model. This work used a 200-nanosecond duration (PEF) treatment with 2.5 kV/cm amplitude delivered during 100 seconds with a 100 Hz frequency. So-called transient spark discharge plasmas, operated in either open atmospheric air or closed air reactors, were used to treat water samples. Changing plasma source operation regime is a first way to prepare plasma-treated water with different RONS compositions, as reported by many authors. More interestingly, Mentheour and Machala [27] added hydrogen peroxide (H_2O_2) to the plasma-treated samples first to achieve different chemical compositions of RONS in the water and second to reach a deeper understanding of the role of these RONS in combination with PEF for bacterial inactivation. It has been demonstrated that plasma-treated solutions enriched with large amounts of both nitrites and hydrogen peroxide have a significant antibacterial effect that is dramatically enhanced by the subsequent application of PEF. Conversely, plasma-treated solutions enriched with large amounts of nitrites but with very low concentrations of hydrogen peroxide have low antibacterial action when combined with PEF delivery, although this action can be significantly enhanced by adding hydrogen peroxide. Detailed chemical analysis revealed that sufficient concentrations of both nitrites and hydrogen peroxide resulted in the efficient generation of the peroxynitrous acid, which caused a strong peroxidation of the cell membranes, leading to cell death. This peroxidation is thus a key pre-treatment, resulting in a much higher vulnerability of the bacteria when exposed to the PEF treatment.

Finally, two studies reported preliminary translation to *in vivo* mouse models combined plasma-PEF treatments in oncology. The pioneering study of Daeschlein et al. [28] compared cold plasma, electrochemotherapy, and combined therapy in a melanoma mouse model. After melanoma implantation into the C57BL/6N mice flanks, two different plasma sources (plasma jet and DBD devices) were applied directly to the tumor surface. Electrochemotherapy was operated in combination with intravenous bleomycin injection at a field strength of 1000 V/cm without or combined with the plasma delivery. Tumor growth, daily tumor volume progression, and survival after treatment were measured during the study for different mouse groups. Plasma treatment with any of the two plasma sources showed a reduction in the tumor growth rate, but a smaller one than that achieved with the PEF protocol. Promising outcomes were documented when non-thermal plasma treatment was combined with electrochemotherapy, leading to significant mouse survival improvement compared to the PEF treatment alone. It was concluded that non-thermal plasma treatment alone or in combination with electrochemotherapy may represent a new option in palliative skin melanoma therapy.

Chung et al. [29] documented a preliminary *in vivo* study of antitumor plasma-electrochemotherapy. Three independent mouse studies were performed, dealing with immunocompetent mice bearing fibrosarcoma tumors and involving control groups, a PEF-treated group (bleomycin electrochemotherapy), and combined groups of plasma-treated PBS and PEF treatments. A single-shot treatment was applied 12 days after tumor cell injection, and tumor volume follow-up was performed for more than two months. The main conclusions of these studies are that:

- Bleomycin, p-PBS, and their combined injection have no antitumor action;
- Electrochemotherapy as a reference therapy is efficient, especially for fast-growing tumors;
- Combined plasma-treated PBS and electrochemotherapy treatment allow for slower growth of fast-growing tumors and lead to a significant increase in the number of tumor regressions; and
- The composition of the plasma-treated PBS solution for different plasma exposures is a unique parameter for fine-tuning the efficiency of the combined antitumor treatment.

8.5 RECENT REPORTS SUGGESTING REDOX CHEMISTRY MAY NOT EXPLAIN THE FULL MODE OF ACTION OF PLASMAS

8.5.1 TISSUE MODEL, CELL PERMEABILIZATION

Szili and coworkers [30] developed a tissue model based on a gelatin substrate to study the delivery of reactive oxygen species following helium plasma jet exposure (see also Chapter 6). They report a counterintuitive result regarding the supposed barrier function of the gelatin sample to hydrogen peroxide or any other reactive nitrogen oxygen species penetration. They set the assumption that a "physical force" should be responsible for such species penetration across a millimeter-thick tight gelatin barrier. Measuring and documenting star-shaped-pattern RONS delivery onto the surface of the gelatin samples and by analogy with well-known Lichtenberg figures described when plasma streamers propagate over dielectric surfaces, they speculate that surface charging and consecutive charge-induced electric field may be the cause of the RONS electric field assisted diffusion across their tissue model. They nevertheless mention that their results were inconclusive and that much further work is needed to understand mechanisms that drive plasma-generated RONS deep into tissues.

He et al. [31] reported on similar experiments using a gelatin tissue model to study the barrier effects for reactive species generated by a surface air DBD. The authors mentioned that the diffusion of RONS in human tissues strongly depends on the tissue depth and plasma-delivery parameters such as treatment time. Interestingly, the authors also report that adding a small but constant electric field with a strength of up to about 20 V/cm considerably enhances the RONS penetration, but they do not fully explain the exact mechanism. They mention that such a low-amplitude electric field may play only a minor role as a physical force to promote RONS penetration but indicate that the electric field may modulate the tissue microstructure similarly to the mode of action for drug uptake in human skin reported by other authors. Low-amplitude electric field may change tissue permeability without requiring any electroporation process. The recent development of plasma skin explants treatment [32] for cosmetic ingredient uptake and the similarity of measured kinetics with those achieved with so-called micro-current applicators may shed new light on and offer tentative hypotheses for RONS penetration across human skin. It is well-known that plasma delivery on almost any surface will result in a transient charge deposition and a decrease in pH. Such plasma-induced surface action may then trigger electro-repulsion and electro-osmosis, processes fundamental to drug-delivery enhancement with micro-current applicators developed for iontophoresis [8]. Marro et al. [33] reported that at pH 7.4, human or pig skin is negatively charged and selectively permeable to cations. Such negative charging can be reduced, neutralized, or even reversed when applying iontophoresis. The reported impacts of pH modulation included the following: at physiological pH, the mannitol transport is cation-selective dominant the first time when the skin is negatively charged, while lowering the pH to 3.5 reverses the electro-osmosis, turning the skin surface to a positive charge and thus rendering transport anion-selective.

8.5.2 Treatment on Cells in Which EF Was Suspected to Play a Role

Vijayrangan et al. published two papers on cell permeabilization when exposed to a limited number of plasma jet pulses and the comparison between plasma-treated liquid (so-called indirect) and direct plasma treatments. In the first one [11], HeLa cells were treated in cell culture medium with a plasma jet capillary to cells distance of 11 mm. Surprisingly, a detachment of the cells from the petri dish was documented as having a circular pattern with a 1 mm in diameter extension. This detachment was consistently observed and was presumed to be caused either by the mechanical stress produced at the plasma impact spot or by the electric field, the amplitude of which exhibited a substantial radial decrease. The cells on the border of the detachment area were highly permeabilized. The fluorescence signal decreased as a function of the distance from the detachment area. This observation suggests that permeabilization is localized around the plasma-treated area. Using the same number of pulses, a 100-second plasma treatment (10 Hz) led to a much higher (3.84 times) propidium iodide (PI) and Dextran uptake compared to the 10-second (100 Hz) plasma treatment. This result suggests that the cell resting time between two pulses is essential for efficient uptake. Moreover, this indicated that reactive species having a lifetime of a few tens of milliseconds may be essential for drug uptake. Nevertheless, the ring shape of the PI and Dextran uptake is challenging to correlate solely with the action of RONS generated by the plasma jet and kept in contact with cells for a long duration (30 min). One hypothesis is that the electric field alone is inefficient for cell molecule penetration. However, it may play a role in combination with long- and short-lifespan reactive species present during plasma delivery on cell membranes. The better efficiency achieved with low repetition rate is likely connected with the combination with intermediate lifetime species. This favorable low repetition rate regime for plasma delivery is a common feature with results obtained with PEFs. A better uptake was indeed achieved with a one-hertz PEFs regime, leading to an optimized matching with the cell membrane dynamics. Cell detachment was also reported following PEF application [34] and connected with induced damages in the HeLa cell cytoskeleton. The actin cytoskeleton plays a significant role in allowing cells to adhere to the extracellular matrix and one another. A way to disturb cell adhesion is to stress this cytoskeleton by external force [35]. Interestingly, besides such an external force pathway, it was also recently suggested that cell adhesion may be compromised also through dipole formation or cell roughness modification associated with the electric field and ROS oxidation effects when exposed to non-thermal plasma [36].

In the second paper [37], plasma-treated medium was prepared and added at the optimal condition for cellular uptake, i.e., with a three-minute delay after incubating with the fluorescent marker (Propidium iodide, PI). In this protocol, no detachment but the same uptake as for direct plasma jet exposure was achieved for HeLa cells, while no significant uptake was measured for the 4T1 breast cancer cells. Exposure to the electric field alone was also reported to be relatively ineffective.

Thus, the authors concluded that the plasma cell "sensitization" triggered by the plasma jet exposure should involve both electric fields and short-living RONS. These results show that, for HeLa cells, electric field sensitization was not mandatory for the drug uptake. However, if PI was added after an optimized time, i.e., three minutes in the investigated experimental conditions for HeLa cells, plasma sensitization would allow for a doubling of the uptake rate. Such a potentially dramatic role of plasma sensitization was revealed to be even more drastic for 4T1 cells, for which plasma-treated medium exposure was shown to be ineffective. Cell sensitization is also involved in pulsed electric field reversible electroporation, where transient electric fields are speculated to lead to so-called "pore" formation before reactive species and drugs may interact with the cell membrane. Lin et al. [38] offered interesting insights about how electric field treatments can stabilize lipid rafts in the plasma membrane and how those rafts are the primary sensing mechanism to external electric fields and regulate downstream protein signaling. The raft reorganization into the lipid bilayer of the cells could possibly prevent the action of plasma-treated medium and reduce molecular uptake, which might explain why the plasma-treated medium applied on the cells after treatment with electric fields alone was ineffective.

8.5.3 MEMBRANE, CELL CHARGING, AND IRREVERSIBLE ELECTROPORATION

Dezest et al. [39] studied *E. coli* bacteria inactivation when exposed to helium or helium-oxygen plasma jet. Like several previous authors, they concluded that the oxidative stress following plasma treatment results in severe damage to *E. coli*. Their biochemical analysis indicated massive intracellular protein oxidation, as measured when such bacteria are PEF-exposed. Interestingly, the comparison between helium only and helium with oxygen admixtures plasma jet exposures led to the conclusion that RONS were not the only species involved in *E. coli* inactivation and that electric field and charge particles could play a significant role during helium-oxygen non-thermal plasma exposition. This interpretation was inferred from a cell membrane depolarization study revealing the likely combined effects of charged particles and electric field. Scanning electron microscopy analysis revealed that membrane pores were generated, and some bacteria experienced morphological changes. Some *E. coli* were no longer rod-shaped but rounded, becoming coccoid and losing their bacillus shape. Such damages have been described in the literature and could be attributed to charged particles or electric fields. Bacteria became round to minimize the electrostatic forces exerted by charges, and if the charge effect or the electric field was too significant, the membranes eventually broke. This is a complementary mode of action to that triggered by RONS, leading to the oxidation of the lipid membrane.

Weaver et al. [40] summarized the mode of action of electric field and chemical species from experiments conducted on artificial bilayer lipid membranes. They mention that a membrane exposed to an external electric field will charge like a capacitor, and subsequently, a transmembrane potential is induced. A short-lived steady-state current across the membrane is established when the membrane is fully charged, inducing the membrane's permeability to hydrophilic molecules. This phenomenon is most fully explained by models involving the formation of transient pores in the membrane due to exposure to the external electric field. Two critical parameters influence the reversibility of this electropermeabilization: the magnitude of the induced transmembrane potential and the duration of exposure to the external electric field. For cells, transmembrane potentials above 1 V and longer pulse times lead to irreversible permeabilization and cell death. The inactivation of bacteria by electrochemical means may involve different mechanisms, such as changes in the membrane potential, which lead to local ion flux imbalances. Death occurs due to either the formation of permanent pores and subsequent cell membrane destabilization or the loss of essential cell components and the destruction of chemical gradients via transport through transient pores. If electrochemically generated oxidants or plasma-generated RONS are present, these pores may allow the oxidants free access to the cell's interior, aiding the inactivation process. This is the irreversible permeabilization of the cell membrane.

8.6 CONCLUSION

Non-thermal plasma developed for biomedical applications delivers reactive nitrogen or oxygen species, whose intricate role in the biological response of cells or tissues has been unambiguously established [41]. As for any plasma, non-thermal plasmas, such as plasma jets, plasma torches, dielectric barrier discharges, and corona discharges, are generated in gas discharge based on the application of an electric field. Other electric fields are produced during plasma propagation out of the electrode gap of the plasma reactor, when moving towards the biological target, or as a result of charge deposition on the exposed sample. All of these transient and intense electric fields, alone or in combination with the reactive species simultaneously produced by plasma devices, may be involved in the interaction with tissues. So far, clear evidence for the role of these electric fields alone has, to the best of our knowledge, not been reported in the context of plasma medicine or plasma agriculture studies. Nevertheless, there is some evidence of their combinative or synergistic action with pulsed electric field developed for *in vitro* and *in vivo* electroporation, electrostimulation, and electrochemotherapy. They are also suspected of being involved in the plasma effects on tissue in many protocols where the plasma-derived reactive species cannot fully explain the measured biological response. Still, much work is needed to design dedicated experiments for establishing their role, in order to optimize the plasma delivery protocol alone or in combination with other relevant therapies. The ultimate goal is to develop innovative medical approaches in critical areas such as oncology, chronic wounds, dermatology, and resistant-pathogen eradication. With this objective, as discussed in this chapter, non-thermal plasmas may offer unique opportunities as not only the electric field but rich redox chemistry can be simultaneously delivered and finely tuned.

REFERENCES

[1] E. Robert, V. Sarron, T. Darny, D. Riès, S. Dozias, J. Fontane, L. Joly and J-M. Pouvesle, *Rare gas flow structuration in plasma jet experiments*, Plasma Sources Sci. Technol., **2014**, 10.1088/0963-0252/23/1/012003

[2] T. Darny, J-M. Pouvesle, J. Fontane, L. Joly, S. Dozias and E. Robert, *Plasma action on helium flow in cold atmospheric pressure plasma jet experiments*, Plasma Sources Sci. Technol., **2017**, 10.1088/1361-6595/aa8877

[3] N. Jiang, J.L. Yang, F. He and Z. Cao, *Interplay of discharge and gas flow in atmospheric pressure plasma jets*, J. Appl. Phys., 109, **2011**, 10.1063/1.358106

[4] Z. Cao, Q. Nie, D.L. Bayliss, J.L. Walsh, C.S. Ren, D.Z. Wang and M.G. Kong, *Spatially extended atmospheric plasma arrays*, Plasma Sources Sci. Technol., **2010**, 10.1088/0963-0252/19/2/025003

[5] E.R.W. Van Doremaele, V.S.S.K. Kondeti and P.J. Bruggeman, *Effect of plasma on gas flow and air concentration in the effluent of a pulsed cold atmospheric pressure helium plasma jet*, Plasma Sources Sci. Technol., **2018**, 10.1088/1361-6595/aadbd3

[6] M. Boselli, V. Colombo, E. Ghedini, M. Gherardi, R. Lorita, A. Sanibondi and A. Stancampiano, *Schlieren high-speed imaging of a nanosecond pulsed atmospheric pressure non-equilibrium plasma jet*, Plasma Chem. Plasma Process, **2014**, 10.1007/s11090-014-9537-1

[7] M.D.V.S. Mussard, O. Guaitella and A. Rousseau, *Propagation of plasma bullets in helium within a dielectric capillary—Influence of the interaction with surfaces*, J. Phys. D: Appl. Phys., **2013**, 10.1088/0022-3727/46/30/302001

[8] R.H. Guy, Y.N. Kalia, M.B. Delgado-Charro, V. Merino, A. López and D. Marro, *Iontophoresis: Electrorepulsion and electroosmosis*, J. Control. Release, **2000**, 10.1016/s0168-3659(99)00132-7

[9] A. Kumada, S. Okabe and K. Hidaka, *Residual charge distribution of positive surface streamer*, J. Phys. D: Appl. Phys., **2009**, 10.1088/0022-3727/42/9/095209

[10] E. Robert, S. Dozias, T. Maho, A. Briolay, S. Milsant and J-M. Pouvesle, *Atmospheric pressure plasma treatment of seeds: Evaluation of plasma component effects*, Workshop on Application of Advanced Plasma Technologies in CE Agriculture, 17th-21th April 2016, WAAPT, Ljubljana, Slovenia.

[11] V. Vijayarangan, A. Delalande, S. Dozias, J.-M. Pouvesle, C. Pichon and E. Robert, *Cold atmospheric plasma parameters investigation for efficient drug delivery in HeLa cells*, IEEE Trans. Radiat. Plasma Med. Sci., **2018**, 10.1109/TRPMS.2017.2759322

[12] M. Jinno, Y. Ikeda, H. Motomura, Y. Kido and S. Satoh, *Investigation of plasma induced electrical and chemical factors and their contribution processes to plasma gene transfection*, Arch. Biochem. Biophys., **2016**, 10.1016/j.abb.2016.04.013

[13] Q. Zhang, J. Zhuang, T. von Woedtke, J.F. Kolb, J. Zhang, J. Fang and K.D. Weltmann, *Synergistic antibacterial effects of treatments with low temperature plasma jet and pulsed electric fields*, Appl. Phys. Lett., **2014**, 10.1063/1.4895731

[14] W. Zhu, S-J. Lee, N.J. Castro, D. Yan, M. Keidar and L.G. Zhang, *Synergistic effect of cold atmospheric plasma and drug loaded core-shell nanoparticles on inhibiting breast cancer cell growth*, Sci. Rep., **2016**, 10.1038/srep21974

[15] Y. Ogawa, N. Morikawa, A. Ohkubo-Suzuki, S. Miyoshi, H. Arakawa, Y. Kita and S. Nishimura, *An epoch-making application of discharge plasma phenomenon to gene-transfer*, Biotechnol. Bioeng., **2005**, 10.1002/bit.20659

[16] Y. Sakai, V. Khajoee, Y. Ogawa, K. Kusuhara, Y. Katayama and T. Hara, *A novel transfection method for mammalian cells using gas plasma*, J. Biotechnol., **2006**, 10.1016/j.jbiotec.2005.08.020

[17] M. Leduc, D. Guay, R.L. Leask and S. Coulombe, *Cell permeabilization using a non-thermal plasma*, New J. Phys., **2009**, 10.1088/1367-2630/11/11/115021

[18] T. Kaneko, S. Sasaki, Y. Hokari, S. Horiuchi, R. Honda and M. Kanzaki, *Improvement of cell membrane permeability using a cell-solution electrode for generating atmospheric-pressure plasma*, Biointerphases, **2015**, 10.1116/1.4921278

[19] T. Polajzer, J. Dermol-Cerne, M. Rebersek, R. O'Connor and D. Miklavcic, *Cancellation effect is present in high-frequency reversible and irreversible electroporation*, Bioelectrochemistry, **2019**, 10.1016/j.bioelechem.2019.107442

[20] E. Robert, T. Darny, S. Dozias, S. Iseni and J-M. Pouvesle, *New insights on the propagation of pulsed atmospheric plasma streams: From single jet to multi jet arrays*, Phys. Plasmas, **2015**, 10.1063/1.4934655

[21] T. Darny, J-M. Pouvesle, V. Puech, C. Douat, S. Dozias and E. Robert, *Analysis of conductive target influence in plasma jet experiments through helium metastable and electric field measurements*, Plasma Sources Sci. Technol., **2017**, 10.1088/1361-6595/aa5b15

[22] N.Y. Babaeva and M.J. Kushner, *Intracellular electric fields produced by dielectric barrier discharge treatment of skin*, J. Phys. D: Appl. Phys., **2010**, 10.1088/0022-3727/43/18/185206

[23] C.M. Wolff, A. Steuer, I. Stoffels, T. von Woedtke, K-D. Weltmann, S. Bekeschus and J.F. Kolb, *Combination of cold plasma and pulsed electric fields—A rationale for cancer patients in palliative care*, Clin. Plasma Med., **2019**, 10.1016/j.cpme.2020.100096

[24] T-H. Chung, A. Stancampiano, K. Sklias, K. Gazeli, F. M André, S. Dozias, C. Douat, J-M. Pouvesle, J. Santos Sousa, E. Robert and L.M. Mir, *Cell electropermeabilisation enhancement by non-thermal-plasma-treated pbs*, Cancers, **2020**, 10.3390/cancers12010219

[25] M. Breton and L.M. Mir, *Microsecond and nanosecond electric pulses in cancer treatments*, Bioelectromagnetics, **2012**, 10.1002/bem.20692

[26] L.M. Mir, J. Gehl, G. Sersa, C.G. Collins, J.R. Garbay, V. Billard, Poul F. Geertsen, Z. Rudolf, G.C. O'Sullivan and M. Marty, *Standard operating procedures of the electrochemotherapy: Instructions for the use of bleomycin or cisplatin administered either systemically or locally and electric pulses delivered by the cliniporatorTM by means of invasive or non-invasive electrodes*, EJC Suppl., **2006**, 10.1016/j.ejcsup.2006.08.003

[27] R. Mentheour and Z. Machala, *Coupled antibacterial effects of plasma-activated water and pulsed electric field*, Front. Phys., **2022**, 10.3389/fphy.2022.895813

[28] G. Daeschlein, S. Scholz, S. Lutze, A. Arnold, S. von Podewils, T. Kiefer, T. Tueting, O. Hardt, H. Haase, O. Grisk, S. Langner, C. Ritter, T. von Woedtke and M. Jünger, *Comparison between cold plasma, electrochemotherapy and combined therapy in a melanoma mouse model*, Exp. Dermatol., **2013**, 10.1111/exd.12201

[29] S Bekeschus, S Emmert, R Clemen and L Boeckmann, *Anticancerous plasma-electro-chemotherapy: first in-vivo studies*, Cancers 13 (18), 4549, Therapeutic ROS and Immunity in Cancer—The TRIC-21 Meeting, contribution by T-H. Chung, K. Sklias, A. Stancampiano, K. Gazeli, T. Darny, F.M. André, S. Dozias, C. Douat, J-M. Pouvesle, J. Santos Sousa, E. Robert, L.M. Mir.

[30] E.J. Szili, J.W. Bradley and R.D. Short, *A 'tissue model' to study the plasma delivery of reactive oxygen species*, J. Phys. D: Appl. Phys., **2014**, 10.1088/0022-3727/47/15/152002

[31] T. He, D. Liu, H. Xu, Z. Liu, D. Xu, D. Li, Q. Li, M. Rong and M.G. Kong, *A 'tissue model' to study the barrier effects of living tissues on the reactive species generated by surface air discharge*, J. Phys. D: Appl. Phys., **2016**, 10.1088/0022-3727/49/20/205204

[32] V. Vijayarangan, S. Dozias, C. Heusèle, O. Jeanneton, C. Nizard, C. Pichon, J-M. Pouvesle and E. Robert, *The development of cold plasma for cosmetics*, 12th International Symposium on Plasma Bioscience (ISPB12), **November 2022**, Kwangwoon University, Seoul (Korea).

[33] D. Marro, R.H. Guy and M. Begoña Delgado-Charro, *Characterization of the iontophoretic perselectivity properties of human and pig skin*, J. Control. Release, **2001**, 10.1016/S0168-3659(00)00350-3

[34] M. Stacey, P. Fox, S. Buescher and J. Kolb, *Nanosecond pulsed electric field induced cytoskeleton, nuclear membrane and telomere damage adversely impact cell survival*, Bioelectrochemistry, **2011**, 10.1016/j.bioelechem.2011.06.002

[35] N.M. Cronin and K.A. DeMali, *Dynamics of the actin cytoskeleton at adhesion complexes*, Biology, **2022**, 10.3390/biology11010052

[36] K. Hajizadeh, H. Mehdian, K. Hajisharifi and E. Robert, *A van der Waals force-based adhesion study of stem cells exposed to cold atmospheric plasma jets*, Sci. Rep., **2022**, 10.1038/s41598-022-16277-1

[37] V. Vijayarangan, A. Delalande, S. Dozias, J-M. Pouvesle, E. Robert and C. Pichon, *New insights on molecular internalization and drug delivery following plasma jet exposures*, Int. J. Pharm., **2020**, 10.1016/j.ijpharm.2020.119874

[38] B. Lin, S. Tsao, A. Chen, S. Hu, L. Chao and P. Chao, *Lipid rafts sense and direct electric field induced migration*, Proc. Nati. Acad. Sci., **2017**, 10.1073/pnas.1702526114

[39] M. Dezest, A-L. Bulteau, D. Quinton, L. Chavatte, M. Le Bechec, J-P. Cambus, S. Arbault, A. Nègre-Salvayre, F. Clément and S. Cousty, *Oxidative modification and electrochemical inactivation of Escherichia coli upon cold atmospheric pressure plasma exposure*, PLoS One, **2017**, 10.1371/journal.pone.0173618

[40] J.C. Weaver and Y.A. Chizmadzhev, *Theory of electroporation: A review*, Bioelectrochem. Bioener., **1996**, 10.1016/S0302-4598(96)05062-3

[41] D.B. Graves, *The emerging role of reactive oxygen and nitrogen species in redox biology and some implications for plasma applications to medicine and biology*, J. Phys. D: Appl. Phys., **2012**, 10.1088/0022-3727/45/26/263001

9 Reactive Species and Redox Processes in Medical Gas Plasma-Promoted Wound Healing

Anke Schmidt, Sander Bekeschus

9.1 INTRODUCTION

Human skin is a multifunctional, highly adaptive organ that protects us from environmental challenges. Improper or impaired wound healing, and even minor tissue damage combined with persistent skin infection, have had life-or-death consequences throughout human history [1]. A pronounced wound management supports sophisticated reparative wound processes that allow the skin to heal efficiently [2]. Pronounced wound management may include techniques such as debridement, irrigation, dressing changes, infection control, and advanced wound care modalities depending on the specific needs of the wound and the individual patient. However, age and common diseases such as diabetes, angiopathy, or immunosuppressive diseases often result in pathological systemic changes and, thus, in non-healing wounds such as diabetic foot ulcers, venous ulcers, and pressure sores [3]. Nowadays, the therapeutic modalities for targeting defective wound healing are numerous [4,5], with a widespread polypragmasia in the type and sequence of therapies [6]. Consequently, wound care management with new concepts, strategies, and treatment options is indispensable.

While wound healing is strongly associated with a balanced content of reactive species in which oxygen-dependent, redox-sensitive signaling represents an essential step in the healing cascade, the concept of redox control was introduced in wound healing [7]. The concept was based on the findings that oxidants serve mainly to kill bacteria and prevent wound infection. Secondly, enlarged generation of reactive oxygen species (ROS) such as hydrogen peroxide and superoxide anions improved wound healing with increased epithelialization and granulation and promoted the release of vascular endothelial growth factor (VEGF), stimulating angiogenesis [8]. The requirements of both, not only ROS but also reactive nitrogen species (RNS), as essential components of wound healing-related processes lead to the hypothesis of redox signaling involvement in wound healing mechanism [9,10]. For decades, wound treatment with ozone [11,12], dimethyl sulfoxide [13], catalase, and iron chelators [14] has been reported to actively intervene in ROS metabolism, with benefits for wound healing. However, pharmacological control of inflammation in wound healing remains challenging [15]. Consequently, novel concepts have emerged in so-called medical gas plasma systems, capable of generating a multitude of gaseous ROS/RNS simultaneously in a localized and time-restricted manner [16–18]. This technology, operated at body temperature, modulates numerous cellular processes related to redox signaling and may help target many specific wound-healing-related pathways [19,20]. This chapter summarizes the progress of medical gas plasma technology in experimental and clinical wound-healing studies.

DOI: 10.1201/9781003328056-11

9.2   SKIN WOUND HEALING AND MEDICAL GAS PLASMA TECHNOLOGY

9.2.1   SKIN WOUND HEALING

The body is protected from infectious agents and environmental stressors by the skin, the largest organ of vertebrate bodies. Structurally complex and compartmentalized, the skin plays multiple roles. The dermis is just beneath the epidermis, containing a dense gland, hair follicle, immune cell, and vasculature network [21]. In life, skin tissue injury is inevitable, and this becomes critically important when microorganisms invade and cause infection. The wound healing process is classically simplified into four main phases [22]: hemostasis, inflammation, proliferation, and maturation with skin remodeling. These continuous, overlapping, and precisely programmed phases resulted in architectural and physiological restoration of the skin tissue (Figure 9.1). Briefly, hemostasis is characterized by platelet aggregation, blood clot and fibrin formation, and vasoconstriction [23]. During the inflammation phase, neutrophils and monocytes infiltrate the wound site, and bacteria are removed by these cells [24]. Re-epithelialization occurs during the proliferation phase, allowing keratinocyte and fibroblast immigration into the wound bed guided by growth factor production and gradients [25]. New blood vessels are also formed (angiogenesis), and collagen is synthesized for new extracellular matrix formation. Finally, the deposited matrix is remodeled further by fibroblasts, myofibroblast-mediated wound contraction, blood vessel maturation, and apoptosis of cells, forming a scar [26].

In contrast, the wounds become chronic if healing is not achieved within weeks. Chronic wounds sometimes have prolonged, incomplete, delayed, or uncoordinated inflammatory and proliferative phases, resulting in a chronification of these processes with poor anatomic, functional, and skin-architecture outcomes [27]. By definition, chronic wounds "failed to progress through the normal

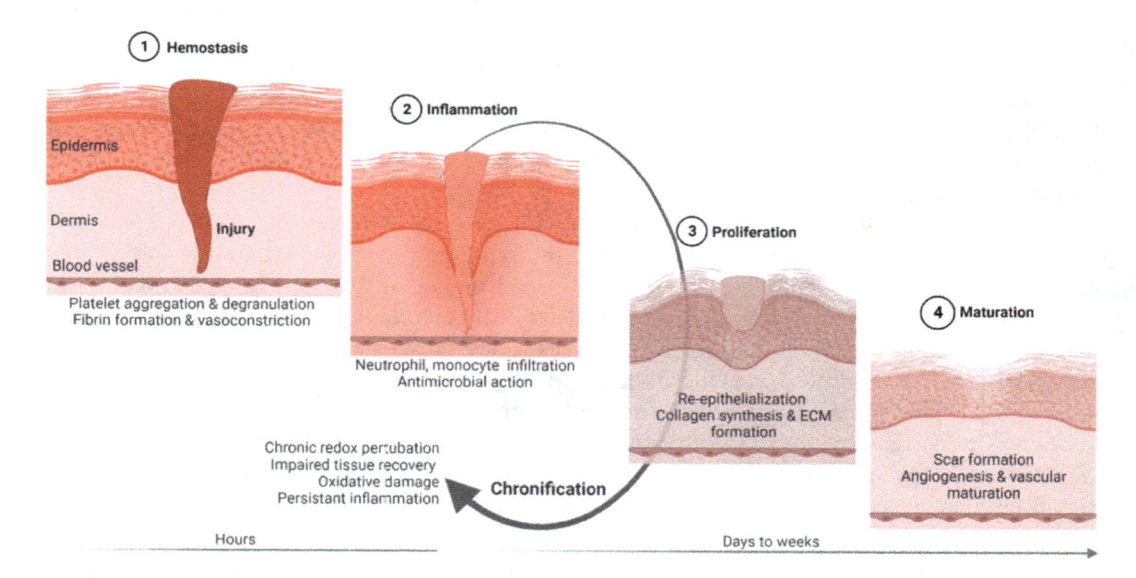

FIGURE 9.1   Stages of wound healing. Hemostasis is characterized by blood clot formation through platelet activation and release of cytokines and growth factors. It is followed by the inflammation phase, characterized by extensive infiltration of immune cells to promote inflammation and pathogen clearance. During the proliferation phase, profound proliferation-promoting processes with tissue granulation are stimulated, leading to keratinocyte and fibroblast immigration into the wound bed. In the maturation phase, the remodeling of epidermis and dermis by fibroblasts and extracellular matrix synthesis is finished and accompanied by contraction and blood vessel formation to restore near-original skin architecture. Non-healing, chronic wounds appear to persist in the inflammatory stage leading to a chronification of wounds linked to oxidative damage, persistent inflammation, and impaired tissue recovery.

stages of healing and therefore enter a state of persistent inflammation" [28], resulting in oxidative damage, impaired tissue recovery, and redox malfunction. Distinct wound areas may be trapped in different healing phases, losing the ideal synchrony of events [29].

9.2.2 Medical Gas Plasma Technology

Wound healing has a redox dimension [7]. ROS/RNS have a role in all wound healing stages, acting as signaling molecules during signal transduction processes. Additionally, ROS/RNS are integral components of antimicrobial defense elicited by phagocytes [30]. Hence, it appears reasonable to utilize ROS/RNS as therapeutics supplied exogenously to the wound microenvironment for supporting healing-mediated signaling and anti-infective action.

Physical plasma as "radiating matter" was used as early as the late 1870s [31]. Later, Langmuir investigated the flow of physical plasma, which reminded him of blood plasma flowing through veins [32], culminating in the concept of the fourth state of matter in physics [33]. More than 99% of all known matter in the universe is in a plasma state, appearing naturally on Earth in the form of lightning, aurora borealis, or fire. Plasma is technically generated for different applications, such as light systems and bulbs, welding, and televisions. Generally, plasmas are generated by supplying energy to gas up to a point at which electrons dissociate from atoms, resulting in a partial or complete ionization of gas. The charged particles, such as electrons and ions plasmas, are conductive while the overall charge remains electrically neutral. The complex physicochemical characteristics of a plasma depend on a multitude of parameters: type and composition of the gas, gas mixture used for plasma generation, applied energy, pressure, and environment. Electric and magnetic fields and light (visible, infrared, ultraviolet) and other particles are generated, including ions, electrons, and reactive atomic and molecular species [34,35]. The availability of techniques for stable and reproducible plasma generation opened up the new field of 'plasma medicine', meaning the direct application of physical gas plasma on or in the human or animal body to achieve therapeutic outcomes. It is important to note that the active plasma zone must be in direct contact with the treatment target.

Different techniques, plasma sources, and engineering concepts of medical gas plasma systems have been extensively described [36,37]. For dielectric barrier discharges (DBDs) and plasma jets, marketed gas plasma systems in Europe were introduced, which comply with the guidelines set out by medical device regulations. Such devices are usually classified as medical devices class IIa operating at body temperature under atmospheric pressure [20]. Medical gas plasma has long been used in electrosurgery, where argon plasma coagulation with targeted thermal necrotization of tissue (cauterization) or tissue removal is achieved [38,39]. The plasma jets, especially the kINPen, are the best-investigated plasma devices from the physical and biomedical points of view [40]. The plasma is generated using a high-frequency alternating voltage to a noble gas, such as argon or helium. The electron flux displays high kinetic energy and is hot. These fast electrons then ionize molecules of the argon or helium feed gas. Compared to electrons, argon or helium molecules are heavyweight and, therefore, slower in the electric field. This prevents acceleration of these molecules. In addition, fast electrons inefficiently transfer their kinetic energy to ions. Therefore, the ions remain cold, and because their temperature determines the overall temperature of a gas, the plasma harbors strongly energized particles without showing elevated temperature increases. Additional cooling of plasma can be achieved using high feed gas fluxes. The feed gas flux drives the plasma and its charged particles to the ambient air containing oxygen and nitrogen, eventually generating ROS/RNS [34]. These are usually mixtures of non-radical species (e.g., hydrogen peroxide, ozone), radical species (e.g., hydroxyl radical, nitric oxide), and several other components (Figure 9.2). Wende et al. [41] recently outlined more information on primary and secondary ROS/RNS that appear in plasma sources relevant for medical applications. In other plasma device concepts in which atmospheric air is used as working gas for plasma generation, ROS/RNS are also generated, which can be seen as a general characteristic of cold atmospheric plasmas. A comprehensive overview of gas plasma

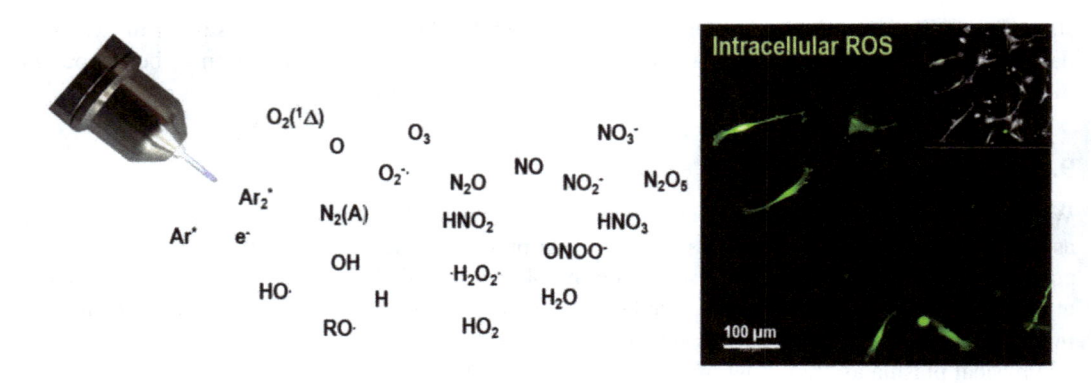

FIGURE 9.2 Several types of ROS/RNS are generated and emitted simultaneously. Among the species generated by plasma devices (here *kINPen* jet) are nitric oxide (NO), peroxynitrite (ONOO⁻), atomic oxygen (O), singlet oxygen ($O_2(^1\Delta)$), superoxide anion (O_2^-), hydrogen peroxide (H_2O_2), ozone, hydroxyl radicals (OH), nitrite (NO_2^-), and others. Intracellular ROS can be detected in skin cells by staining with 2′, 7′-dichlorodihydrofluorescein diacetate (H_2DCF-DA, green) fluorescent probe, which may react with several ROS, including hydrogen peroxide, hydroxyl radicals, and peroxynitrite.

devices and operation principles, antimicrobial effects, and *in vitro* findings in eukaryotic cells can be found in literature [42–49].

9.3 TARGETS POTENTIALLY AFFECTED DURING GAS PLASMA-STIMULATED WOUND HEALING

Medical gas plasma systems are operated at atmospheric pressure and body temperature, expelling ROS/RNS cocktails locally to the exposed site and reacting with biomolecules and wound resident cells. Several cell types are involved in healing and orchestrating wound healing phases. Sequential immigration of neutrophils, macrophages, and lymphocytes supports the extinction of infectious factors, along with regulating inflammation, cell migration, and antimicrobial immune responses. While the exact antimicrobial mechanism of action of gas plasma treatment is beyond the scope of this work, it is worth mentioning that killing bacteria via ROS is an old and conserved and, therefore, effective evolutionary strategy of higher organisms [50,51]. Beyond the antiseptic effectivity of gas plasma, there were several early hints from preclinical and clinical observations that plasma application can stimulate regeneration of impaired tissue. Additionally, macrophages play an essential role in angiogenesis by aiding microvascular endothelial cell migration [2]. Fibroblasts respond to the release of growth factors and other signaling molecules from endothelial cells and are essential—together with keratinocytes—for extracellular matrix deposition and tissue remodeling for wound closure. Chemokine and cytokine release with new blood vessel formation and angiogenesis were further observed in murine skin ear wounds (Figure 9.3).

9.3.1 ANTIMICROBIAL ACTION OF GAS PLASMA TREATMENT

Wound healing can be compromised by infection with microorganisms [52]. Although bacteria play a role in improper wound healing and chronic wounds [53], the killing of bacteria is not always associated with accelerated healing responses [54]. Hence, an infection can be both cause and consequence of hampered wound closure. In healing wounds, inflammatory processes effectively kill microbial contaminations. While this process may have deteriorated in ischemic, hypoxic, and devitalized tissue with chronic inflammation, and inflammation mediated by neutrophils cannot be resolved [55], neutrophil presence is self-limited in acute healing three to four days after wounding [56]. Consequently, these cells are present in senescent states throughout the

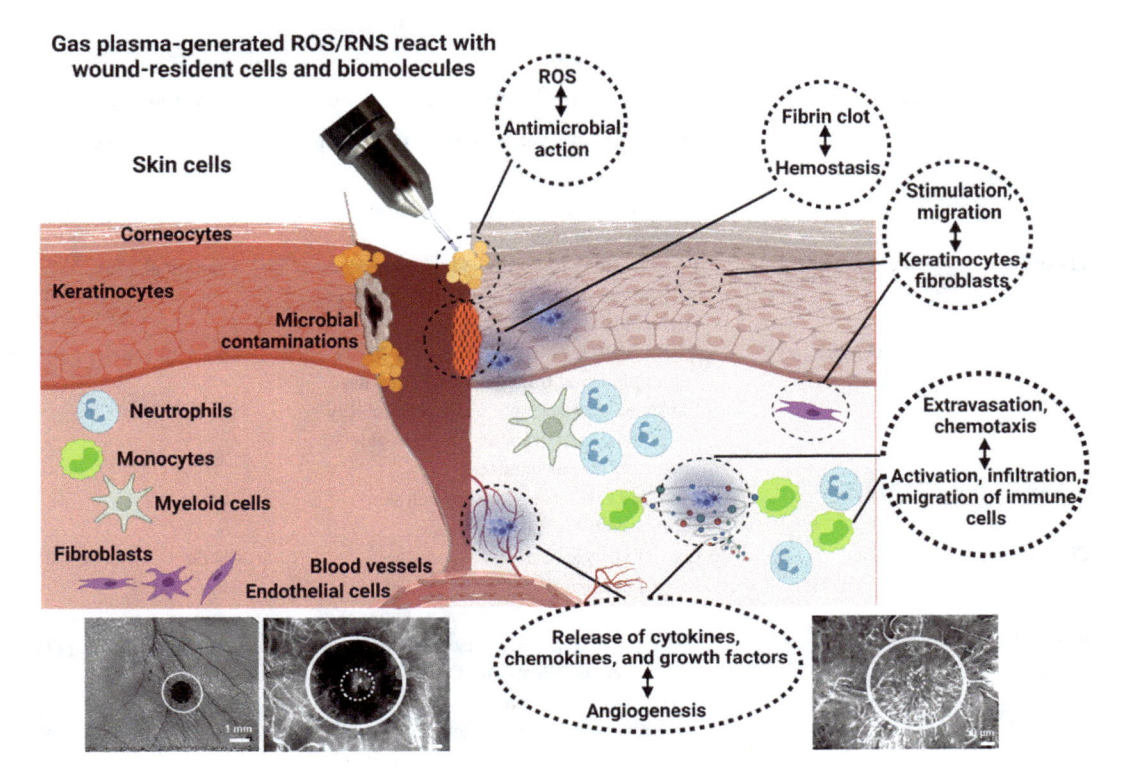

FIGURE 9.3 Wound-relevant cells and targets affected by medical gas plasma treatment. Many different cell types, such as epidermal keratinocytes, dermal fibroblasts, immune cells (*e.g.*, neutrophils, macrophages, lymphocytes), and endothelial cells (blood vessels), are involved in the healing process. Microbial and viral contaminations are reduced immediately after medical gas plasma treatment. Gas plasma exposure induced keratinocyte and fibroblast stimulation and migration, macrophage activation, infiltration, and migration of neutrophils and lymphocytes along a chemotactic gradient towards the site of tissue injury, and chemokine and cytokine release with the formation of new blood vessels. Angiogenesis was detected by intravital microscopy in murine skin ear wounds (white circle; left = before plasma treatment; middle = untreated control; right = after plasma treatment).

stagnated healing process in chronic wounds [57]. In the 1990s, promising antimicrobial effects were observed following gas plasma treatment [58], and the term "plasma medicine" was born [59]. In the following decades, several non-medical gas plasma systems were found to inactivate most microbial species, including *Pseudomonas aeruginosa* [60], *Streptococcus mutants* [61], *Escherichia coli* [62], and *Staphylococcus aureus* [63]. More than a hundred clinical isolates of bacteria, including drug-resistant strains, were killed by gas plasma treatment in wounds [64]. Bacterial resistance development has not been observed so far [65,66]. There is evidence that the bacterial burden was decreased with gas plasma treatment, fostering healing in chronic wounds. Moreover, a treatment time window exists between antimicrobial activity and tissue damage caused by gas plasma [17].

9.3.2 Gas Plasma Treatment of Non-Wounded Intact Mouse Skin

Various reports have outlined the effects of gas plasma on intact skin, predominantly in mouse models (Table 9.1).

The analysis of non-wounded skin generated a better understanding of plasma-induced effects and mechanisms. While an increased nitric oxide deposition in the DBD-treated healthy

TABLE 9.1
Medical Gas Plasma-Stimulated Effects in Non-Wounded or Lesioned Skin Mouse Models.

Mouse strain	Plasma (gas)	Findings in the gas plasma-treated group	Reference
Balb/c	**DBD (Air)**	**NO deposition was increased; no inflammation or changes of collagen deposition in skin**	**Rajas et al. [67]**
129Sv/Ev	**Plasma torch (Ar)**	**Proliferation of keratinocytes; no migration or changes in apoptosis; upregulation of β defensins**	**Arndt et al. [70]**
HRM-2	**DBD (Air)**	**TGFβ, VEGF, GM-CSF, and EGF levels and epidermal thickness were increased**	Choi et al. [68]
KM	DBD (Air)	Lidocaine drug absorption and permeability were increased; the latter was reversible 30 min after plasma treatment	Xin et al. [71]
HRM-2	Jet (Ar)	Increased EGF absorption through skin; decreased E-cadherin	Choi et al. [74]
CBA	Jet (Ar)	Lack of allergic response in skin	von der Linde et al. [72]
Balb/c	APPJ jet (He)	Several feed gas fluxes and treatment times damaged the skin; the jet reached up to 90°C	Kos et al. [73]
Balb/c	Jet (Ar)	Local treatment through intact diabetic mice skin; systemic effects on AOPP, MDA, oxLDL, and cytokines	Rezaeinezhad et al. [77]
HRS	Jet (Ar)	No dermal degeneration in intact skin; elevated catalase and Nrf2 expression; lack of follicular atrophy	Pasqual-Melo et al. [76]
SKH1	Jet (Ar)	Increased tissue oxygenation and microcirculation as well as uptake of a model drug; changes in junctional protein expression and composition of lipid bilayers	Schmidt et al. [75]
C57BL/6J	Jet (N_2)	Psoriasis-like inflammation triggered via Imiquimod; decreased epidermal thickness, leukocyte infiltrate, and chemokine/cytokine expression (IL6, IL17, IL22, CXCL1, CCL20)	Lee et al. [82]
Balb/c	DBD (Ar/N_2/O_2/ humidity)	Allergic contact dermatitis; addition of water to argon plasma decreased disease severity greater than add of N_2 or O_2	Xiong et al. [83]
SKH1	Jet (Ar)	Model of atopic dermatitis; decrease of symptoms and catalase activity; increase of SOD activity	Ara et al. [85]
Balb/c	Jet (He/O_2)	Imiquimod-induced psoriatic lesions with ameliorated morphological manifestation and reduced proliferation of keratinocytes	Gan et al. [80]
NC/Nga	N_2 DBD	Model of atopic dermatitis-like allergic skin inflammation; less recruited eosinophils and mast cell; decreased thickness of epidermis and T_H2 differentiation	Lee et al. [84]
C57BL/6J	Jet (He/Ar/N_2/air)	Effectiveness of plasma treatment in a psoriasis model	Ma et al. [81]

mouse skin was found, other side effects such as permanently over-shooting collagen production or increased inflammation were not identified [67]. Other experiments using an argon-driven DBD found increased levels of cytokines and chemokines, such as tumor necrosis factor (TGF) β, vascular endothelial growth factor (VEGF), granulocyte/macrophage colony-stimulating factor (GM-CSF), and epidermal growth factor (EGF), along with enhanced epidermal thickness [68]. Arndt and colleagues treated 129Sv/Ev mice with a prototype of the commercially available SteriPlas device, an experimental argon plasma torch [69]. They found an increase in β-defensins, which are host defense peptides with potent antimicrobial activity and capable of immune stimulation [70]. In addition, changes in keratinocyte proliferation, migration, and apoptosis were not observed, leading the authors to conclude that the device is safe. Enhanced uptake rates of the anesthetic lidocaine were determined in gas plasma-treated (air DBD) skin [71].

The kINPen has often been used to treat murine skin for gas plasma jets. When compared to allergy-inducing agents as a positive control, gas plasma treatment showed no allergic effects, which can cause discomfort in patients with hypersensitive skin [72]. In contrast, cumulative topical treatment with a helium plasma jet can lead to skin damage [73]. The treatment with the kINPen jet decreased E-cadherin expression and increased EGF absorption [74], an outcome corroborated in a different study that found an increased uptake of the model drug curcumin. Simultaneously, there was altered expression of junctional genes and proteins, elevated tissue oxygenation, and changes in microcirculatory parameters as measured using hyperspectral imaging [75]. Interestingly, in layers of the stratum corneum, this study also reported for the first time changes in the lipid composition of this upper layer following gas plasma treatment and using sophisticated lipidomics approaches. Moreover, in this same study and using an immuno-competent mouse model, the kINPen treatment activated (Nuclear factor erythroid 2-related factor 2) Nrf2 signaling and increased the catalase levels in the epidermis and dermis without any dermal degeneration [76]. Another study on diabetic Balb/c mice found that local long-term skin treatment changed malondialdehyde levels and the oxidation of protein products in the blood circulation [77]. These results are debatable, however, and we could not confirm the data for an argon plasma jet [76]. However, intact murine skin exposed to gas plasma showed changed lipid and protein compositions, which points to improved drug uptake, and the treatment was well-tolerated, as reported a decade ago in pig skin [78,79], suggesting a promising new application for gas plasma in dermatology.

Another possible area of gas plasma utilization is treating pathological skin diseases, such as psoriasis, an autoimmune disease. Using a helium/oxygen plasma jet, epidermal thickness and psoriatic skin lesions decreased [80], confirming results for nitrogen and argon plasma jet treatments [81]. Additionally, an abolished expression of psoriasis-associated inflammatory cytokines, such as interleukin (IL) 17 and IL22, and decreased leukocyte infiltrates were found [82]. Further skin diseases were investigated for gas plasma therapy effects, such as atopic dermatitis, an inflammation of the skin typically characterized by itchiness, redness, and rash. A humidified DBD treatment with argon decreased disease severity much more than nitrogen or oxygen plasma did in Balb/c mice [83]. Another preclinical study examined disease immuno-infiltrates and reported reduced epidermal thickness along with lower eosinophils and mast cells, which are strongly associated with disease severity [84]. A study on plasma-treated water also demonstrated an active reduction of the disease burden [85]. Since only long-lived reactive species such as hydrogen peroxide and nitrate are present (at physiological pH) in plasma-treated liquids [41], the findings could likely be achieved using chemically enriched liquids without a gas plasma treatment. These data showed gas plasma treatment to be safe and well-tolerated and suggested possible dermatological uses of plasma technology apart from wounds.

9.3.3 Preclinical Studies on Gas Plasma Treatment in Animal Wound Models

9.3.3.1 Mouse models

Multifaceted preclinical studies have shown favorable effects that promote wound healing following gas plasma exposure to clinically approved devices and experimental gas plasma prototypes (Table 9.2). A generally accepted, human-like chronic wound animal model has not been available [86,87], making animal experiments challenging to transfer to human conditions. The first study on experimental gas plasma sources, published in 2011, used helium plasma with different admixtures of oxygen and nitrogen in mice and confirmed the healing-promoting effects of gas plasma exposure on wound size reduction, granulation, epithelialization, and neovascularization [88]. These results were repeated in an experimental study that found expanded early inflammation, increased cellular proliferation and epithelialization, and elevated nitric oxide deposition in skin tissue, together with accelerated angiogenesis [89]. Moreover, helium plasma needle-induced wound healing was attributed to RNS (e.g., nitric oxide), whereas argon plasma predominantly caused blood coagulation [90]. The authors of the studies recommend that the ROS/RNS quantities and gas admixtures of plasma sources can be optimized for targeting therapies by favoring the accumulation of certain species over others. However, the absolute plasma treatment time and the number of treatment cycles also affect wound healing. Three treatments outperformed single exposures regarding re-epithelialization, blood flow, nitric oxide deposition into wound tissue, wound resistance, and laminin production in murine laser-induced wounds [91]. A helium plasma jet accelerated vascularization and angiogenesis of newly formed blood vessels in ear-wounded SKH1-hr mice [92]. However, a homemade helium jet did not affect acute wound healing after direct and indirect plasma treatment using plasma-activated water [93]. Using a mouse model of a third-degree burn wound, the authors showed that gas plasma treatment was characterized by enhanced angiogenesis, *in vivo* wound healing, cellular proliferation, and production of endothelial nitric oxide synthase [94]. In several diabetic mouse models, accelerated wound healing was obtained depending on the gas plasma exposure time [95]. Even more intriguingly, a complementary study demonstrated that improved wound healing depends on the type of feed gas and the percentage of admixture. It was shown that 0.1% of a helium/oxygen plasma had better wound healing effects than 1% helium or helium/oxygen conditions. The former was also associated with higher VEGF and bFGF expression levels [96].

In other studies, a former version of the commercially available SteriPlas system was applied in 129 Sv/Ev mice that had received punch biopsies on the back. Accelerated wound closure was associated with increased angiogenesis, macrophage and neutrophil infiltration, and elevated expression levels of fibroblast growth factor 2, monocyte-chemoattractant protein 1, collagen type I, and IL6 [97,98]. Apart from beneficial wound healing responses, the gas plasma treatment with a helium-based DBD was associated with reduced scar formation [99] in concert with decreased scar thickness and cell death [100].

Several studies investigated the effects of argon plasma jets on wound healing, mainly using kINPen. Other studies repeated and extended these findings to accelerated healing [101], re-epithelialization and angiogenesis, along with elevated collagen fiber and keratin production, granulation, inflammation, macrophage and neutrophil immigration, and antioxidant (Nrf2) and proliferative (p53) responses in ear wounds of SKH1-hr mice [102,103]. The latter study also revealed that hydrogen peroxide treatment did not fully support the findings identified after gas plasma exposure. Additionally, for several results such as re-epithelialization and expression levels of IL1β, IL6, TGFβ, Nrf2, heme oxygenase 1 (HMOX1), NAD(P)H dehydrogenase [quinone] 1 (NQO1), SOD1 (superoxide dismutase 1), catalase, and KGF (keratinocyte growth factor), differences between short- and long-term exposure times were identified, suggesting differential redox regulation depending on the quantity of ROS/RNS in the wound area. Notably, the evaluation of plasma-treated tissues one year later did not show any adverse long-term effects regarding carcinogenesis or skin-architecture deficiencies [102]. Up to two weeks after gas plasma exposure, changes in matrix remodeling, including, e.g., matrix metalloproteinases and tissue inhibitors of

TABLE 9.2

Animal Studies on Gas Plasma-Stimulated Effects in Wounded Skin in Mice, Medium-Sized Rodents (Rats, Rabbits), and Larger Animals, Including Pet Patients.

Size (location)	Plasma (gas)	Findings in the gas plasma-treated group	Reference
		Mouse wound models	
20 x 5 mm scar burn (dorsum)	DBD (Air/He)	Two scars were exposed two weeks after wounding with gas plasma, which resulted in less scar tissue and thickness as well as vascularization	Hoon Lee et al. [100]
6 mm punch (dorsum)	Jet ($He/O_2/N_2$)	Improved epithelalization and neovascularization; decreased microbial burden of natural wound flora	Yu et al. [88]
8 mm punch (dorsum)	*SteriPlas* (Ar)	Accelerated wound closure; increased macrophage and neutrophil infiltration; increased MCP1, IL6, and collagen type I production	Arndt et al. [98]
6–10 mm (rear leg)	Jet (Ar or He)	Accelerated wound healing; faster hemostasis	Garcia-Alcantara et al. [90]
10x15 mm skin removal (dorsum)	Jet (He)	Accelerated wound closure with intermediate gas plasma treatment times	Jacofsky et al. [95]
2 mm punch (dorsum)	Jet (Ar)	Accelerated wound healing and re-epithelialization	Nasruddin et al. [114]
$2^{nd}\,°$ burn (dorsum)	Jet (Ar/N_2)	Increased healing rates, blood flow, angiogenesis, epithelialization, wound contraction, and formation of secondary ROS/RNS in wound tissue	Ngo Thi et al. [89]
5 mm punch (dorsum)	Jet (Ar, Ar/Air)	Accelerated wound closure; better in Ar+Air over single Ar, together with more IL6 mRNA in wounds	Kim et al. [108]
4 mm punch (dorsum)	Jet (Ar)	Accelerated wound healing and increased myofibroblasts in humidified wounds	Nasruddin et al. [115]
5 mm burn (dorsum)	Jet (N_2/air)	Improved wound healing and contraction	Nguyen et al. [231]
4 mm punch (dorsum)	Jet (Ar)	Accelerated healing with short and intermediate but not long exposure times	Xu et al. [112]
1.5 mm punch (ear)	Jet (He)	Increased angiogenesis and vascularization early after gas plasma exposure	Kim et al. [92]
3x20 mm laser (dorsum)	Jet (Ar/N_2)	Improved wound healing after repeated (3x) over once treatment; increased blood flow, RNS release into tissue, wound strength, and laminin; decreased matrix metalloproteinases 3 expression	Shao et al. [91]
4 mm punch (dorsum)	Jet (Ar) + honey	Lack of accelerated wound healing and re-epithelialization in combination therapy	Nasruddin et al. [116]
2–3 mm skin removal (ear)	*kINPen* jet (Ar)	Improved wound healing and angiogenesis	Schmidt et al. [101,102]
6 mm punch (dorsum)	*SteriPlas* (Ar)	Accelerated angiogenesis and FGF2 expression	Arndt et al. [97]
6 mm punch (dorsum)	Jet (He)	Direct gas plasma exposure and gas plasma-treated liquid did not improve wound healing	Duchesne et al. [93]
6 mm punch (dorsum)	Jet (He)	Accelerated wound healing across several exposure durations	Shahbazi Rad et al. [111]
4 mm punch (dorsum)	Jet (Ar)	Treatment time-dependent changes of collagen and vimentin	Shi et al. [232]

(Continued)

TABLE 9.2 *(Continued)*
Animal Studies on Gas Plasma-Stimulated Effects in Wounded Skin in Mice, Medium-Sized Rodents (Rats, Rabbits), and Larger Animals, Including Pet Patients.

Size (location)	Plasma (gas)	Findings in the gas plasma-treated group	Reference
4 mm punch (dorsum)	Jet (Ar)	Lack of beneficial effect of combination therapy with medical honey and hydrocolloid dressings	Wahyuningtyas et al. [117]
4 mm punch (dorsum)	Jet (Ar)	Heating of tissue (>50°C) after prolonged exposure to direct contact-style treatment; longer distances promoted wound healing and re-epithelialization	Darmawati et al. [110]
10 mm burn (dorsum)	Jet (He)	Skin grafts on wounds with plasma-elevated angiogenesis and hemoglobin expression; increased CD31, VEGFR2, PDGFRβ, and endothelial nitric oxide synthase; decreased TSP-1 expression	Duchesne et al. [94]
4 mm punch (dorsum)	Jet (Ar/Air)	Accelerated healing with dual-frequency over single-frequency; two treatment cycles were more beneficial than one, three, and four cycles	Lee et al. [109]
6 mm burn wound (dorsum)	Jet (Ar/O$_2$ oil)	Accelerated wound healing	Pan et al. [233]
6 mm punch (dorsum)	Jet (Ar/air or Ar/O$_2$ oil)	Improved wound closure with oil application pretreated with either air or O$_2$ plasma	Xu et al. [234]
2–3 mm skin removal (ear)	*kINPen* jet (Ar)	Accelerated wound healing, re-epithelialization, and angiogenesis; increased keratin production, collagen fibers, Nrf2 and p53 activation, inflammation, macrophage infiltration, and granulation	Schmidt et al. [103]
4 mm punch (dorsum)	APPJ jet (Ar)	Improved wound healing with plasma alone, but Binahong extract combined with gas plasma exposure impedes wound healing	Darmawati et al. [118]
6 mm punch (dorsum)	Jet (He, He/O$_2$)	Accelerated wound healing and highest VEGF and bFGF with He/0.1% O$_2$ over He and He/1% O$_2$	Pan et al. [96]
6 mm punch (dorsum)	DBD, Jet (He; Ar)	Improved wound healing with DBD; better wound closure with Ar and He gas plasma jet treatment	Shahbazi Rad et al. [113]
2 cm ICR (dorsum)	DBD (Air-treated water	Wound healing was accelerated; antimicrobial efficacy was improved; wound microbiome was changed	Xu et al. [235]
4 mm punch (dorsum)	Jet (Ar)	Infection model; gas plasma jet-to-wound contact decreased infectious burden; remote gas plasma jet exposure was more beneficial for wound healing stimulation	Darmawati et al. [236]
1 cm burn (dorsum)	Jet (Ar)	Sterile and infected burn wounds showed improved wound healing with plasma treatment; decreased TNFα levels; unchanged bacterial burden	Dang et al. [107]
5 mm punch (dorsum)	RADIX-0502 (Ar)	Wound healing was accelerated; keratin and granular layer formation was improved; collagen and αSMA deposition were increased	Jung et al. [106]
1.5 x 3 cm^2 skin flap (dorsum)	DBD (Ar/N$_2$)	Gas plasma-conditioned solutions were added to hydrogels, which accelerated wound healing in random-pattern skin-flap full-thickness wounds	Lee et al. [237]
2–3 mm skin removal (ear)	*kINPen* jet (Ar)	Increased tissue oxygenation and skin perfusion; changed phosphorylation of signaling molecules; affected cell surface adhesion receptors, structural proteins, differentiation markers (FMT, EMT), and matrix metabolism (*e.g.*, matrix metalloproteinases)	Schmidt et al. [104]

TABLE 9.2 *(Continued)*

Animal Studies on Gas Plasma-Stimulated Effects in Wounded Skin in Mice, Medium-Sized Rodents (Rats, Rabbits), and Larger Animals, Including Pet Patients.

Size (location)	Plasma (gas)	Findings in the gas plasma-treated group	Reference
2–3 mm skin removal (ear)	*kINPen* jet (Ar)	In superficial and deep layers, increased wound tissue oxygenation, tissue hemoglobin, and tissue water index at late time points	Schmidt et al. [105]
2 cm punch (dorsum)	DBD (Air)	Improved wound healing; enhanced pro- and anti-inflammatory cytokines and growth factors	Wang et al. [238]
		Medium-sized rodent models (rats, rabbits)	
300 mm² skin removal (back)	Plasmatron, NO-generator (Air),	Pronounced wound healing in aseptic and infected wounds	Shekhter et al. [129]
300 mm² skin removal (back)	Plasmatron, NO-generator (Air)	Improved angiogenesis, wound healing, granulation, and shortened inflammation in both sterile and infected (1×10^9 *S. aureus*) wounds	Shekhter et al. [131]
2 mm punch (dorsum)	Jet (He)	Accelerated wound healing in both diabetic and non-diabetic rats; more acute inflammation and neovascularization	Fathollah et al. [121]
8 mm punch (dorsum)	Jet (He)	Accelerated wound closure; mRNA expression changes (increased IL6, nitric oxide synthase 2, and Ptgs2; decreased Nfκb and Sod1); effects only at day 7 but not at days 3 and 10	Kubinova et al. [123]
6 mm punch (dorsum)	*kINPen* jet (Ar)	Accelerated wound closure and re-epithelialization; decreased fibrosis; more acute inflammation	Breathnach et al. [119]
10 mm pressure ulcers (dorsum)	Jet (He)	Improved wound healing; higher force resistance and elastic stiffness; increased angiogenesis, re-epithelialization and inflammation	Chatraie et al. [122]
8 mm punch (dorsum)	Jet (Ar)	Accelerated wound closure in both healthy and type I/II diabetic rats; at late stages, improved re-epithelialization and fewer neutrophils and T cells in diabetic wounds; increased Sod, catalase, and glutathione peroxidases expression	Cheng et al. [120]
300 mm² skin removal (back)	Plason device (Air)	Accelerated wound healing; decreased long-term inflammation; improved collagen I/collagen III ratio	Shekhter et al. [130]
Burn wounds (dorsum)	Jet (Ar)	Decelerated burn wound injury progression regarding necrosis; lower neutrophil influx in rats	Lee et al. [127]
13x13 mm skin removal (dorsum)	Jet (He)	Accelerated wound healing, angiogenesis, epithelialization, and wound contraction; increased TNFα and IL1β	Zhang et al. [128]
17 mm punch (dorsum)	Jet (He; He/Ar)	Accelerated wound healing and formation of granulation tissue; He/Ar performed better than He gas plasma; short gas plasma treatment times performed best	Lou et al. [124]
2nd-degree burn (dorsum)	Jet (Ar)	Increased early leukocyte influx, MPO release, free thiols, and angiogenesis	Souza et al. [126]
(spine and two leg joints)	Jet (He)	Accelerated wound healing; less scar formation; increased TGFβ/pSmad2/3 expression, αSMA positive cells, and collagen deposition	Wang et al. [99]

(Continued)

TABLE 9.2 *(Continued)*
Animal Studies on Gas Plasma-Stimulated Effects in Wounded Skin in Mice, Medium-Sized Rodents (Rats, Rabbits), and Larger Animals, Including Pet Patients.

Size (location)	Plasma (gas)	Findings in the gas plasma-treated group	Reference
1.2 x1.2 cm (dorsum)	Jet (He)	Accelerated skin wound healing; decreased evidence of fungal infection in diabetic wounds	Choi et al. [239]
3.5 cm² (spine, leg joints)	DBD (Air)	Accelerated wound healing in diabetic rats; improved epidermal thickness, re-epithelialization, collagen deposition, angiogenesis, proliferation, TGFβ, and fibroblast counts; T cell counts were decreased	Guo et al. [125]
2 cm punch (dorsum)	DBD (Air)	Elevated adhesion of polypropylene to absorb betaine hydrochloride after gas plasma treatment; improved wound healing in diabetic rats	Zahedi et al. [132]
1 x 1 cm² (ear)	DBD (Air)	Lower leukocyte infiltration and less inflammation early after gas plasma treatment in New Zealand white rabbits	Ding et al. [133]
2 mm (tendon)	Jet (Ar)	Alteration of the tender structure in skin joints in New Zealand white rabbits	Amini et al. [134]
(Not stated)	DBD (He)	Removal of MRSA in infected wounds; accelerated re-epithelialization and healing; modulation of cytokine secretion and inflammation in New Zealand white rabbits	Li et al. [135]
Mammals and pet patients			
Pigs	Jet (Ar)	Good healing of superficial dermal wounds and deep wounds; no side effects; fewer immune cells visible in fully healed gas plasma-treated wounds	Kramer et al. [137]
Pigs	DBD (Air)	Long exposure times severely damaged intact skin; the energy deposition was moderate; tissue was not harmed by short exposure times; plasma-mediated hemostasis was observed	Wu et al. [136]
Bergamasca sheep	Plasma torch (He)	Wound healing was accelerated; microbial wound burden was decreased; proliferation and VEGF release during early wound healing phases were improved	Martines et al. [138]
Bergamasca sheep	Plasma torch (He)	Improved wound healing efficacy in combination with mesenchymal stem cells	Melotti et al. [139]
Dogs and cats	Jet (Ar)	Chronic wounds in pets, and summary of 8 case studies; healing was complete in 7 of 8 patients	Bender et al. [240]
Dogs, cats, Guinea pigs	Jet (Ar)	Summary of 12 case studies on pets suffering from incomplete wound healing; complete remission after repeated plasma treatment in all 12 patients	Bender et al. [141]
Dog (case report)	Jet (Ar)	Cutaneous infection with *Alternaria spp.* was decreased; immunosuppressive drug administration could be suspended; complete clinical remission	Classen et al., [142]
Dogs	Jet (Ar)	40 dogs with canine bite wounds without superior antimicrobial efficacy; lack of side effects	Winter et al. [144]
Dogs	Jet (Ar)	85 dogs with canine bite wounds without superior antimicrobial efficacy; lack of side effects	Norff et al. [143]

metalloproteinases (TIMPs), and regulation of fibrillary and integrin adhesion complexes were identified in plasma-treated wounds [104]. Adding to the molecular findings and safety aspects, gas plasma treatment improved wound tissue oxygenation in superficial and deep layers and changed tissue water and hemoglobin indices [105]. Leveled gas plasma exposure supported alpha-smooth muscle actin (αSMA) and collagen 1 and 3 depositions [106]. Another study found that once-daily plasma flux administration for seven days promoted the healing process in burn wounds with or without infection and reduced cytokines (tumor necrosis factor (TNF) a, IL6) in the tissue [107]. In SKH1-hr mice, low air admixtures to an argon plasma showed improved wound healing compared to the lack of air admixture [108]. Investigating several gas plasma conditions demonstrated that the excitation frequency of the gas plasma source affected the extent of wound healing progress [109]. As mentioned, the gas plasma system must be operated at body temperature, as higher temperatures harm wound healing [110]. Furthermore, the gas plasma treatment duration [111,112] and the type of device (DBD vs. jet) were found to be critical for improved wound healing [113]. Nasruddin and colleagues not only identified improved healing responses using an argon plasma jet [114] but also determined in a follow-up study that moisturized wounds markedly increased the extent of gas plasma-promoted wound healing and myofibroblast formation [115]. Combination with other methods, such as medical honey, was less successful [116,117]. Gas plasma treatment using Binahong leaf extract also tended to impede wound healing [118]. Such wound pre-conditioning has received little attention in plasma medicine wound studies so far.

9.3.3.2 Medium-sized rodent models

The effects and conditions of the gas plasma treatment were confirmed in punch-wounds in Sprague Dawley rats. An increased wound closure, re-epithelialization, and acute inflammation, together with reduced fibrosis, were demonstrated in argon gas plasma-exposed wound tissues [119]. A more extensive study confirmed the beneficial effects of gas plasma in healing responses accompanied by elevated re-epithelialization and Sod, catalase, and glutathione peroxidase expression in healthy rats or rats conditioned with type I or type II diabetes [120]. Overall, similar results were obtained for helium plasma-treated wounds [121]. In a model of dorsal pressure ulcers, wound closure, angiogenesis, re-epithelialization, inflammation, force resistance, and elastic stiffness were accelerated [122]. In Wistar rats with punch-biopsy wounds, accelerated healing and elevated IL6 and nitric oxide synthase 2 mRNA levels were identified on day 7 [123]. Interestingly, a helium and argon gas mixture was found to outperform He alone in optimal granulation tissue formation and accelerated wound healing, but only for short treatment times [124]. The helium plasma treatment accelerated wound healing and decreased collagen deposition and scar formation [99]. One study used DBD to investigate wound healing in Wistar rats, finding that not only was wound healing improved, but also epidermal thickness, re-epithelialization, collagen deposition, TGFβ release, proliferation, angiogenesis, and fibroblast numbers were increased in diabetic rats [125].

Studies using burn wounds focused on the inflammatory consequences of gas plasma treatment in Wistar rats. As a mechanism, one report found increases in free thiols and neutrophil granulocyte activity, as seen in increased cellular influx, myeloperoxidase and VEGF levels, and collagen I/collagen III ratios in wounds. Blood vessel formation and TGFβ levels were elevated throughout healing, and IL10 was down-regulated [126]. Together with other reports showing decelerated injury progression regarding necrosis and neutrophil influx [127] and elevated levels of TNFα and IL1β [128], these findings underline that gas plasma treatment increases early inflammatory processes. Curiously, this effect appears to simultaneously enhance antioxidant and anti-inflammatory processes, ultimately supporting matrix remodeling, tissue formation, and angiogenesis [126]. Therefore, it can be hypothesized that gas plasma treatment significantly increases existing biological tissue formation pathways connected with redox signaling and inflammatory pathways. This finding is confirmed by accelerated healing and mitigated long-term inflammation in aseptic and infected wounds in rats, using an air-plasma nitric oxide donor device together with NADPH oxidase (NOX)-derived superoxide anions, which might also have amplified peroxynitrite in the wound microenvironment [129–131]. In these technical setups, ozone and nitric oxide generators are not directly categorized as plasma medicine devices since the gas plasma is ignited within a zone and not directly in contact with the target. Nevertheless, the findings emphasize the importance and potential of gas plasma technology in wound healing along with applied redox medicine. Moreover, gas plasma is also used as a so-called enabling technology for functionalizing surfaces, such as polypropylene particles, to elevate the uptake of substances such as betaine hydrochloride, which was shown to be beneficial in diabetic rat wound healing [132].

In another medium-sized animal model with New Zealand white rabbits, wounds were treated with DBD plasma. It was found that the infiltration of leukocytes was significantly enhanced shortly after the treatment and wound analysis. This indicates that ROS/RNS derived from gas plasma exposure may be able to act as chemo-attractants for leukocytes, which in turn may accelerate pathogen clearance and healing [133]. In another report using diabetic rabbits, it was found that skin exposure to a gas plasma jet changed animal joints' tender structure [134]. However, it has not been established how short-lived agents such as ROS/RNS or other gas plasma products can diffuse to such tissue depths to exert such effects. Regarding the reduction of microbial burden, it was shown in rabbit wounds that after DBD treatment, decreased contamination with methicillin-resistant *Staphylococcus aureus* (MRSA) was present, concomitant with accelerated re-epithelialization and healing [135].

9.3.3.3 Veterinary medicine in larger animals and pet patients

Fundamental studies have investigated the effects of gas plasma exposure on intact and wounded skin in larger animal models and in veterinary patients, mainly cats and dogs. DBD plasma was safely applied to intact porcine skin for up to two minutes, and histological analysis revealed no significant damage in the tissue. Another benefit of gas plasma exposure to fresh wounds is the pro-coagulative properties, which promote faster wound sealing and clotting and protect underlying tissue [136]. Good healing of superficial dermal wounds and deep wounds without side effects and with fewer inflammatory cells was found in a porcine wound healing model following kINPen gas plasma exposure [137]. Apart from this report, a newly designed helium plasma torch with veterinary applications in mind was investigated in Bergamasca sheep. The advantage of this plasma torch is the large surface area, which decreases treatment times required for larger wounds. A lower microbial burden was observed, as well as accelerated healing of gas plasma-treated wounds along with higher cell proliferation and VEGF presence during early wound healing phases [138]. The findings were confirmed in a follow-up study, which added an extra experimental group receiving mesenchymal stem cells. Combining the two treatment types was optimal for wound healing [139]. The findings may be transferred into daily life in small sheep farms with wool production.

Repeated kINPen plasma exposure yielded complete wound remission in seven cats and dogs suffering from therapy-resistant wounds for months to years [140,141]. Partial remission was achieved in eight veterinary patients. The noteworthy findings of remission were critical and demonstrated the concept of gas plasma-provided wound healing stimulation for the first time. A second case-cohort with 12 pet patients suffering from insufficient wound healing confirmed these findings of complete healing [140]. After several gas plasma therapy sessions using the kINPenVET, the only medical product accredited for veterinary medicine, complete clinical remission was identified in a dog that experienced cutaneous *Alternaria* species infection. Consequently, immunosuppressive drug dosing could be reduced, improving quality of life [142]. However, a dedicated gas plasma-mediated antimicrobial action could not be identified in canine bite wounds when comparing healing to polyhexanide usage in more comprehensive studies, including of 40 and 85 dogs [143,144]. Although these studies were the first to investigate gas plasma performance in veterinary sciences within a prospective, randomized clinical trial, the wounds were not characterized as chronic per se, which could have affected the results. Yet gas plasma seems to show promise for problematic wounds that need therapeutic attend beyond standard-of-care regimens.

Promising data from preclinical studies indicates that gas plasma-induced wound healing is independent of the animal species and present in both wild type and diabetic rodents. The studies mainly investigated acute wounds (punch biopsy, surgical removal of epidermis and dermis, laser-induced wound, heat-induced burn) and not experimentally infected wounds. Notably, the effects of gas plasma-derived ROS/RNS on the wound cells and micro-environmental milieu are hypothesized to be the primary mechanism of action, bringing this area into the center of attention in redox medicine [49]. Most impressive, the treatment procedure was critical when eliciting improved wound healing, including several conditions such as the treatment time, the number of exposure cycles, excitation frequency, feed gas admixtures and subsequent ROS/RNS cocktails, and the type of plasma source. Rather than DBD systems, plasma jets were primarily used, benefiting from improved penetration of micro-cavities abundant in wounded tissues. Biological reactions of gas plasma treatment were found in mice during all wound healing stages. ROS/RNS signaling molecules are the first to be released upon tissue injury, initially evoking leukocyte chemo attraction and inflammation [145]. Reactive species were also found in gas plasma-treated wounds accompanied by increased bacterial inactivation, antioxidant defense responses, growth factor release, angiogenesis, stimulation, and migration, culminating in re-epithelialization and remodeling (Figure 9.3). Nevertheless, the studies' conformity and treatment conditions had low congruency regarding the use of plasma device, feed gas, gas plasma treatment time and cycles, endpoints of wound tissue analysis, wound type, animal model, and investigated biological factors. However, similar trends are observable across all studies. More standardization and harmonization of study designs are needed to identify key wound pathways responsible for the accelerated healing responses and optimized gas plasma ROS/RNS cocktails critical for efficient wound healing.

9.3.4 Gas Plasma-Treated Human Skin

Several conclusive studies on gas plasma-mediated wound healing have been conducted in humans. However, relatively few research facilities with only a handful of plasma devices have been tested for treating experimental or clinical wounds in humans. This is because clinical studies are essential for plasma devices to be accredited as a medical device for clinical use and sold for this purpose (Table 9.3).

First, few studies on gas plasma effects in clinically relevant acute wounds are available. The kINPen plasma treatment promoted objective healing responses in laser-induced acute wounds on the forearms of five volunteers [146] without negative short-term (e.g., pain) or long-term effects (e.g., scars) [147]. These findings were confirmed using a similar study setup, where no increase or decrease in melanin production was observed in twelve patients [148]. In a randomized placebo-controlled trial in 40 patients, the SteriPlas system was successfully applied to support the healing of skin graft donor sites [149]. Significantly accelerated healing was also observed in kINPen-treated vacuum-generated wounds in volunteers [150] and helium plasma-treated second-degree burn wounds in one patient [151].

Secondly and more importantly, gas plasma treatment was investigated in non-healing, chronic wounds in humans. Chronic wounds have no single etiology, but most ulcers are caused by ischemia, diabetes mellitus, venous stasis, and pressure [152]. In the late 1990s, some studies on gas plasma-generated nitric oxide therapy were performed in Russia. Beneficial effects on wound healing were reported in 68 patients [129], following a study series in 318 patients [153] and a case study in 2018 [154] using a nitric oxide-generating device called Plasotron/Plazon. While the study descriptions are insufficient to retrieve further information, accelerated wound healing was reported and confirmed in 113 patients in 2001 [155]. As outlined above, such nitric oxide generators do not put the active plasma zone in direct contact with the target and are not classified as medical plasma devices. Nevertheless, these devices reflect the potential of this technology in therapeutic redox biology and medicine.

More detailed studies are available in Europe on the clinical experience of accredited medical plasma devices. In the first randomized, prospective clinical trial in plasma medicine, earlier versions of the SteriPlas device significantly reduced microbial burden in chronic wounds in 36 patients [156], independent of the type of microorganism and plasma device modifications [157]. Subsequently, a significant reduction of particular wound parameters was observed in 70 patients in a follow-up trial [158]. A significant bacterial load and wound size reduction was also observed in a randomized prospective clinical trial in 14 patients [159]. Using the same PlasmaDerm device,

TABLE 9.3

Gas Plasma Treatment in Experimental and Clinical Wounds and Skin Diseases in Patients (d = days; w = weeks, ctrl = control).

# Patients	Conditions	Plasma device	Consequences in gas plasma-exposed wounds	Reference
Acute wounds				
5 (volunteers)	Laser-induced acute wounds	*kINPen* (Ar); 10 s, 30 s, 10 s once for 3 days vs. ctrl	Repeated short-term (10 s) treatment with best wound healing; well-tolerated	Metelmann et al. [146]
40 (randomized)	Acute skin graft wounds	*SteriPlas* (Ar); 2 min daily (except weekends) for days to weeks	Accelerated healing and positive effects on re-epithelialization; well-tolerated; few fibrin layers, and blood crusts	Heinlin et al. [149]
5 (volunteers)	Laser-induced acute wounds	*kINPen* (Ar); 10 s, 30 s, 10 s once for 3 days vs. control	Increased inflammation at d6; less post-traumatic disorders at d180; no long-term side effects after one year	Metelmann et al. [147]
6 (volunteers)	Vacuum-generated acute wounds	*kINPen* (Ar); 60 s once per week	Significantly increased healing; well-tolerated	Vandersee et al. [150]
1 (case report)	Acute 2nd-degree burn	Jet (He); 3 min per wound (79 cm^2, 9 cm^2) twice per day once per week	Lack of increase in inflammation; well-tolerated	Betancourt-Angeles et al. [151]
12 (volunteers)	Laser-induced wounds	*kINPen* (Ar); 60 s/cm^2 single application	Minimal improvement of wound healing; lack of effect on melanin production	Nishijama et al. [148]
Chronic wounds				
68 (case collection)	Chronic wounds	NO-generator (Air); 5–12 s every or every other day, *no direct plasma use*	Improved tissue hemodynamics and healing responses	Shekhter et al. [129]
113 (case collection)	Purulent and long-standing wounds and trophic ulcers	Low-energy plasma device (unnamed)	Local antimicrobial regenerating response of purulent and long-standing wounds and trophic ulcers	Khrupkin et al. [155]
36 (randomized)	Chronic wounds	*SteriPlas* (Ar); 5 min daily (except weekends) for days to weeks	Significantly reduced bacterial load; well-tolerated	Isbary et al. [156]
24 (randomized)	Chronic wounds	*SteriPlas* (Ar); 2 min daily (except weekends) for days to weeks	Significantly reduced bacterial load; well-tolerated	Isbary et al. [157]
70 (randomized)	Chronic wounds	*SteriPlas* (Ar); 3–7 min daily (except weekends) for days to weeks	Non-significant wound size reduction; significant wound width reduction; well-tolerated	Isbary et al. [158]
14 (randomized)	Chronic venous leg ulcers	*PlasmaDerm* (DBD); 2x45 s/cm^2 three times per week for 8 weeks	Significantly reduced bacterial load; wound size reduction; well-tolerated	Brehmer et al. [159]
34 (case collection)	Chronic wounds	*kINPen* (Ar); 60 s/cm^2	Significantly reduced wound exudation and microbial burden in combination with antiseptics; changed wound microbiome; well-tolerated	Klebes et al. [163]

TABLE 9.3 *(Continued)*

Gas Plasma Treatment in Experimental and Clinical Wounds and Skin Diseases in Patients (d = days; w = weeks, ctrl = control).

# Patients	Conditions	Plasma device	Consequences in gas plasma-exposed wounds	Reference
16 (case collection)	Chronic leg ulcers	*kINPen* (Ar); 60 s/cm^2 3x per week over 2 weeks	Significantly higher microbial reduction with complete removal of *P. aeruginosa*; well-tolerated	Ulrich et al. [162]
50 (randomized)	Chronic pressure ulcers	Microbeam (DBD, Ar); 1 min/cm^2 once per week for 8 weeks	Significantly improved healing; reduced wound exudate and microbial load; well-tolerated	Chuangsuwanich [164]
4 (case collection)	Chronic skin graft forearm wounds	*kINPen* (Ar); 30–60 s/cm^2 single application	Lack of increase of inflammation or infection; complete healing of all wounds; well-tolerated	Hartwig et al. [161]
6 (case collection)	Chronic wounds after surgery	*PlasmaDerm* (DBD); 90 s/cm^2 twice within 1 week	Lack of increase of inflammation or infection; complete healing of all wounds; well-tolerated	Hartwig et al. [160]
4 (case collection)	Chronic diabetic foot ulcers	Jet; 1 min every second day (7x in total)	Elevated levels of TGFβ, EGF, and KGF in wounds	Naderi et al. [172]
1 (case report)	Diabetic foot ulcer	Jet (He); 3x per day for 26 days, twice a day after that until remission	Complete healing of previously non-healing ulcer	López-Callejas et al. [171]
1 (case report)	Chronic wound	NO-generator (Air); 5 min and a total of 33 treatments, *no direct plasma application*	Complete wound healing of 2.5-year therapy-resistant wound	Pekshev et al. [154]
32 (case collection)	Chronic ulcers	DBD (He); 30 s/cm^2 once per day until patient suspension	100% (neuropathic), 75% (chronic), and 59% (mixed) ulcer healing rates; well-tolerated	González-Mendoza et al. [170]
44 (randomized)	Chronic diabetic foot ulcers	Jet (He); 5 min 3x per week for 3 weeks	Decreased IL1, IL8, IFNγ, TNFα independent of plasma; less antimicrobial burden; clinical benefit modest; well-tolerated	Amini et al. [168]
37 (randomized)	Chronic wounds	*SteriPlas* (Ar); 2 min 1 to 3 times per week	Improvement of various parameters of wound healing; once-a-week treatment suggested as optimal	Moelleken et al. [165]
44 (randomized)	Chronic diabetic foot ulcers	Jet (He); 5 min/3x per week for 3 consecutive weeks	Significantly accelerated wound closure; short-term antimicrobial effects	Mirpour et al. [169]
45 (randomized, multicentric)	Chronic diabetic foot ulcers	*kINPen* (Ar); 30 s/cm^2 per day for 5 consecutive days followed by 3x every 2nd day	Accelerated healing speed and wound reduction independent from background infection; well-tolerated	Stratmann et al. [167]

(Continued)

TABLE 9.3 *(Continued)*

Gas Plasma Treatment in Experimental and Clinical Wounds and Skin Diseases in Patients (d = days; w = weeks, ctrl = control).

# Patients	Conditions	Plasma device	Consequences in gas plasma-exposed wounds	Reference
		Other skin injuries		
5 (case series)	Actinic keratosis	Custom-made device, 1–2 min once per week	Upregulation of immune response; induction of apoptosis	Friedmann et al. [174]
7 (case series)	Actinic keratosis	*kINPen* (Ar); 120 s twice per week	Reduction of hyperpigmentation, keratosis, and scarring	Wirtz et al. [175]
12 (case series)	Actinic keratosis	*SteriPlas* (Ar); 120 s 2x/ week over max. 7 weeks	Increase of dermal thickness; improvement of chronic photodamage	Arisi et al. [177]
1 (case report)	Actinic keratosis	*Jet* (Ar); 60 s once	Complete cure of AK	Daeschlein et al [176]
6 patients	Psoriasis vulgaris	*kINPen* (Ar); 15 min 3–5 x per week	Reduction of redness, infiltration, and bacterial colonization	Klebes et al. [178]
2 (case report)	Psoriasis vulgaris	*DBD* (Air); 5x per week for 10 min per lesion or 3x per week for 5 min for a total of 8 treatments	Lesions cleared up almost entirely; no sign of recurrence	Zheng et al. [179]
14 patients	Androgenetic alopecia	*Nitrogen jet-type*	Improvement of hair loss; well-tolerated treatment	Khan et al. [241]
2 (cases)	Warts	FPG10–01NM10 pulse generator, 2–3 treatments for 2 min/lesion	Resolution of lesions; induction of apoptosis	Friedman et al. [180]
5 (case series)	Warts	*DBD* (Ar); 2–20 min once to fourth treatments over 4–16 weeks	Effective treatment for warts; well-tolerated; easy to administer	Friedman et al. [181]

surgical wounds in six patients were successfully treated [160]. Using the kINPen jet, the possibility of promoted healing was outlined in four patients with chronic skin graft forearm wounds [161]. With the kINPen plasma treatment, chronic leg ulcer treatment significantly reduced microbial burden in 16 patients [162] and wound exudation in 34 patients suffering from poor wound healing [163]. Interestingly, gas plasma treatment did not change the wound microbiome in contrast to antiseptics, highlighting the deadly effect of gas plasma on different bacterial species. Using an argon gas DBD treatment, researchers found not only a reduction of the bacterial load but also a significantly better score for pressure ulcer scale for healing [164].

Additionally, a significant wound area reduction was detectable in a randomized clinical pilot study in 37 patients with therapy-refractory chronic wounds using the SteriPlas device. Surprisingly, there was no difference in wound healing after treatments one or three times per week. A differentiation between antiseptic plasma effects and direct stimulation of tissue regeneration was not possible because patients with systemic or topical antibiotic treatment were excluded [165]. Venous insufficiencies and neuropathy make diabetic patients especially prone to defective wound healing [166]. Recently, the wound healing-promoting properties of the kINPen plasma treatment were reiterated in a clinical trial in patients suffering from diabetic foot ulcers [167]. Strikingly, gas plasma treatment was performed in addition to wound antiseptics, suggesting the mode of action of gas plasma treatment to be independent or at least less dependent on its antimicrobial efficacy.

Clinical observations have also been reported for plasma devices without clinical accreditation in western countries. A helium plasma exposure reduced the antimicrobial burden and production of IL1, IL8, IFNγ, and TNFα, while clinical benefit in 44 patients suffering from diabetic foot ulcers was relatively modest in a randomized trial in Iran [168]. This finding suggested a plasma-induced inflammation decrease, possibly due to decreased infection. However, a similar randomized study identified a transient antimicrobial efficacy with significantly improved healing effects in 44 patients using a different type of helium plasma jet [169], confirming observations for the kINPen treatment (unpublished observation). In a randomized trial with 50 patients, significantly improved healing and reduced microbial contamination were also found with a microbeam DBD (Bio-Plasma Cell modulation) in Thailand. In contrast, the microbial composition was unchanged in the wound area [164]. Additionally, 32 patients with chronic ulcers benefited from helium-operated DBD plasma treatment to a low-to-medium-scale extent, depending on ulcer type [170]. Other case reports provided findings on safe application in wound healing using a helium plasma jet [151,171]. Interestingly, elevated levels of TGFβ, EGF, and KGF were found in gas plasma-treated wound exudates of four patients, as shown in a mechanistic study using 2D gel electrophoresis [172].

Thirdly, in several studies, authors report successes in gas plasma treatment of skin diseases such as actinic keratosis (AK), psoriasis vulgaris, or warts. Intra-epidermal keratinocytic dysplasia may transform into skin cancer [173]. Gas plasma treatment using a custom-made device may serve as an effective, well-tolerated, curative, and economical alternative for current treatment options for AK [174], which was also shown after repeated kINPen treatments in seven patients [175] and one patient [176]. Additionally, the SteriPlas application effectively reduced the cumulative AK area and the number of AKs in 12 patients [177]. Since patients with psoriasis vulgaris often experience therapy-resistant lesions, these promising findings make gas plasma a valuable option in the therapy of those patients. The first clinical case reports on treating psoriatic plaques were only partially encouraging and showed no significant advantage over conventional therapies [178], or only in two patients with inverse psoriasis positive therapeutic effects [179]. Additionally, inhibiting effects on warts were partially found in gas plasma-treated patients, especially children [180,181]. Several parameters can be changed to support the treatment of these diseases, including the selection of gas plasma devices and feed gas admixtures, the treatment time and frequency, and the combination therapy. Finally, the gas plasma exposure was overall well-tolerated with no severe adverse events observed in hundreds of clinical gas plasma studies. Another important conclusion is that the plasma technology is based on safety guidelines, which were previously described [182], and/or an accreditation as medical device has been performed [20,183].

9.4 GAS PLASMA-PROMOTED WOUND HEALING AND POTENTIAL MECHANISMS

Originally, it was stated in the literature that the antimicrobial effects following gas plasma treatment were mainly responsible for enhanced healing. With increasing knowledge, studies dealing with sterile wounds in vertebrates [103] or human patients receiving both gas plasma treatment and antiseptic agents [167] propose a mode of action independent of microbicide activity. However, the compromised wound-healing response in immunosuppressed patients remains unclear, especially with the burden of antimicrobial-resistant bacterial strains. The microbial load can be not only the cause of defective wound healing but also its consequence. In line with the data of several preclinical trials (Table 9.1, Table 9.2) and one clinical trial [167], it can be hypothesized that gas plasma promotes endogenous wound healing independent of its proven antimicrobial effect. The hypothesis is supported by the fact that microbial growth is reduced in the wound only temporarily following gas plasma treatment and returns to the elevated baseline values one day later, suggesting that the gas plasma-induced microbial burden reduction may not be durable [17].

The fundamental insight that reactive species might be essential components in wound healing inspired plasma medicine research focusing on redox biology to explain and interpret several biological effects in gas plasma-stimulated skin [31]. Preclinical and, more critically, clinical studies (Table 9.3) underline the efficient and well-tolerated healing responses in gas plasma-stimulated tissues described above. It is unsurprising that no single factor or mechanism can be identified to explain the effect of gas plasma in wound healing. Wound therapy depends on the healing phase and context. Therefore, a multitude of responsible mediators can be defined [184,185]. Medical gas plasma systems are assembled of multiple components such as neutral, charged, and reactive particles, and accurately identifying all single components simultaneously or separating some types of reactive species in the plasma gas phase is difficult to impossible. While the plasma gas phase has been investigated very well, less is known about the mechanisms of action, thanks to the short half-life of ROS/RNS and their quick reaction with biomolecules, cells, or each other, and the technical difficulties of defining the contribution of single types of ROS/RNS appearing simultaneously on a quantitative scale [47]. Hence, many mechanisms and hypotheses are proposed as contributing to gas plasma-induced wound healing. From the hundreds of studies, it can be concluded that such mechanisms are independent of the species, meaning that gas plasma-derived ROS/RNS drive an evolutionarily older tissue renewal and regeneration program.

Twenty years ago, redox-based strategies were suggested to treat non-healing wounds, because wound healing is subject to redox control [7]. Today, there is strong evidence that gas plasma therapy fulfills this criterion. Thus, gas plasma treatment of the murine wound margins around a scab induced wound healing below the scab, which suggests paracrine redox signaling to distant tissue sites with relayed ROS/RNS components [98]. It should be emphasized that gas plasma-derived ROS/RNS (e.g., superoxide anions, nitric oxide, peroxynitrite, hypochlorous acid, hydrogen peroxide, nitrite, nitrate, hydroxyl radical, singlet oxygen) are part of the reactive species physiologically generated by enzymes (e.g., NOX, MPO, nitric oxide synthase) and therefore activate existing redox signaling pathways. Both *in vitro* [186] and *in vivo* [103] data identified a transient activation of the antioxidant nuclear transcription factor Nrf2 pathway (Figure 9.4), which was described as a precious target for pharmacological intervention and wound healing [187]. The gas plasma exposure enhanced the levels of free thiols in the wound tissue, suggesting the activation of the Nrf2 pathway upon plasma-generated ROS/RNS exposure in a rat burn wound model [127]. Similarly, gas plasma exposure led to increased cellular proliferation and levels of Nrf2 and catalase in intact murine skin [76], which was also confirmed in ex vivo gas plasma-treated human skin tissue. However, proliferation decreased for longer treatment times [188], in congruence with observations of a healing-supporting effect of short (3 s) over intermediate (20 s) treatment times in murine ear wounds [103]. Helmut Sies proposed the dual and hormetic role of gas plasma-derived ROS/RNS acting as signaling agents at low concentrations and damaging agents at higher concentrations as oxidative eustress and distress paradigm [189]. To counteract the oxidative stress, keratinocytes stimulated by gas plasma repeatedly over three months continuously expressed high levels of heat-shock protein 27 *in vitro* [190]. Generally, wound therapies often target inflammation. Several cell types of skin tissue release dozens of chemokines and cytokines during the different healing stages. However, many details about their interplay and straightforward solutions for controlling inflammation towards optimal healing have yet to be identified [191]. Chemokine and cytokine generation and release are a consequence of signaling and a hallmark of chronic wounds [192,193]. While a few therapies aim at adding or removing distinct agents from the wound bed with some success [194–196], a gas plasma-targeted wound therapy would address the cause of such pathological release. Therefore, it was hypothesized that macrophages are a primary cell type responding to gas plasma-generated reactive species. Proper healing strongly depends on these cells, allowing an appropriate inflammation level, removing infections, supporting keratinocyte migration and epithelial-to-mesenchymal transition for re-epithelialization of the wound bed, and matrix reorganization [197]. A hypothesis is a gas plasma-induced Nrf2 activation in macrophages, leading to pro-healing M2 polarization, anti-inflammatory cytokine release, increased angiogenesis, and improved matrix remodeling in wounds [198].

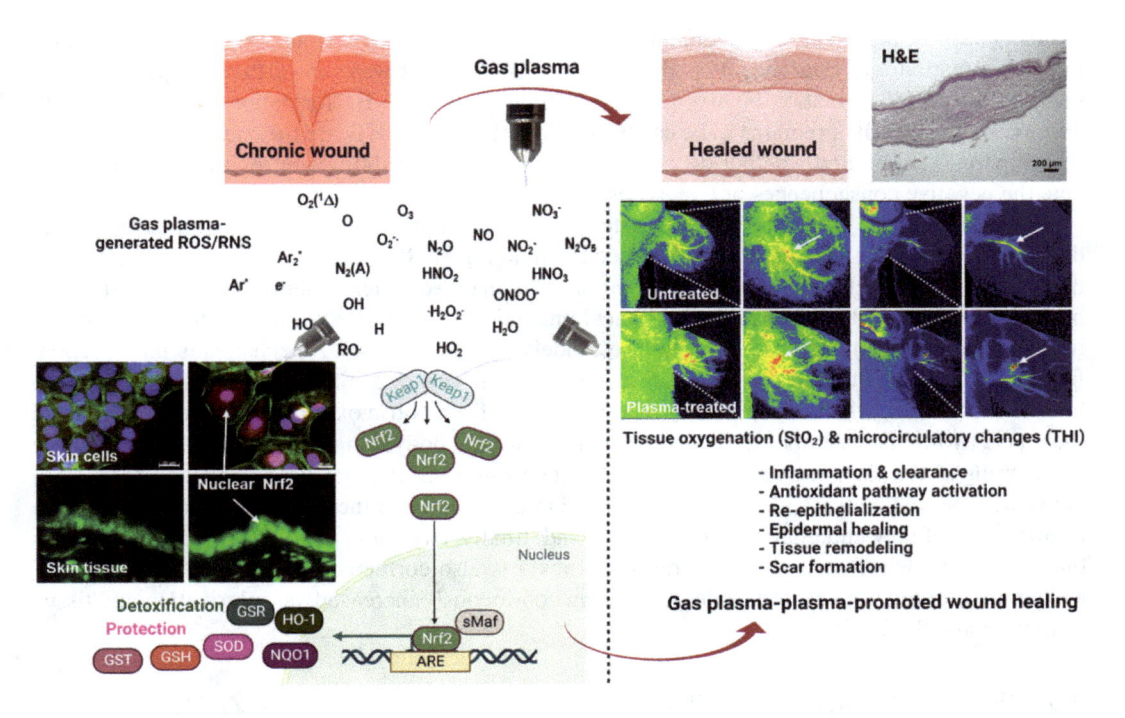

FIGURE 9.4 Biological consequences of medical gas plasma-applied ROS/RNS. Gas plasma-generated ROS/RNS target chronic wounds, which fail to heal, and react with cells and biomolecules. The treatment leads to biological consequences with the activation of redox signaling during wound healing. These include antioxidant Nrf2 pathway activation responses with detoxification and protection signaling, inflammation and clearance, and tissue remodeling. In preclinical studies, accelerated healing was found together with enlarged re-epithelialization (H&E staining of ear wounds after healing), granulation, and collagen fiber production (PSR staining). Beneficial effects on microcirculatory parameters such as tissue oxygenation (StO2;) and tissue hemoglobin index (THI) were observed by hyperspectral imaging. Keap1 = Kelch-like ECH-associated protein 1; sMaf = Adipogenin; ARE = antioxidative response elements; GST = glutathione S-transferase; GSH = glutathione; SOD = superoxide dismutates; NQO1 = NAD(P)H dehydrogenase [quinone] 1.

However, *in vitro* and preclinical models reflect the clinical situation only partially, since they have no severe underlying disease or excessive infection that induces the chronification of wounds in patients. Wound healing is sometimes complicated but always functional in preclinical models. In contrast, healing in humans is often stopped completely. Consequently, the question is not how to promote wound healing but how to get it started in the first place, a process often referred to as turning a chronic wound into an acute wound. First, chronic wounds are subject to repeated micro-injuries and bleeding, especially during debridement and cleaning of the wound from necrotic material. As demonstrated in liver injury, gas plasma treatment was strongly hemostatic, along with platelet activation-dependent coagulation in mice [199]. Via secretion of platelet-derived and epidermal growth factors [200], platelet activation supports the wound-healing process. Platelet activation has been thoroughly investigated as biological therapy in clinical trials to promote wound healing [201]. This mode of action was confirmed in gas plasma-treated human blood platelets that are strongly activated via hemolysis [202]. Interestingly, autologous platelet gel therapy [203] and in situ electric pulse stimulation [204] have been used for wound therapy, inducing hemostasis, local infection control, and platelet activation. Secondly, non-healing wounds are caused by several factors, such as hypoxia, bacterial infections, and senescent cells [205]. Apoptosis through gas plasma-induced oxidative distress in senescent fibroblasts [206] and neutrophils [207] might be a mechanism of action, a hypothesis that could be tested in clinical plasma research. Neutrophilic products such as neutrophil extracellular traps (NETs) are present in acute wounds [208], promoting inflammation and clearance of pathogens. In this regard, it was shown that gas plasma treatment induces NET formation *in vitro* [209]. Although excessive NET formation is associated with chronic wounds [210], it remains to be established whether such NETs are the cause or consequence of hampered healing and whether de novo NET formation from dying neutrophils might outcompete the negative consequences of live, senescent neutrophils.

Finally, one factor that probably contributes to gas plasma-mediated wound healing is the immediate tissue responses detected by hyperspectral imaging (HSI). Thus, increased superficial and deep tissue microcirculation, tissue oxygenation, and changed water content were observed in gas plasma-treated patient wounds [211] and intact human skin [212–214]. Similar results were observed in gas plasma-treated ear wounds in mouse models [104,105]. Several mechanisms may account for these findings. Plasma devices such as the kINPen generate nitric oxide [215,216], a well-known vasodilator, which subsequently increases blood flow. Nitric oxide generation could even be increased by the intrinsic thermal energy of plasma around body temperature (37–40°C), resulting in warming of the plasma-treated skin (30°C) and encouraging endogenous nitric oxide production along with blood flow [217]. Moreover, enhanced microcirculation increases nutrient transport, the immigration of new and unprimed leukocytes, and, finally, oxygen to counteract hypoxia [218]. As one of the key drivers of wound ulceration, hypoxia is also corrected by increased angiogenesis, as identified in preclinical studies demonstrating continuously increased vascularization and tissue oxygenation [89,97,103].

9.5   ISSUES AND CHALLENGES LINKED TO GAS PLASMA UTILIZATION

Successful preclinical and clinical studies on plasma-assisted wound healing or skin therapies support the utilization of gas plasma technology in routine clinical practice. Yet despite the solid experimental evidence of plasma-aided wound healing, several un-examined issues, challenges, and opportunities remain in the field of plasma medicine.

Because of its multimodal ROS/RNS nature, gas plasma technology is attractive for clinical use in dermatology and redox biology. However, plasma diagnostics need to be further developed. Due to their diverse reactivity, hundreds of chemical reactions occur in the gas phase parallel with individual spatiotemporal concentrations of individual ROS/RNS along the effluent-air-gradients [219–221]. A variety of ROS/RNS are described as having biological functions. Among them are hydroxyl radicals, atomic and singlet oxygen, superoxide anions, hydrogen peroxide, ozone, triplet oxygen, nitric oxide, nitrogen dioxide, peroxynitrite, hydroperoxyl radicals [222], and secondary species such as organic alkoxyl radicals and organic peroxides [41]. However, determining the main biologically active component in the plasma-generated ROS/RNS mixture is a complex technical task. Such reactive species can be precisely quantified in the plasma gas phase [40]. Current methods, techniques, and model systems face several limitations in unambiguously identifying the kind and levels of each individual ROS/RNS in tissues [104]. Considering the dose-dependent action of ROS/RNS, additional *in vivo* parameter studies would shed more light on essential aspects, such as the duration and frequency of gas plasma treatment. This research allows the identification of ROS/RNS concentrations and molecular biomarkers that determine the boundary between supporting healing (i.e., oxidative eustress) and hampering healing (i.e., distress). Due to the multiple factors that can influence gas plasma effects, we emphasize that further improvement of plasma devices for patients must be made under natural treatment conditions. The knowledge gap regarding how ROS/RNS enter the gas plasma-treated cells, change communication between cells, and activate intracellular signaling has not yet been closed. Since it is not yet entirely clear how ROS/RNS interact with tissue, or how deeply various species penetrate and alter a complex environment such as tissue or the target cell composition, there is a great need for more *in vivo* studies. In this regard, mouse models with functional knockouts of various oxidative genes or immune cells would be of great interest. Experiments related to NET formation [208,222], senescent cell apoptosis, and treatment-time-dependent nitrotyrosine formation [223] in gas plasma-treated experimental wounds may allow better comprehension of how to exploit redox systems therapeutically. Gas plasma-induced inflammation is an attractive option for mimicking inflammatory redox environments because many wound healing processes involve multiple redox chemistry pathways [222]. To understand redox control in the wound environment, chemokine and cytokine quantification of selected targets in wound material samples is one economical way of understanding the mechanisms of gas plasma exposure. A more insightful but technically challenging way is mass spectrometry analysis of wound material [208].

Depending on the primary trigger, chronic wounds form a heterogeneous group of skin injuries, including venous leg ulcers, pressure ulcers, and diabetic foot ulcers. To develop medical devices for topical ROS/RNS gas plasma therapies in such diseases, human trials are inevitably needed to optimize towards specific wound applications. In particular, jet-generated plasma systems can be operated with different feed gas compositions that produce distinct ROS/RNS mixtures, as shown for many animal models [88,91,96,108,124]. Preclinical evidence on optimizing healing responses via changing the ROS/RNS composition suggests an excellent chance of increasing the already promising promotion of wound healing by gas plasma therapy. Besides essential wound factors such as underlying disease, size, or volume, the location of the wound is also critical for deciding which plasma source is used. Jet-assisted point treatments of non-flat body tissues will deliver the gas plasma appropriately, unlike flat DBD devices. In contrast, plasma jets are less helpful in treating larger areas, a feature of chronic diabetic wounds. A relatively recent concept is using gas plasma multi-jet-array systems. These are scalable to facilitate both small- and large-area treatments, enabling rapid care of large-sized wounds. No research has been conducted on whether there is an optimal time for gas plasma therapy within existing standard treatment regimens. Above all, it is necessary to determine the factors that are decisive for the length of treatment, which ranges from seconds to minutes per square centimeter, and the start and end of gas plasma treatment. While some companies are making recommendations based on relatively little evidence, there is often no consensus on the frequency of treatment (e.g., single vs. repeated), how many times per week, what treatment breaks should be applied, and whether wound areas (e.g., the margins or heavily infected areas) should be preferentially exposed to gas plasma. The complexity of gas plasmas, the structural differences in devices, and missing key biomarkers complicate the definition of a 'plasma dose'. In order to make devices and treatment intensities comparable, the definition of a unit to describe plasma dose must be introduced worldwide. Wound age, stage, and degree of chronification, which influence traditional wound care [223,224], as well as immune status and co-morbidities, continue to be of therapeutic importance [225]. Other parameters related to the wound colonization signatures and antimicrobial resistance that hinder proper wound healing must also be considered [226]. The combination of plasma devices with technical facilities for continuous real-time monitoring of both plasma properties and the biological target parallel to the exposure may facilitate the development of a novel generation of gas plasma-based therapy systems [227]. Additionally, intensive work is being done on portable, battery-powered plasma devices, which could revolutionize wound care for immobile patients in home care.

Despite these challenges, selecting appropriate gas plasma treatment conditions is vital to maximize the clinical efficacy of gas plasma-assisted wound care. Hundreds of plasma units (mainly two or three types from two or three companies) are used daily in European clinical centers. Numerous preclinical studies have examined safety standards and potential healthy risk of gas plasma treatment in animals [228], but there are few structured studies and reports on clinical experiences in plasma-treated humans. From the clinical perspective, gas plasma treatment show a low incidence of adverse side effects or fundamental complications such as mutations, genotoxicity, tumor formation, and cell metastasis [229,230]. However, more parameter studies are needed for gas plasma therapy in humans to iterate and understand the influence of plasma source selection, treatment times and cycles, bacterial contaminations, and local or systemic wound responses. Additionally, the standard of care for wound management varies worldwide, making capturing large cohorts of patients with similar wound characteristics difficult. The first steps of gas plasma use in humans were done. Standardizing medical plasma devices with the CE certification of category IIa guarantees safety and therapeutic aid for European patient care. International guidelines, standardized sampling protocols, and databases should be introduced and continuously developed to support effective clinical gas plasma therapy and to drive standardization procedures worldwide.

The clinical application on a global scale of gas plasma therapy as consensus-guided wound treatment is still in its infancy but promises to be an essential step in developing improved wound care. Several hypotheses should be addressed and tested in wound centers to better understand gas plasma-assisted wound care: Are certain microorganisms that strongly impede wound healing more readily killed by specific gas plasma-derived ROS/RNS mixtures? Do aged wounds require a different treatment frequency? Is the wound size linearly related to treatment time for optimal results, or should certain wound regions be given priority? Given the breakthrough clinical results that gas plasma therapy has produced to date, such questions are not only an academic endeavor but also have practical implications for patients worldwide, especially those for whom standard therapies have failed.

9.6 CONCLUSION

Medical gas plasma devices can inactivate microorganisms, enabling their medical use in defective wound healing. Conclusions from preclinical studies on wound healing-supporting effects of the gas plasma-derived ROS/RNS suggested the exciting paradigm of biological redox regulation through reactive gaseous species. The effects are independent of antimicrobial action, modestly to significantly independent of the technology (jet vs. DBD), and effective across several species and wound types (e.g., acute and/or chronic), etiologies, and underlying diseases (e.g., diabetes). Dependent on several gas plasma treatment parameters (e.g., exposure time, feed gas mixture, ROS/RNS cocktail, excitation frequency, number of exposures) and device types, gas plasma treatment is associated with increased leukocyte infiltration and generation of inflammatory mediators (chemokines and cytokines) followed by accelerated antioxidant defense responses, growth factor release, angiogenesis, tissue oxygenation, re-epithelialization, and tissue remodeling. Not only preclinical but also clinical studies demonstrated impressive healing effects in gas plasma-treated skin tissue. A decrease in wound size is accompanied by decreased microbial burden without changing the composition of the microbial community. Gas plasma treatment is mainly applied more than once as adjuvant therapy rather than in place of standard-of-care. However, specific clinical recommendations are hard to make, given the wide variety of systems and applications, and the evidence-based results, while hopeful, are still inadequate.

REFERENCES

1. Erdem, H.; Tetik, A.; Arun, O.; Besirbellioglu, B.A.; Coskun, O.; Eyigun, C.P. *War and infection in the pre-antibiotic era: The third ottoman army in 1915*. Scandinavian Journal of Infectious Diseases **2011**, 10.3109/00365548.2011.577801.
2. Wilkinson, H.N.; Hardman, M.J. *Wound healing: Cellular mechanisms and pathological outcomes*. Open Biology **2020**, 10.1098/rsob.200223.
3. Richmond, N.A.; Maderal, A.D.; Vivas, A.C. *Evidence-based management of common chronic lower extremity ulcers*. Dermatologic Therapy **2013**, 10.1111/dth.12051.
4. Frykberg, R.G.; Banks, J. *Challenges in the treatment of chronic wounds*. Advances in Wound Care (New Rochelle) **2015**, 10.1089/wound.2015.0635.
5. Cutting, K.F.; White, R.J.; Legerstee, R. *Evidence and practical wound care—An all-inclusive approach*. Wound Medicine **2017**, 10.1016/j.wndm.2017.01.005.
6. Kramer, A.; Dissemond, J.; Kim, S.; Willy, C.; Mayer, D.; Papke, R.; Tuchmann, F.; Assadian, O. *Consensus on wound antisepsis: Update 2018*. Skin Pharmacology and Physiology **2018**, 10.1159/000481545.
7. Sen, C.K. *The general case for redox control of wound repair*. Wound Repair and Regeneration **2003**.

8. Sen, C.K.; Khanna, S.; Babior, B.M.; Hunt, T.K.; Ellison, E.C.; Roy, S. *Oxidant-induced vascular endothelial growth factor expression in human keratinocytes and cutaneous wound healing.* Journal of Biological Chemistry **2002**, 10.1074/jbc.M203391200.

9. Sen, C.K.; Roy, S. *Redox signals in wound healing.* Biochimica et Biophysica Acta **2008**, 10.1016/j.bbagen.2008.01.006.

10. Sen, C.K. *Wound healing essentials: Let there be oxygen.* Wound Repair and Regeneration **2009**, 10.1111/j.1524-475X.2008.00436.x.

11. Martinez-Sanchez, G.; Al-Dalain, S.M.; Menendez, S.; Re, L.; Giuliani, A.; Candelario-Jalil, E.; Alvarez, H.; Fernandez-Montequin, J.I.; Leon, O.S. *Therapeutic efficacy of ozone in patients with diabetic foot.* European Journal of Pharmacology **2005**, 10.1016/j.ejphar.2005.08.020.

12. Valacchi, G.; Fortino, V.; Bocci, V. *The dual action of ozone on the skin.* British Journal of Dermatology **2005**, 10.1111/j.1365-2133.2005.06939.x.

13. Patil, M. *Pharmacology and clinical use of dimethyl sulfoxide (dmso): A review.* International Journal of Molecular Veterinary Research **2013**, 10.5376/ijmvr.2013.03.0006.

14. Wright, J.A.; Richards, T.; Srai, S.K. *The role of iron in the skin and cutaneous wound healing.* Frontiers in Pharmacology **2014**, 10.3389/fphar.2014.00156.

15. Shukla, S.K.; Sharma, A.K.; Gupta, V.; Yashavarddhan, M.H. *Pharmacological control of inflammation in wound healing.* Journal of Tissue Viability **2019**, 10.1016/j.jtv.2019.09.002.

16. Kramer, A.; Lindequist, U.; Weltmann, K.D.; Wilke, C.; von Woedtke, T. *Plasma medicine—Its perspective for wound therapy.* GMS Krankenhhyg Interdiszip **2008**.

17. Bekeschus, S.; von Woedtke, T.; Emmert, S.; Schmidt, A. *Medical gas plasma-stimulated wound healing: Evidence and mechanisms.* Redox Biology **2021**, 10.1016/j.redox.2021.102116.

18. Bekeschus, S.; Kramer, A.; Schmidt, A. *Gas plasma-augmented wound healing in animal models and veterinary medicine.* Molecules **2021**, 10.3390/molecules26185682.

19. Barton, A.; Wende, K.; Bundscherer, L.; Hasse, S.; Schmidt, A.; Bekeschus, S.; Weltmann, K.-D.; Lindequist, U.; Masur, K. *Nonthermal plasma increases expression of wound healing related genes in a keratinocyte cell line.* Plasma Medicine **2013**, 10.1615/PlasmaMed.2014008540.

20. Bekeschus, S.; Schmidt, A.; Weltmann, K.-D.; von Woedtke, T. *The plasma jet kinpen—A powerful tool for wound healing.* Clinical Plasma Medicine **2016**, 10.1016/j.cpme.2016.01.001.

21. Shaw, T.J.; Martin, P. *Wound repair at a glance.* Journal of Cell Science **2009**, 10.1242/jcs.031187.

22. El Ayadi, A.; Jay, J.W.; Prasai, A. *Current approaches targeting the wound healing phases to attenuate fibrosis and scarring.* International Journal of Molecular Sciences **2020**, 10.3390/ijms21031105.

23. Broos, K.; Feys, H.B.; De Meyer, S.F.; Vanhoorelbeke, K.; Deckmyn, H. *Platelets at work in primary hemostasis.* Blood Reviews **2011**, 10.1016/j.blre.2011.03.002.

24. Broughton, G., 2nd; Janis, J.E.; Attinger, C.E. *The basic science of wound healing.* Plastic and Reconstructive Surgery **2006**, 10.1097/01.prs.0000225430.42531.c2.

25. Guo, S.; Dipietro, L.A. *Factors affecting wound healing.* Journal of Dental Research **2010**, 10.1177/0022034509359125.

26. Werner, S.; Krieg, T.; Smola, H. *Keratinocyte-fibroblast interactions in wound healing.* Journal of Investigative Dermatology **2007**, 10.1038/sj.jid.5700786.

27. Wilkinson, H.N.; Hardman, M.J. *Senescence in wound repair: Emerging strategies to target chronic healing wounds.* Frontiers in Cell and Developmental Biology **2020**, 10.3389/fcell.2020.00773.

28. Lazarus, G.S.; Cooper, D.M.; Knighton, D.R.; Margolis, D.J.; Pecoraro, R.E.; Rodeheaver, G.; Robson, M.C. *Definitions and guidelines for assessment of wounds and evaluation of healing.* Archives of Dermatology **1994**, 10.1001/archderm.130.4.489.

29. Loots, M.A.M.; Kenter, S.B.; Au, F.L.; van Galen, W.J.M.; Middelkoop, E.; Bos, J.D.; Mekkes, J.R. *Fibroblasts derived from chronic diabetic ulcers differ in their response to stimulation with egf, igf-i, bfgf and pdgf-ab compared to controls.* European Journal of Cell Biology **2002**, 10.1078/0171-9335-00228.

30. Winterbourn, C.C.; Kettle, A.J. *Reactions of superoxide with myeloperoxidase and its products.* Japanese Journal of Infectious Diseases **2004**.

31. Crookes, W. *On radiant matter.* Journal of the Franklin Institute **1879**.

32. Langmuir, I. *Oscillations in ionized gases.* Proceedings of the National Academy of Sciences of the United States of America **1928**, 10.1073/pnas.14.8.627.

33. Fridman, A. *Plasma chemistry.* Cambridge University Press: Philadelphia, USA, 2008.

34. Graves, D.B. *The emerging role of reactive oxygen and nitrogen species in redox biology and some implications for plasma applications to medicine and biology.* Journal of Physics D-Applied Physics **2012**, 10.1088/0022-3727/45/26/263001.

35. Weltmann, K.D.; von Woedtke, T. *Plasma medicine-current state of research and medical application.* Plasma Physics and Controlled Fusion **2017**, 10.1088/0741-3335/59/1/014031.
36. Winter, J.; Brandenburg, R.; Weltmann, K.D. *Atmospheric pressure plasma jets: An overview of devices and new directions.* Plasma Sources Science & Technology **2015**, 10.1088/0963-0252/24/6/064001.
37. Brandenburg, R. *Dielectric barrier discharges: Progress on plasma sources and on the understanding of regimes and single filaments.* Plasma Sources Science and Technology **2017**, 10.1088/1361-6595/aa6426.
38. Raiser, J.; Zenker, M. *Argon plasma coagulation for open surgical and endoscopic applications: State of the art.* Journal of Physics D-Applied Physics **2006**, 10.1088/0022-3727/39/16/S10.
39. Canady, J.; Wiley, K.; Ravo, B. *Argon plasma coagulation and the future applications for dual-mode endoscopic probes.* Reviews in Gastroenterological Disorders **2006**.
40. Reuter, S.; von Woedtke, T.; Weltmann, K.D. *The kinpen-a review on physics and chemistry of the atmospheric pressure plasma jet and its applications.* Journal of Physics D-Applied Physics **2018**, 10.1088/1361-6463/aab3ad.
41. Wende, K.; von Woedtke, T.; Weltmann, K.D.; Bekeschus, S. *Chemistry and biochemistry of cold physical plasma derived reactive species in liquids.* Biological Chemistry **2018**, 10.1515/hsz-2018-0242.
42. Khlyustova, A.; Labay, C.; Machala, Z.; Ginebra, M.P.; Canal, C. *Important parameters in plasma jets for the production of rons in liquids for plasma medicine: A brief review.* Frontiers of Chemical Science and Engineering **2019**, 10.1007/s11705-019-1801-8.
43. Laroussi, M.; Lu, X.; Keidar, M. *Perspective: The physics, diagnostics, and applications of atmospheric pressure low temperature plasma sources used in plasma medicine.* Journal of Applied Physics **2017**, 10.1063/1.4993710.
44. Setsuhara, Y. *Low-temperature atmospheric-pressure plasma sources for plasma medicine.* Archives of Biochemistry and Biophysics **2016**, 10.1016/j.abb.2016.04.009.
45. Kramer, A.; Bekeschus, S.; Matthes, R.; Bender, C.; Stope, M.B.; Napp, M.; Lademann, O.; Lademann, J.; Weltmann, K.-D.; Schauer, F. *Cold physical plasmas in the field of hygiene-relevance, significance, and future applications.* Plasma Processes and Polymers **2015**, 10.1002/ppap.201500170.
46. Kramer, A.; Schauer, F.; Papke, R.; Bekeschus, S. Plasma application for hygienic purposes in medicine, industry, and biotechnology: Update 2017. In *Comprehensive clinical plasma medicine*, Metelmann, H.-R.; von Woedtke, T.; Weltmann, K.-D., Eds. Springer International Publishing: Cham, 2018; pp. 253–281.
47. Privat-Maldonado, A.; Schmidt, A.; Lin, A.; Weltmann, K.D.; Wende, K.; Bogaerts, A.; Bekeschus, S. *Ros from physical plasmas: Redox chemistry for biomedical therapy.* Oxidative Medicine and Cellular Longevity **2019**, 10.1155/2019/9062098.
48. von Woedtke, T.; Oehmigen, K.; Brandenburg, R.; Hoder, T.; Wilke, C.; Hähnel, M.; Weltmann, K.D. *Plasma-liquid interactions: Chemistry and antimicrobial effects.* Plasma for Bio-Decontamination, Medicine and Food Security 2012.
49. von Woedtke, T.; Schmidt, A.; Bekeschus, S.; Wende, K.; Weltmann, K.D. *Plasma medicine: A field of applied redox biology.* In Vivo **2019**, 10.21873/invivo.11570.
50. Ulfig, A.; Leichert, L.I. *The effects of neutrophil-generated hypochlorous acid and other hypohalous acids on host and pathogens.* Cellular and Molecular Life Sciences **2021**, 10.1007/s00018-020-03591-y.
51. Hazen, S.L.; Hsu, F.F.; Mueller, D.M.; Crowley, J.R.; Heinecke, J.W. *Human neutrophils employ chlorine gas as an oxidant during phagocytosis.* Journal of Clinical Investigation **1996**, 10.1172/JCI118914.
52. Bowler, P.G. *Wound pathophysiology, infection and therapeutic options.* Annals of Medicine **2002**.
53. Renner, R.; Sticherling, M.; Ruger, R.; Simon, J. *Persistence of bacteria like pseudomonas aeruginosa in non-healing venous ulcers.* European Journal of Dermatology **2012**, 10.1684/ejd.2012.1865.
54. Sibbald, R.G.; Coutts, P.; Woo, K.Y. *Reduction of bacterial burden and pain in chronic wounds using a new polyhexamethylene biguanide antimicrobial foam dressing-clinical trial results.* Advances in Skin & Wound Care **2011**, 10.1097/01.ASW.0000394027.82702.16.
55. Bauer, S.M.; Bauer, R.J.; Velazquez, O.C. *Angiogenesis, vasculogenesis, and induction of healing in chronic wounds.* Vascular and Endovascular Surgery **2005**, 10.1177/153857440503900401.
56. Menke, N.B.; Ward, K.R.; Witten, T.M.; Bonchev, D.G.; Diegelmann, R.F. *Impaired wound healing.* Clinics in Dermatology **2007**, 10.1016/j.clindermatol.2006.12.005.
57. Diegelmann, R.F. *Excessive neutrophils characterize chronic pressure ulcers.* Wound Repair and Regeneration **2003**.
58. Fridman, G.; Friedman, G.; Gutsol, A.; Shekhter, A.B.; Vasilets, V.N.; Fridman, A. *Applied plasma medicine.* Plasma Processes and Polymers **2008**, 10.1002/ppap.200700154.
59. Laroussi, M. *The biomedical applications of plasma: A brief history of the development of a new field of research.* Ieee Transactions on Plasma Science **2008**, 10.1109/tps.2008.917167.

60. Matthes, R.; Hubner, N.O.; Bender, C.; Koban, I.; Horn, S.; Bekeschus, S.; Weltmann, K.D.; Kocher, T.; Kramer, A.; Assadian, O. *Efficacy of different carrier gases for barrier discharge plasma generation compared to chlorhexidine on the survival of pseudomonas aeruginosa embedded in biofilm in vitro.* Skin Pharmacology and Physiology **2014**, 10.1159/000353861.

61. Sladek, R.E.; Filoche, S.K.; Sissons, C.H.; Stoffels, E. *Treatment of streptococcus mutans biofilms with a nonthermal atmospheric plasma.* Letters in Applied Microbiology **2007**, 10.1111/j.1472-765X.2007.02194.x.

62. Boxhammer, V.; Morfill, G.E.; Jokipii, J.R.; Shimizu, T.; Klampfl, T.; Li, Y.F.; Koritzer, J.; Schlegel, J.; Zimmermann, J.L. *Bactericidal action of cold atmospheric plasma in solution.* New Journal of Physics **2012**, 10.1088/1367-2630/14/11/113042.

63. Matthes, R.; Bekeschus, S.; Bender, C.; Koban, I.; Hubner, N.O.; Kramer, A. *Pilot-study on the influence of carrier gas and plasma application (open resp. Delimited) modifications on physical plasma and its antimicrobial effect against pseudomonas aeruginosa and staphylococcus aureus.* GMS Krankenhhyg Interdiszip **2012**, 10.3205/dgkh000186.

64. Daeschlein, G.; Napp, M.; von Podewils, S.; Lutze, S.; Emmert, S.; Lange, A.; Klare, I.; Haase, H.; Gumbel, D.; von Woedtke, T.; Junger, M. *In vitro susceptibility of multidrug resistant skin and wound pathogens against low temperature atmospheric pressure plasma jet (appj) and dielectric barrier discharge plasma (dbd).* Plasma Processes and Polymers **2014**, 10.1002/ppap.201300070.

65. Zimmermann, J.L.; Shimizu, T.; Schmidt, H.U.; Li, Y.F.; Morfill, G.E.; Isbary, G. *Test for bacterial resistance build-up against plasma treatment.* New Journal of Physics **2012**, 10.1088/1367-2630/14/7/073037.

66. Matthes, R.; Assadian, O.; Kramer, A. *Repeated applications of cold atmospheric pressure plasma does not induce resistance in staphylococcus aureus embedded in biofilms.* GMS Hygiene and Infection Control **2014**, 10.3205/dgkh000237.

67. Rajasekaran, P.; Oplander, C.; Hoffmeister, D.; Bibinov, N.; Suschek, C.V.; Wandke, D.; Awakowicz, P. *Characterization of dielectric barrier discharge (dbd) on mouse and histological evaluation of the plasma-treated tissue.* Plasma Processes and Polymers **2011**, 10.1002/ppap.201000122.

68. Choi, B.B.R.; Choi, J.H.; Ji, J.; Song, K.W.; Lee, H.J.; Kim, G.C. *Increment of growth factors in mouse skin treated with non-thermal plasma.* International Journal of Medical Sciences **2018**, 10.7150/ijms.26342.

69. Herbst, F.; van Schalkwyk, J.; McGovern, M. Microplaster and steriplas. In *Comprehensive clinical plasma medicine*, Metelmann, H.-R.; von Woedtke, T.; Weltmann, K.-D., Eds. Springer International Publishing: Cham, 2018; pp 503–509.

70. Arndt, S.; Landthaler, M.; Zimmermann, J.L.; Unger, P.; Wacker, E.; Shimizu, T.; Li, Y.F.; Morfill, G.E.; Bosserhoff, A.K.; Karrer, S. *Effects of cold atmospheric plasma (cap) on ss-defensins, inflammatory cytokines, and apoptosis-related molecules in keratinocytes in vitro and in vivo.* PloS One **2015**, 10.1371/journal.pone.0120041.

71. Xin, Y.; Wen, X.; Hamblin, M.R.; Jiang, X. *Transdermal delivery of topical lidocaine in a mouse model is enhanced by treatment with cold atmospheric plasma.* Journal of Cosmetic Dermatology **2021**, 10.1111/jocd.13581.

72. van der Linde, J.; Liedtke, K.R.; Matthes, R.; Kramer, A.; Heidecke, C.-D.; Partecke, L.I. *Repeated cold atmospheric plasma application to intact skin does not cause sensitization in a standardized murine model.* Plasma Medicine **2017**, 10.1615/PlasmaMed.2017019167.

73. Kos, S.; Blagus, T.; Cemazar, M.; Filipic, G.; Sersa, G.; Cvelbar, U. *Safety aspects of atmospheric pressure helium plasma jet operation on skin: In vivo study on mouse skin.* PloS One **2017**, 10.1371/journal.pone.0174966.

74. Choi, J.H.; Nam, S.H.; Song, Y.S.; Lee, H.W.; Lee, H.J.; Song, K.; Hong, J.W.; Kim, G.C. *Treatment with low-temperature atmospheric pressure plasma enhances cutaneous delivery of epidermal growth factor by regulating e-cadherin-mediated cell junctions.* Archives for Dermatological Research. Archiv für Dermatologische Forschung **2014**, 10.1007/s00403-014-1463-9.

75. Schmidt, A.; Liebelt, G.; Striesow, J.; Freund, E.; von Woedtke, T.; Wende, K.; Bekeschus, S. *The molecular and physiological consequences of cold plasma treatment in murine skin and its barrier function.* Free Radical Biology and Medicine **2020**, 10.1016/j.freeradbiomed.2020.09.026.

76. Pasqual-Melo, G.; Nascimento, T.; Sanches, L.J.; Blegniski, F.P.; Bianchi, J.K.; Sagwal, S.K.; Berner, J.; Schmidt, A.; Emmert, S.; Weltmann, K.D.; von Woedtke, T.; Gandhirajan, R.K.; Cecchini, A.L.; Bekeschus, S. *Plasma treatment limits cutaneous squamous cell carcinoma development in vitro and in vivo.* Cancers **2020**, 10.3390/cancers12071993.

77. Rezaeinezhad, A.; Eslami, P.; Mirmiranpour, H.; Ghomi, H. *The effect of cold atmospheric plasma on diabetes-induced enzyme glycation, oxidative stress, and inflammation; in vitro and in vivo.* Scientific Reports **2019**, 10.1038/s41598-019-56459-y.

78. Lademann, O.; Richter, H.; Kramer, A.; Patzelt, A.; Meinke, M.C.; Graf, C.; Gao, Q.; Korotianskiy, E.; Ruhl, E.; Weltmann, K.D.; Lademann, J.; Koch, S. *Stimulation of the penetration of particles into the skin by plasma tissue interaction.* Laser Physics Letters **2011**, 10.1002/lapl.201110055.

79. Lademann, O.; Richter, H.; Meinke, M.C.; Patzelt, A.; Kramer, A.; Hinz, P.; Weltmann, K.D.; Hartmann, B.; Koch, S. *Drug delivery through the skin barrier enhanced by treatment with tissue-tolerable plasma.* Experimental Dermatology **2011**, 10.1111/j.1600-0625.2010.01245.x.

80. Gan, L.; Duan, J.; Zhang, S.; Liu, X.; Poorun, D.; Liu, X.; Lu, X.; Duan, X.; Liu, D.; Chen, H. *Cold atmospheric plasma ameliorates imiquimod-induced psoriasiform dermatitis in mice by mediating antiproliferative effects.* Free Radical Research **2019**, 10.1080/10715762.2018.1564920.

81. Ma, S.; Lee, M.-H.; Kang, S.U.; Lee, Y.S.; Kim, C.-H.; Kim, K. *Development of an atmospheric non-thermal multineedle dielectric barrier discharge jet for large area treatment of skin diseases.* Current Applied Physics **2021**, 10.1016/j.cap.2021.02.003.

82. Lee, Y.S.; Lee, M.H.; Kim, H.J.; Won, H.R.; Kim, C.H. *Non-thermal atmospheric plasma ameliorates imiquimod-induced psoriasis-like skin inflammation in mice through inhibition of immune responses and up-regulation of pd-l1 expression.* Scientific Reports **2017**, 10.1038/s41598-017-15725-7.

83. Xiong, Q.; Wang, X.; Yin, R.; Xiong, L.; Chen, Q.; Zheng, M.-X.; Xu, L.; Huang, Q.-H.; Hamblin, M.R. *Surface treatment with non-thermal humid argon plasma as a treatment for allergic contact dermatitis in a mouse model.* Clinical Plasma Medicine **2018**, 10.1016/j.cpme.2018.09.002.

84. Lee, M.H.; Lee, Y.S.; Kim, H.J.; Han, C.H.; Kang, S.U.; Kim, C.H. *Non-thermal plasma inhibits mast cell activation and ameliorates allergic skin inflammatory diseases in nc/nga mice.* Scientific Reports **2019**, 10.1038/s41598-019-49938-9.

85. Ara, J.; Bajgai, J.; Sajo, M.E.J.; Fadriquela, A.; Kim, C.S.; Kim, S.K.; Lee, K.J. *The immunological and oxidative stress regulation of non-thermal plasma-aided water on atopic dermatitis-like lesion in dinitrochlorobenzene-induced skh-1 hairless mice.* Molecular & Cellular Toxicology **2019**, 10.1007/s13273-019-0023-y.

86. Falanga, V. *Wound healing and its impairment in the diabetic foot.* Lancet **2005**, 10.1016/S0140-6736(05)67700-8.

87. Pastar, I.; Liang, L.; Sawaya, A.P.; Wikramanayake, T.C.; Glinos, G.D.; Drakulich, S.; Chen, V.; Stojadinovic, O.; Davis, S.C.; Tomic-Canic, M. Preclinical models for wound-healing studies. In *Skin tissue models.* Elsevier: Amsterdam, 2018; pp 223–253.

88. Yu, Y.; Tan, M.; Chen, H.; Wu, Z.; Xu, L.; Li, J.; Cao, J.; Yang, Y.; Xiao, X.; Lian, X.; Lu, X.; Tu, Y. *Non-thermal plasma suppresses bacterial colonization on skin wound and promotes wound healing in mice.* Journal of Huazhong University of Science and Technology Medical Sciences **2011**, 10.1007/s11596-011-0387-2.

89. Ngo Thi, M.-H.; Shao, P.-L.; Liao, J.-D.; Lin, C.-C.K.; Yip, H.-K. *Enhancement of angiogenesis and epithelialization processes in mice with burn wounds through ros/rns signals generated by non-thermal n2/ar micro-plasma.* Plasma Processes and Polymers **2014**, 10.1002/ppap.201400072.

90. Garcia-Alcantara, E.; Lopez-Callejas, R.; Morales-Ramirez, P.R.; Pena-Eguiluz, R.; Fajardo-Munoz, R.; Mercado-Cabrera, A.; Barocio, S.R.; Valencia-Alvarado, R.; Rodriguez-Mendez, B.G.; Munoz-Castro, A.E.; de la Piedad-Beneitez, A.; Rojas-Olmedo, I.A. *Accelerated mice skin acute wound healing in vivo by combined treatment of argon and helium plasma needle.* Archives of Medical Research **2013**, 10.1016/j.arcmed.2013.02.001.

91. Shao, P.L.; Liao, J.D.; Wong, T.W.; Wang, Y.C.; Leu, S.; Yip, H.K. *Enhancement of wound healing by non-thermal n2/ar micro-plasma exposure in mice with fractional-co2-laser-induced wounds.* PloS One **2016**, 10.1371/journal.pone.0156699.

92. Kim, D.W.; Park, T.J.; Jang, S.J.; You, S.J.; Oh, W.Y. *Plasma treatment effect on angiogenesis in wound healing process evaluated in vivo using angiographic optical coherence tomography.* Applied Physics Letters **2016**, 10.1063/1.4967375.

93. Duchesne, C.; Frescaline, N.; Lataillade, J.-J.; Rousseau, A. *Comparative study between direct and indirect treatment with cold atmospheric plasma on in vitro and in vivo models of wound healing.* Plasma Medicine **2018**, 10.1615/PlasmaMed.2019028659.

94. Duchesne, C.; Banzet, S.; Lataillade, J.J.; Rousseau, A.; Frescaline, N. *Cold atmospheric plasma modulates endothelial nitric oxide synthase signalling and enhances burn wound neovascularisation.* Journal of Pathology **2019**, 10.1002/path.5323.

95. Jacofsky, M.C.; Lubahn, C.; McDonnell, C.; Seepersad, Y.; Fridman, G.; Fridman, A.A.; Dobrynin, D. *Spatially resolved optical emission spectroscopy of a helium plasma jet and its effects on wound healing rate in a diabetic murine model.* Plasma Medicine **2014**, 10.1615/PlasmaMed.2015012190.

96. Pan, S.; Zhang, S.; Chen, H. *Low temperature plasma promotes the healing of chronic wounds in diabetic mice.* Journal of Physics D: Applied Physics **2020**, 10.1088/1361-6463/ab7514.

97. Arndt, S.; Unger, P.; Berneburg, M.; Bosserhoff, A.K.; Karrer, S. *Cold atmospheric plasma (cap) activates angiogenesis-related molecules in skin keratinocytes, fibroblasts and endothelial cells and improves wound angiogenesis in an autocrine and paracrine mode.* Journal of Dermatological Science **2018**, 10.1016/j.jdermsci.2017.11.008.

98. Arndt, S.; Unger, P.; Wacker, E.; Shimizu, T.; Heinlin, J.; Li, Y.F.; Thomas, H.M.; Morfill, G.E.; Zimmermann, J.L.; Bosserhoff, A.K.; Karrer, S. *Cold atmospheric plasma (cap) changes gene expression of key molecules of the wound healing machinery and improves wound healing in vitro and in vivo.* PloS One **2013**, 10.1371/journal.pone.0079325.

99. Wang, X.F.; Fang, Q.Q.; Jia, B.; Hu, Y.Y.; Wang, Z.C.; Yan, K.P.; Yin, S.Y.; Liu, Z.; Tan, W.Q. *Potential effect of non-thermal plasma for the inhibition of scar formation: A preliminary report.* Scientific Reports **2020**, 10.1038/s41598-020-57703-6.

100. Hoon Lee, D.; Lee, J.-O.; Jeon, W.; Choi, I.-G.; Kim, J.-S.; Hoon Jeong, J.; Kang, T.-C.; Hoon Seo, C. *Suppression of scar formation in a murine burn wound model by the application of non-thermal plasma.* Applied Physics Letters **2011**, 10.1063/1.3662040.

101. Schmidt, A.; Bekeschus, S.; Wende, K.; Vollmar, B.; von Woedtke, T. *A cold plasma jet accelerates wound healing in a murine model of full-thickness skin wounds.* Experimental Dermatology **2017**, 10.1111/exd.13156.

102. Schmidt, A.; Woedtke, T.V.; Stenzel, J.; Lindner, T.; Polei, S.; Vollmar, B.; Bekeschus, S. *One year follow-up risk assessment in skh-1 mice and wounds treated with an argon plasma jet.* International Journal of Molecular Sciences **2017**, 10.3390/ijms18040868.

103. Schmidt, A.; von Woedtke, T.; Vollmar, B.; Hasse, S.; Bekeschus, S. *Nrf2 signaling and inflammation are key events in physical plasma-spurred wound healing.* Theranostics **2019**, 10.7150/thno.29754.

104. Schmidt, A.; Liebelt, G.; Niessner, F.; von Woedtke, T.; Bekeschus, S. *Gas plasma-spurred wound healing is accompanied by regulation of focal adhesion, matrix remodeling, and tissue oxygenation.* Redox Biology **2021**, 10.1016/j.redox.2020.101809.

105. Schmidt, A.; Niesner, F.; von Woedtke, T.; Bekeschus, S. *Hyperspectral imaging of wounds reveals augmented tissue oxygenation following cold physical plasma treatment in vivo.* IEEE Transactions on Radiation and Plasma Medical Sciences **2021**, 10.1109/trpms.2020.3009913.

106. Jung, J.M.; Yoon, H.K.; Jung, C.J.; Jo, S.Y.; Hwang, S.G.; Lee, H.J.; Lee, W.J.; Chang, S.E.; Won, C.H. *Cold plasma treatment promotes full-thickness healing of skin wounds in murine models.* The International Journal of Lower Extremity Wounds **2021**, 10.1177/15347346211002144.

107. Dang, C.P.; Weawseetong, S.; Charoensappakit, A.; Sae-Khow, K.; Thong-Aram, D.; Leelahavanichkul, A. *Non-thermal atmospheric pressure argon-sourced plasma flux promotes wound healing of burn wounds and burn wounds with infection in mice through the anti-inflammatory macrophages.* Applied Sciences **2021**, 10.3390/app11125343.

108. Kim, H.Y.; Kang, S.K.; Park, S.M.; Jung, H.Y.; Choi, B.H.; Sim, J.Y.; Lee, J.K. *Characterization and effects of ar/air microwave plasma on wound healing.* Plasma Processes and Polymers **2015**, 10.1002/ppap.201500017.

109. Lee, J.; Kim, W.S.; Bae, K.B.; Yun, G.; Lee, J.K. *Variation of physical parameters and anomalous healing observation in plasma wound healing.* Ieee Transactions on Plasma Science **2019**, 10.1109/tps.2019.2923750.

110. Darmawati, S.; Rohmani, A.; Nurani, L.H.; Prastiyanto, M.E.; Dewi, S.S.; Salsabila, N.; Wahyuningtyas, E.S.; Murdiya, F.; Sikumbang, I.M.; Rohmah, R.N.; Fatimah, Y.A.; Widiyanto, A.; Ishijima, T.; Sugama, J.; Nakatani, T.; Nasruddin, N. *When plasma jet is effective for chronic wound bacteria inactivation, is it also effective for wound healing?* Clinical Plasma Medicine **2019**, 10.1016/j.cpme.2019.100085.

111. Shahbazi Rad, Z.; Abbasi Davani, F.; Etaati, G. *Determination of proper treatment time for in vivo blood coagulation and wound healing application by non-thermal helium plasma jet.* Australasian Physical and Engineering Sciences in Medicine **2018**, 10.1007/s13246-018-0686-z.

112. Xu, G.M.; Shi, X.M.; Cai, J.F.; Chen, S.L.; Li, P.; Yao, C.W.; Chang, Z.S.; Zhang, G.J. *Dual effects of atmospheric pressure plasma jet on skin wound healing of mice.* Wound Repair and Regeneration **2015**, 10.1111/wrr.12364.

113. Shahbazi Rad, Z.; Abbasi Davani, F. *Measurements of the electrical parameters and wound area for investigation on the effect of different non-thermal atmospheric pressure plasma sources on wound healing time.* Measurement **2020**, 10.1016/j.measurement.2020.107545.

114. Nasruddin; Nakajima, Y.; Mukai, K.; Rahayu, H.S.E.; Nur, M.; Ishijima, T.; Enomoto, H.; Uesugi, Y.; Sugama, J.; Nakatani, T. *Cold plasma on full-thickness cutaneous wound accelerates healing through promoting inflammation, re-epithelialization and wound contraction.* Clinical Plasma Medicine **2014**, 10.1016/j.cpme.2014.01.001.

115. Nasruddin; Nakajima, Y.; Mukai, K.; Komatsu, E.; Rahayu, H.S.E.; Nur, M.; Ishijima, T.; Enomoto, H.; Uesugi, Y.; Sugama, J.; Nakatani, T. *A simple technique to improve contractile effect of cold plasma jet on acute mouse wound by dropping water.* Plasma Processes and Polymers **2015**, 10.1002/ppap.201400236.

116. Nasruddin; Putri, I.K.; Kamal, S.; Esti Rahayu, H.S.; Lutfiyati, H.; Pribadi, P.; Kusuma, T.M.; Muhlisin, Z.; Nur, M.; Nurani, L.H.; Santosa, B.; Ishijima, T.; Nakatani, T. *Evaluation the effectiveness of combinative treatment of cold plasma jet, indonesian honey, and micro-well dressing to accelerate wound healing.* Clinical Plasma Medicine **2017**, 10.1016/j.cpme.2017.03.001.

117. Wahyuningtyas, E.S.; Iswara, A.; Sari, Y.; Kamal, S.; Santosa, B.; Ishijima, T.; Nakatani, T.; Putri, I.K.; Nasruddin, N. *Comparative study on manuka and indonesian honeys to support the application of plasma jet during proliferative phase on wound healing.* Clinical Plasma Medicine **2018**, 10.1016/j.cpme.2018.08.001.

118. Darmawati, S.; Nasruddin, N.; Kurniasiwi, P.; Mukaromah, A.H.; Iswara, A.; Putri, G.S.A.; Rahayu, H.S.E.; Wahyuningtyas, E.S.; Lutfiyati, H.; Kartikadewi, A.; Rejeki, S.; Ishijima, T.; Nakatani, T.; Sugama, J. *Plasma jet effectiveness alteration in acute wound healing by binahong (anredera cordifolia) extract.* Plasma Medicine **2020**, 10.1615/PlasmaMed.2021037264.

119. Breathnach, R.; McDonnell, K.A.; Chebbi, A.; Callanan, J.J.; Dowling, D.P. *Evaluation of the effectiveness of kinpen med plasma jet and bioactive agent therapy in a rat model of wound healing.* Biointerphases **2018**, 10.1116/1.5046489.

120. Cheng, K.Y.; Lin, Z.H.; Cheng, Y.P.; Chiu, H.Y.; Yeh, N.L.; Wu, T.K.; Wu, J.S. *Wound healing in streptozotocin-induced diabetic rats using atmospheric-pressure argon plasma jet.* Scientific Reports **2018**, 10.1038/s41598-018-30597-1.

121. Fathollah, S.; Mirpour, S.; Mansouri, P.; Dehpour, A.R.; Ghoranneviss, M.; Rahimi, N.; Safaie Naraghi, Z.; Chalangari, R.; Chalangari, K.M. *Investigation on the effects of the atmospheric pressure plasma on wound healing in diabetic rats.* Scientific Reports **2016**, 10.1038/srep19144.

122. Chatraie, M.; Torkaman, G.; Khani, M.; Salehi, H.; Shokri, B. *In vivo study of non-invasive effects of non-thermal plasma in pressure ulcer treatment.* Scientific Reports **2018**, 10.1038/s41598-018-24049-z.

123. Kubinova, S.; Zaviskova, K.; Uherkova, L.; Zablotskii, V.; Churpita, O.; Lunov, O.; Dejneka, A. *Non-thermal air plasma promotes the healing of acute skin wounds in rats.* Scientific Reports **2017**, 10.1038/srep45183.

124. Lou, B.S.; Hsieh, J.H.; Chen, C.M.; Hou, C.W.; Wu, H.Y.; Chou, P.Y.; Lai, C.H.; Lee, J.W. *Helium/argon-generated cold atmospheric plasma facilitates cutaneous wound healing.* Frontiers in Bioengineering Biotechnology **2020**, 10.3389/fbioe.2020.00683.

125. Guo, P.; Liu, Y.; Li, J.; Zhang, N.; Zhou, M.; Li, Y.; Zhao, G.; Wang, N.; Wang, A.; Wang, Y.; Wang, F.; Huang, L. *A novel atmospheric-pressure air plasma jet for wound healing.* International Wound Journal **2021**, 10.1111/iwj.13652.

126. Souza, L.B.; Silva, J.I.S.; Bagne, L.; Pereira, A.T.; Oliveira, M.A.; Lopes, B.B.; Amaral, M.; de Aro, A.A.; Esquisatto, M.A.M.; Santos, G.; Andrade, T.A.M. *Argon atmospheric plasma treatment promotes burn healing by stimulating inflammation and controlling the redox state.* Inflammation **2020**, 10.1007/s10753-020-01305-x.

127. Lee, Y.; Ricky, S.; Lim, T.H.; Jang, K.-S.; Kim, H.; Song, Y.; Kim, S.-Y.; Chung, K.-S. *Wound healing effect of nonthermal atmospheric pressure plasma jet on a rat burn wound model: A preliminary study.* Journal of Burn Care & Research **2019**, 10.1093/jbcr/irz120.

128. Zhang, J.-P.; Guo, L.; Chen, Q.-L.; Zhang, K.-Y.; Wang, T.; An, G.-Z.; Zhang, X.-F.; Li, H.-P.; Ding, G.-R. *Effects and mechanisms of cold atmospheric plasma on skin wound healing of rats.* Contributions to Plasma Physics **2019**, 10.1002/ctpp.201800025.

129. Shekhter, A.B.; Kabisov, R.K.; Pekshev, A.V.; Kozlov, N.P.; Perov, Y.L. *Experimental and clinical validation of plasmadynamic therapy of wounds with nitric oxide.* Bulletin of Experimental Biology and Medicine **1998**, 10.1007/Bf02446923.

130. Shekhter, A.B.; Pekshev, A.V.; Vagapov, A.B.; Telpukhov, V.I.; Panyushkin, P.V.; Rudenko, T.G.; Fayzullin, A.L.; Sharapov, N.A.; Vanin, A.F. *Physicochemical parameters of no-containing gas flow affect wound healing therapy: An experimental study.* European Journal of Pharmaceutical Sciences **2018**, 10.1016/j.ejps.2018.11.034.

131. Shekhter, A.B.; Serezhenkov, V.A.; Rudenko, T.G.; Pekshev, A.V.; Vanin, A.F. *Beneficial effect of gaseous nitric oxide on the healing of skin wounds.* Nitric Oxide **2005**, 10.1016/j.niox.2005.03.004.

132. Zahedi, L.; Ghourchi Beigi, P.; Shafiee, M.; Zare, F.; Mahdikia, H.; Abdouss, M.; Abdollahifar, M.A.; Shokri, B. *Development of plasma functionalized polypropylene wound dressing for betaine hydrochloride controlled drug delivery on diabetic wounds.* Scientific Reports **2021**, 10.1038/s41598-021-89105-7.

133. Ding, C.; Huang, P.; Feng, L.; Jin, T.; Zhou, Y.; He, Y.; Wu, Z.; Liu, Y. *Immediate intervention effect of dielectric barrier discharge on acute inflammation in rabbit's ear wound.* AIP Advances **2020**, 10.1063/1.5139953.

134. Amini, M.; Momeni, M.; Jahandideh, A.; Ghoranneviss, M.; Soudmand, S.; Yousefi, P.; Khandan, S.; Amini, M. *Tendon repair by plasma jet treatment.* Journal of Diabetes & Metabolic Disorders **2021**, 10.1007/s40200-021-00789-0.

135. Li, G.; Li, D.; Li, J.; Jia, Y.-N.; Zhu, C.; Zhang, Y.; Li, H.-P. *Promotion of the wound healing of in vivo rabbit wound infected with methicillin-resistant staphylococcus aureus treated by a cold atmospheric plasma jet.* Ieee Transactions on Plasma Science **2021**, 10.1109/tps.2021.3092946.

136. Wu, A.S.; Kalghatgi, S.; Dobrynin, D.; Sensenig, R.; Cerchar, E.; Podolsky, E.; Dulaimi, E.; Paff, M.; Wasko, K.; Arjunan, K.P.; Garcia, K.; Fridman, G.; Balasubramanian, M.; Ownbey, R.; Barbee, K.A.; Fridman, A.; Friedman, G.; Joshi, S.G.; Brooks, A.D. *Porcine intact and wounded skin responses to atmospheric nonthermal plasma.* Journal of Surgical Research **2013**, 10.1016/j.jss.2012.02.039.

137. Kramer, A.; Lademann, J.; Bender, C.; Sckell, A.; Hartmann, B.; Münch, S.; Hinz, P.; Ekkernkamp, A.; Matthes, R.; Koban, I.; Partecke, I.; Heidecke, C.D.; Masur, K.; Reuter, S.; Weltmann, K.D.; Koch, S.; Assadian, O. *Suitability of tissue tolerable plasmas (ttp) for the management of chronic wounds.* Clinical Plasma Medicine **2013**, 10.1016/j.cpme.2013.03.002.

138. Martines, E.; Brun, P.; Cavazzana, R.; Cordaro, L.; Zuin, M.; Martinello, T.; Gomiero, C.; Perazzi, A.; Melotti, L.; Maccatrozzo, L.; Patruno, M.; Iacopetti, I. *Wound healing improvement in large animals using an indirect helium plasma treatment.* Clinical Plasma Medicine **2020**, 10.1016/j.cpme.2020.100095.

139. Melotti, L.; Martinello, T.; Perazzi, A.; Martines, E.; Zuin, M.; Modenese, D.; Cordaro, L.; Ferro, S.; Maccatrozzo, L.; Iacopetti, I.; Patruno, M. *Could cold plasma act synergistically with allogeneic mesenchymal stem cells to improve wound skin regeneration in a large size animal model?* Research in Veterinary Science **2021**, 10.1016/j.rvsc.2021.01.019.

140. Bender, C.; Kramer, A. *Therapy of wound healing disorders in pets with atmospheric pressure plasma.* Tieraerztliche Umschau **2016**.

141. Bender, C.P.; Hübner, N.-O.; Weltmann, K.-D.; Scharf, C.; Kramer, A. In *Tissue tolerable plasma and polihexanide: Are synergistic effects possible to promote healing of chronic wounds? In vivo and in vitro results*, Dordrecht, 2012; Springer Netherlands: Dordrecht, pp 321–334.

142. Classen, J.; Dengler, B.; Klinger, C.J.; Bettenay, S.V.; Rickerts, V.; Mueller, R.S. *Cutaneous alternariosis in an immunocompromised dog successfully treated with cold plasma and cessation of immunosuppressive medication.* Tierarztl Prax Ausg K Kleintiere Heimtiere **2017**, 10.15654/TPK-160851.

143. Nolff, M.C.; Winter, S.; Reese, S.; Meyer-Lindenberg, A. *Comparison of polyhexanide, cold atmospheric plasma and saline in the treatment of canine bite wounds.* Journal of Small Animal Practice **2019**, 10.1111/jsap.12971.

144. Winter, S.; Nolff, M.C.; Reese, S.; Meyer-Lindenberg, A. *Comparison of the antibacterial efficacy of polyhexanide, cold atmospheric argon plasma and saline in the treatment of canine bite wounds.* Tierarztl Prax Ausg K Kleintiere Heimtiere **2018**, 10.15654/TPK-170713.

145. Niethammer, P.; Grabher, C.; Look, A.T.; Mitchison, T.J. *A tissue-scale gradient of hydrogen peroxide mediates rapid wound detection in zebrafish.* Nature **2009**, 10.1038/nature08119.

146. Metelmann, H.-R.; von Woedtke, T.; Bussiahn, R.; Weltmann, K.-D.; Rieck, M.; Khalili, R.; Podmelle, F.; Waite, P.D. *Experimental recovery of co2-laser skin lesions by plasma stimulation.* The American Journal of Cosmetic Surgery **2012**, 10.5992/ajcs-d-11-00042.1.

147. Metelmann, H.-R.; Vu, T.T.; Do, H.T.; Le, T.N.B.; Hoang, T.H.A.; Phi, T.T.T.; Luong, T.M.L.; Doan, V.T.; Nguyen, T.T.H.; Nguyen, T.H.M.; Nguyen, T.L.; Le, D.Q.; Le, T.K.X.; von Woedtke, T.; Bussiahn, R.; Weltmann, K.-D.; Khalili, R.; Podmelle, F. *Scar formation of laser skin lesions after cold atmospheric pressure plasma (cap) treatment: A clinical long term observation.* Clinical Plasma Medicine **2013**, 10.1016/j.cpme.2012.12.001.

148. Nishijima, A.; Fujimoto, T.; Hirata, T.; Nishijima, J. *Effects of cold atmospheric pressure plasma on accelerating acute wound healing: A comparative study among 4 different treatment groups.* Modern Plastic Surgery **2019**, 10.4236/mps.2019.91004.

149. Heinlin, J.; Zimmermann, J.L.; Zeman, F.; Bunk, W.; Isbary, G.; Landthaler, M.; Maisch, T.; Monetti, R.; Morfill, G.; Shimizu, T.; Steinbauer, J.; Stolz, W.; Karrer, S. *Randomized placebo-controlled human pilot study of cold atmospheric argon plasma on skin graft donor sites.* Wound Repair and Regeneration **2013**, 10.1111/wrr.12078.

150. Vandersee, S.; Richter, H.; Lademann, J.; Beyer, M.; Kramer, A.; Knorr, F.; Lange-Asschenfeldt, B. *Laser scanning microscopy as a means to assess the augmentation of tissue repair by exposition of wounds to tissue tolerable plasma.* Laser Physics Letters **2014**, 10.1088/1612-2011/11/11/115701.

151. Betancourt-Angeles, M.; Pena-Eguiluz, R.; Lopez-Callejas, R.; Dominguez-Cadena, N.A.; Mercado-Cabrera, A.; Munoz-Infante, J.; Rodriguez-Mendez, B.G.; Valencia-Alvarado, R.; Moreno-Tapia, J.A. *Treatment in the healing of burns with a cold plasma source.* International Journal of Burns and Trauma **2017**.

152. Hermans, M.H. *Wounds and ulcers: Back to the old nomenclature.* Wounds **2010**.

153. Krylov, A.Yu., Shulutko, A.M., Chirikova, E.G., and Shekhter, A.B., Ross. Beneficial effects on wound healing in 318 patients using a nitric oxide-generating device Plasotron. Med. Zh., **2002**, no. 2, p. 23.

154. Pekshev, A.V.; Shekhter, A.B.; Vagapov, A.B.; Sharapov, N.A.; Vanin, A.F. *Study of plasma-chemical no-containing gas flow for treatment of wounds and inflammatory processes.* Nitric Oxide **2018**, 10.1016/j.niox.2017.06.002.

155. Khrupkin, V.I.; Zudilin, A.V.; Pisarenko, L.V.; Pekshev, A.V.; Vagapov, A.B.; Nastich Iu, N.; Pokrovskaia, M.S. *Local application of low-energy aerial and argon plasma in the treatment of suppurative wounds and trophic ulcers.* Vestnik Khirurgii Imeni I. I. Grekova **2001**.

156. Isbary, G.; Morfill, G.; Schmidt, H.U.; Georgi, M.; Ramrath, K.; Heinlin, J.; Karrer, S.; Landthaler, M.; Shimizu, T.; Steffes, B.; Bunk, W.; Monetti, R.; Zimmermann, J.L.; Pompl, R.; Stolz, W. *A first prospective randomized controlled trial to decrease bacterial load using cold atmospheric argon plasma on chronic wounds in patients.* British Journal of Dermatology **2010**, 10.1111/j.1365-2133.2010.09744.x.

157. Isbary, G.; Heinlin, J.; Shimizu, T.; Zimmermann, J.L.; Morfill, G.; Schmidt, H.U.; Monetti, R.; Steffes, B.; Bunk, W.; Li, Y.; Klaempfl, T.; Karrer, S.; Landthaler, M.; Stolz, W. *Successful and safe use of 2 min cold atmospheric argon plasma in chronic wounds: Results of a randomized controlled trial.* British Journal of Dermatology **2012**, 10.1111/j.1365-2133.2012.10923.x.

158. Isbary, G.; Stolz, W.; Shimizu, T.; Monetti, R.; Bunk, W.; Schmidt, H.U.; Morfill, G.E.; Klämpfl, T.G.; Steffes, B.; Thomas, H.M.; Heinlin, J.; Karrer, S.; Landthaler, M.; Zimmermann, J.L. *Cold atmospheric argon plasma treatment may accelerate wound healing in chronic wounds: Results of an open retrospective randomized controlled study in vivo.* Clinical Plasma Medicine **2013**, 10.1016/j.cpme.2013.06.001.

159. Brehmer, F.; Haenssle, H.A.; Daeschlein, G.; Ahmed, R.; Pfeiffer, S.; Gorlitz, A.; Simon, D.; Schon, M.P.; Wandke, D.; Emmert, S. *Alleviation of chronic venous leg ulcers with a hand-held dielectric barrier discharge plasma generator (plasmaderm(r) vu-2010): Results of a monocentric, two-armed, open, prospective, randomized and controlled trial (nct01415622).* Journal of the European Academy of Dermatology and Venereology **2015**, 10.1111/jdv.12490.

160. Hartwig, S.; Preissner, S.; Voss, J.O.; Hertel, M.; Doll, C.; Waluga, R.; Raguse, J.D. *The feasibility of cold atmospheric plasma in the treatment of complicated wounds in cranio-maxillo-facial surgery.* Journal of Cranio-Maxillo-Facial Surgery **2017**, 10.1016/j.jcms.2017.07.008.

161. Hartwig, S.; Doll, C.; Voss, J.O.; Hertel, M.; Preissner, S.; Raguse, J.D. *Treatment of wound healing disorders of radial forearm free flap donor sites using cold atmospheric plasma: A proof of concept.* Journal of Oral Maxillofacial Surgery **2017**, 10.1016/j.joms.2016.08.011.

162. Ulrich, C.; Kluschke, F.; Patzelt, A.; Vandersee, S.; Czaika, V.A.; Richter, H.; Bob, A.; von Hutten, J.; Painsi, C.; Hugel, R.; Kramer, A.; Assadian, O.; Lademann, J.; Lange-Asschenfeldt, B. *Clinical use of cold atmospheric pressure argon plasma in chronic leg ulcers: A pilot study.* Journal of Wound Care **2015**, 10.12968/jowc.2015.24.5.196.

163. Klebes, M.; Ulrich, C.; Kluschke, F.; Patzelt, A.; Vandersee, S.; Richter, H.; Bob, A.; von Hutten, J.; Krediet, J.T.; Kramer, A.; Lademann, J.; Lange-Asschenfeld, B. *Combined antibacterial effects of tissue-tolerable plasma and a modern conventional liquid antiseptic on chronic wound treatment.* Journal of Biophotonics **2015**, 10.1002/jbio.201400007.

164. Chuangsuwanich, A.; Assadamongkol, T.; Boonyawan, D. *The healing effect of low-temperature atmospheric-pressure plasma in pressure ulcer: A randomized controlled trial.* International Journal of Lower Extremity Wounds **2016**, 10.1177/1534734616665046.

165. Moelleken, M.; Jockenhofer, F.; Wiegand, C.; Buer, J.; Benson, S.; Dissemond, J. *Pilot study on the influence of cold atmospheric plasma on bacterial contamination and healing tendency of chronic wounds.* Journal der Deutschen Dermatologischen Gesellschaft **2020**, 10.1111/ddg.14294.

166. Arya, A.K.; Tripathi, R.; Kumar, S.; Tripathi, K. *Recent advances on the association of apoptosis in chronic non healing diabetic wound.* World Journal of Diabetes **2014**, 10.4239/wjd.v5.i6.756.

167. Stratmann, B.; Costea, T.C.; Nolte, C.; Hiller, J.; Schmidt, J.; Reindel, J.; Masur, K.; Motz, W.; Timm, J.; Kerner, W.; Tschoepe, D. *Effect of cold atmospheric plasma therapy vs standard therapy placebo on wound healing in patients with diabetic foot ulcers: A randomized clinical trial.* JAMA Netwwork Open **2020**, 10.1001/jamanetworkopen.2020.10411.

168. Amini, M.R.; Sheikh Hosseini, M.; Fatollah, S.; Mirpour, S.; Ghoranneviss, M.; Larijani, B.; Mohajeri-Tehrani, M.R.; Khorramizadeh, M.R. *Beneficial effects of cold atmospheric plasma on inflammatory phase of diabetic foot ulcers; a randomized clinical trial.* Journal of Diabetes & Metabolic Disorders **2020**, 10.1007/s40200-020-00577-2.

169. Mirpour, S.; Fathollah, S.; Mansouri, P.; Larijani, B.; Ghoranneviss, M.; Mohajeri Tehrani, M.; Amini, M.R. *Cold atmospheric plasma as an effective method to treat diabetic foot ulcers: A randomized clinical trial.* Scientific Reports **2020**, 10.1038/s41598-020-67232-x.

170. González-Mendoza, B.; López-Callejas, R.; Rodríguez-Méndez, B.G.; Eguiluz, R.P.; Mercado-Cabrera, A.; Valencia-Alvarado, R.; Betancourt-Ángeles, M.; Reyes-Frías, M.d.L.; Reboyo-Barrios, D.; Chávez-Aguilar, E. *Healing of wounds in lower extremities employing a non-thermal plasma.* Clinical Plasma Medicine **2019**, 10.1016/j.cpme.2020.100094.

171. López-Callejas, R.; Peña-Eguiluz, R.; Valencia-Alvarado, R.; Mercado-Cabrera, A.; Rodríguez-Méndez, B.G.; Serment-Guerrero, J.H.; Cabral-Prieto, A.; González-Garduño, A.C.; Domínguez-Cadena, N.A.; Muñoz-Infante, J.; Betancourt-Ángeles, M. *Alternative method for healing the diabetic foot by means of a plasma needle.* Clinical Plasma Medicine **2018**, 10.1016/j.cpme.2018.01.001.

172. Naderi, N.; Zaefizadeh, M. *Expression of growth factors in re-epithelialization of diabetic foot ulcers after treatment with non-thermal plasma radiation.* Biomedical Research-India **2017**.

173. Vegter, S.; Tolley, K. *A network meta-analysis of the relative efficacy of treatments for actinic keratosis of the face or scalp in europe.* PloS One **2014**, 10.1371/journal.pone.0096829.

174. Friedman, P.C.; Miller, V.; Fridman, G.; Lin, A.; Fridman, A. *Successful treatment of actinic keratoses using nonthermal atmospheric pressure plasma: A case series.* Journal of the American Academy of Dermatology **2017**, 10.1016/j.jaad.2016.09.004.

175. Wirtz, M.; Stoffels, I.; Dissemond, J.; Schadendorf, D.; Roesch, A. *Actinic keratoses treated with cold atmospheric plasma.* Journal of the European Academy of Dermatology and Venereology **2018**, 10.1111/jdv.14465.

176. Daeschlein, G.; Arnold, A.; Lutze, S.; Napp, M.; Aly, F.; von Podewils, S.; Sicher, C.; Juenger, M.; Schumacher, U. *Treatment of recalcitrant actinic keratosis (ak) of the scalp by cold atmospheric plasma.* Cogent Medicine **2017**, 10.1080/2331205x.2017.1412903.

177. Arisi, M.; Soglia, S.; Guasco Pisani, E.; Venturuzzo, A.; Gelmetti, A.; Tomasi, C.; Zane, C.; Rossi, M.; Lorenzi, L.; Calzavara-Pinton, P. *Cold atmospheric plasma (cap) for the treatment of actinic keratosis and skin field cancerization: Clinical and high-frequency ultrasound evaluation.* Dermatology and Therapy **2021**, 10.1007/s13555-021-00514-y.

178. Klebes, M.; Lademann, J.; Philipp, S.; Ulrich, C.; Patzelt, A.; Ulmer, M.; Kluschke, F.; Kramer, A.; Weltmann, K.D.; Sterry, W.; Lange-Asschenfeldt, B. *Effects of tissue-tolerable plasma on psoriasis vulgaris treatment compared to conventional local treatment: A pilot study.* Clinical Plasma Medicine **2014**, 10.1016/j.cpme.2013.11.002.

179. Zheng, L.; Gao, J.; Cao, Y.; Yang, X.; Wang, N.; Cheng, C.; Yang, C. *Two case reports of inverse psoriasis treated with cold atmospheric plasma.* Dermatologic Therapy **2020**, 10.1111/dth.14257.

180. Friedman, P.C.; Miller, V.; Fridman, G.; Fridman, A. *Use of cold atmospheric pressure plasma to treat warts: A potential therapeutic option.* Clinical and Experimental Dermatology **2018**, 10.1111/ced.13790.

181. Friedman, P.C.; Fridman, G.; Fridman, A. *Using cold plasma to treat warts in children: A case series.* Pediatric Dermatology **2020**, 10.1111/pde.14180.

182. Mann, M.S.; Tiede, R.; Gavenis, K.; Daeschlein, G.; Bussiahn, R.; Weltmann, K.-D.; Emmert, S.; Woedtke, T.V.; Ahmed, R. *Introduction to din-specification 91315 based on the characterization of the plasma jet kinpen® med.* Clinical Plasma Medicine **2016**, 10.1016/j.cpme.2016.06.001.

183. Gan, L.; Jiang, J.; Duan, J.W.; Wu, X.J.Z.; Zhang, S.; Duan, X.R.; Song, J.Q.; Chen, H.X. *Cold atmospheric plasma applications in dermatology: A systematic review.* Journal of Biophotonics **2021**, 10.1002/jbio.202000415.

184. Eming, S.A.; Martin, P.; Tomic-Canic, M. *Wound repair and regeneration: Mechanisms, signaling, and translation.* Science Translational Medicine **2014**, 10.1126/scitranslmed.3009337.

185. Eming, S.A.; Wynn, T.A.; Martin, P. *Inflammation and metabolism in tissue repair and regeneration.* Science **2017**, 10.1126/science.aam7928.

186. Schmidt, A.; Dietrich, S.; Steuer, A.; Weltmann, K.D.; von Woedtke, T.; Masur, K.; Wende, K. *Non-thermal plasma activates human keratinocytes by stimulation of antioxidant and phase ii pathways.* Journal of Biological Chemistry **2015**, 10.1074/jbc.M114.603555.

187. Victor, P.; Sarada, D.; Ramkumar, K.M. *Pharmacological activation of nrf2 promotes wound healing.* European Journal of Pharmacology **2020**, 10.1016/j.ejphar.2020.173395.

188. Hasse, S.; Hahn, O.; Kindler, S.; Woedtke, T.V.; Metelmann, H.-R.; Masur, K. *Atmospheric pressure plasma jet application on human oral mucosa modulates tissue regeneration.* Plasma Medicine **2014**, 10.1615/PlasmaMed.2014011978.

189. Sies, H. *Oxidative stress: A concept in redox biology and medicine.* Redox Biology **2015**, 10.1016/j.redox.2015.01.002.

190. Schmidt, A.; von Woedtke, T.; Bekeschus, S. *Periodic exposure of keratinocytes to cold physical plasma: An in vitro model for redox-related diseases of the skin.* Oxidative Medicine and Cellular Longevity **2016**, 10.1155/2016/9816072.

191. Eming, S.A.; Krieg, T.; Davidson, J.M. *Inflammation in wound repair: Molecular and cellular mechanisms.* Journal of Investigative Dermatology **2007**, 10.1038/sj.jid.5700701.

192. Barrientos, S.; Stojadinovic, O.; Golinko, M.S.; Brem, H.; Tomic-Canic, M. *Growth factors and cytokines in wound healing.* Wound Repair and Regeneration **2008**, 10.1111/j.1524-475X.2008.00410.x.

193. Bekeschus, S.; Schmidt, A.; Napp, M.; Kramer, A.; Kerner, W.; von Woedtke, T.; Wende, K.; Hasse, S.; Masur, K. *Distinct cytokine and chemokine patterns in chronic diabetic ulcers and acute wounds.* Experimental Dermatology **2017**, 10.1111/exd.13215.

194. Barrientos, S.; Brem, H.; Stojadinovic, O.; Tomic-Canic, M. *Clinical application of growth factors and cytokines in wound healing.* Wound Repair and Regeneration **2014**, 10.1111/wrr.12205.

195. Behm, B.; Babilas, P.; Landthaler, M.; Schreml, S. *Cytokines, chemokines and growth factors in wound healing.* Journal of the European Academy of Dermatology and Venereology **2012**, 10.1111/j.1468-3083.2011.04415.x.

196. Judith, R.; Nithya, M.; Rose, C.; Mandal, A.B. *Application of a pdgf-containing novel gel for cutaneous wound healing.* Life Sciences **2010**, 10.1016/j.lfs.2010.05.003.

197. Mahdavian Delavary, B.; van der Veer, W.M.; van Egmond, M.; Niessen, F.B.; Beelen, R.H. *Macrophages in skin injury and repair.* Immunobiology **2011**, 10.1016/j.imbio.2011.01.001.

198. Brune, B.; Dehne, N.; Grossmann, N.; Jung, M.; Namgaladze, D.; Schmid, T.; von Knethen, A.; Weigert, A. *Redox control of inflammation in macrophages.* Antioxidants and Redox Signaling **2013**, 10.1089/ars.2012.4785.

199. Bekeschus, S.; Brüggemeier, J.; Hackbarth, C.; von Woedtke, T.; Partecke, L.-I.; van der Linde, J. *Platelets are key in cold physical plasma-facilitated blood coagulation in mice.* Clinical Plasma Medicine **2017**, 10.1016/j.cpme.2017.10.001.

200. Golebiewska, E.M.; Poole, A.W. *Platelet secretion: From haemostasis to wound healing and beyond.* Blood Reviews **2015**, 10.1016/j.blre.2014.10.003.

201. Hollinger, J.O.; Hart, C.E.; Hirsch, S.N.; Lynch, S.; Friedlaender, G.E. *Recombinant human platelet-derived growth factor: Biology and clinical applications.* Journal of Bone and Joint Surgery (American Volume) **2008**, 10.2106/JBJS.G.01231.

202. Bekeschus, S.; Poschkamp, B.; van der Linde, J. *Medical gas plasma promotes blood coagulation via platelet activation.* Biomaterials **2021**, 10.1016/j.biomaterials.2020.120433.

203. Piccin, A.; Di Pierro, A.M.; Canzian, L.; Primerano, M.; Corvetta, D.; Negri, G.; Mazzoleni, G.; Gastl, G.; Steurer, M.; Gentilini, I.; Eisendle, K.; Fontanella, F. *Platelet gel: A new therapeutic tool with great potential.* Blood Transfusion: Trasfusione del Sangue **2017**, 10.2450/2016.0038-16.

204. Torres, A.S.; Caiafa, A.; Garner, A.L.; Klopman, S.; LaPlante, N.; Morton, C.; Conway, K.; Michelson, A.D.; Frelinger, A.L., 3rd; Neculaes, V.B. *Platelet activation using electric pulse stimulation: Growth factor profile and clinical implications.* The Journal of Trauma and Acute Care Surgery **2014**, 10.1097/TA.0000000000000322.

205. Telgenhoff, D.; Shroot, B. *Cellular senescence mechanisms in chronic wound healing.* Cell Death & Differentiation **2005**, 10.1038/sj.cdd.4401632.

206. Wilkinson, H.N.; Clowes, C.; Banyard, K.L.; Matteucci, P.; Mace, K.A.; Hardman, M.J. *Elevated local senescence in diabetic wound healing is linked to pathological repair via cxcr2.* Journal of Investigative Dermatology **2019**, 10.1016/j.jid.2019.01.005.

207. Wetzler, C.; Kampfer, H.; Stallmeyer, B.; Pfeilschifter, J.; Frank, S. *Large and sustained induction of chemokines during impaired wound healing in the genetically diabetic mouse: Prolonged persistence of neutrophils and macrophages during the late phase of repair.* Journal of Investigative Dermatology **2000**, 10.1046/j.1523-1747.2000.00029.x.

208. Bekeschus, S.; Lackmann, J.W.; Gumbel, D.; Napp, M.; Schmidt, A.; Wende, K. *A neutrophil proteomic signature in surgical trauma wounds.* International Journal of Molecular Sciences **2018**, 10.3390/ijms19030761.

209. Bekeschus, S.; Winterbourn, C.C.; Kolata, J.; Masur, K.; Hasse, S.; Broker, B.M.; Parker, H.A. *Neutrophil extracellular trap formation is elicited in response to cold physical plasma.* Journal of Leukocyte Biology **2016**, 10.1189/jlb.3A0415-165RR.

210. Larouche, J.; Sheoran, S.; Maruyama, K.; Martino, M.M. *Immune regulation of skin wound healing: Mechanisms and novel therapeutic targets.* Advances in Wound Care (New Rochelle) **2018**, 10.1089/wound.2017.0761.

211. Rutkowski, R.; Schuster, M.; Unger, J.; Seebauer, C.; Metelmann, H.R.; Woedtke, T.v.; Weltmann, K.D.; Daeschlein, G. *Hyperspectral imaging for in vivo monitoring of cold atmospheric plasma effects on microcirculation in treatment of head and neck cancer and wound healing.* Clinical Plasma Medicine **2017**, 10.1016/j.cpme.2017.09.002.

212. Borchardt, T.; Ernst, J.; Helmke, A.; Tanyeli, M.; Schilling, A.F.; Felmerer, G.; Viol, W. *Effect of direct cold atmospheric plasma (dicap) on microcirculation of intact skin in a controlled mechanical environment.* Microcirculation **2017**, 10.1111/micc.12399.

213. Kisch, T.; Helmke, A.; Schleusser, S.; Song, J.; Liodaki, E.; Stang, F.H.; Mailaender, P.; Kraemer, R. *Improvement of cutaneous microcirculation by cold atmospheric plasma (cap): Results of a controlled, prospective cohort study.* Microvascular Research **2016**, 10.1016/j.mvr.2015.12.002.

214. Kisch, T.; Schleusser, S.; Helmke, A.; Mauss, K.L.; Wenzel, E.T.; Hasemann, B.; Mailaender, P.; Kraemer, R. *The repetitive use of non-thermal dielectric barrier discharge plasma boosts cutaneous microcirculatory effects.* Microvascular Research **2016**, 10.1016/j.mvr.2016.02.008.

215. Pipa, A.V.; Bindemann, T.; Foest, R.; Kindel, E.; Ropcke, J.; Weltmann, K.D. *Absolute production rate measurements of nitric oxide by an atmospheric pressure plasma jet (appj).* Journal of Physics D-Applied Physics **2008**, 10.1088/0022-3727/41/19/194011.

216. Iseni, S.; Zhang, S.; van Gessel, A.F.H.; Hofmann, S.; van Ham, B.T.J.; Reuter, S.; Weltmann, K.D.; Bruggeman, P.J. *Nitric oxide density distributions in the effluent of an rf argon appj: Effect of gas flow rate and substrate.* New Journal of Physics **2014**, 10.1088/1367-2630/16/12/123011.

217. Kellogg, D.L., Jr.; Zhao, J.L.; Wu, Y. *Roles of nitric oxide synthase isoforms in cutaneous vasodilation induced by local warming of the skin and whole body heat stress in humans.* Journal of Applied Physiology (1985) **2009**, 10.1152/japplphysiol.00690.2009.

218. Ramasastry, S.S. *Chronic problem wounds.* Clinics in Plastic Surgery **1998**.

219. Schmidt-Bleker, A.; Winter, J.; Bosel, A.; Reuter, S.; Weltmann, K.D. *On the plasma chemistry of a cold atmospheric argon plasma jet with shielding gas device.* Plasma Sources Science and Technology **2016**, 10.1088/0963-0252/25/1/015005.

220. Schmidt-Bleker, A.; Winter, J.; Iseni, S.; Dunnbier, M.; Weltmann, K.D.; Reuter, S. *Reactive species output of a plasma jet with a shielding gas device-combination of ftir absorption spectroscopy and gas phase modelling.* Journal of Physics D-Applied Physics **2014**, 10.1088/0022-3727/47/14/145201.

221. Winter, J.; Sousa, J.S.; Sadeghi, N.; Schmidt-Bleker, A.; Reuter, S.; Puech, V. *The spatio-temporal distribution of he (23s1) metastable atoms in a mhz-driven helium plasma jet is influenced by the oxygen/nitrogen ratio of the surrounding atmosphere.* Plasma Sources Science and Technology **2015**.

222. Hanschmann, E.M.; Godoy, J.R.; Berndt, C.; Hudemann, C.; Lillig, C.H. *Thioredoxins, glutaredoxins, and peroxiredoxins—Molecular mechanisms and health significance: From cofactors to antioxidants to redox signaling.* Antioxidants and Redox Signaling **2013**, 10.1089/ars.2012.4599.

223. Agren, M.S.; Taplin, C.J.; Woessner, J.F., Jr.; Eaglstein, W.H.; Mertz, P.M. *Collagenase in wound healing: Effect of wound age and type.* Journal of Investigative Dermatology **1992**, 10.1111/1523-1747. ep12614202.

224. Zehtabchi, S.; Tan, A.; Yadav, K.; Badawy, A.; Lucchesi, M. *The impact of wound age on the infection rate of simple lacerations repaired in the emergency department.* Injury **2012**, 10.1016/j.injury.2012.02.018.

225. Beyene, R.T.; Derryberry, S.L., Jr.; Barbul, A. *The effect of comorbidities on wound healing.* Surgical Clinics of North America **2020**, 10.1016/j.suc.2020.05.002.

226. Nahid, M.A.; Griffin, J.M.; Lustik, M.B.; Hayes, J.J.; Fong, K.S.K.; Horseman, T.S.; Menguito, M.; Snesrud, E.C.; Barnhill, J.C.; Washington, M.A. *A longitudinal evaluation of the bacterial pathogens colonizing chronic non-healing wound sites at a united states military treatment facility in the pacific region.* Infection and Drug Resistance **2021**, 10.2147/IDR.S260708.

227. von Woedtke, T.; Emmert, S.; Metelmann, H.-R.; Rupf, S.; Weltmann, K.-D. *Perspectives on cold atmospheric plasma (cap) applications in medicine.* Physics of Plasmas **2020**, 10.1063/5.0008093.

228. Wende, K.; Schmidt, A.; Bekeschus, S. Safety aspects of non-thermal plasmas. In *Comprehensive clinical plasma medicine*, Metelmann, H.-R.; von Woedtke, T.; Weltmann, K.-D., Eds. Springer International Publishing: Cham, 2018; pp 83–109.

229. Schuster, M.; Rutkowski, R.; Hauschild, A.; Shojaei, R.K.; von Woedtke, T.; Rana, A.; Bauer, G.; Metelmann, P.; Seebauer, C. *Side effects in cold plasma treatment of advanced oral cancer—Clinical data and biological interpretation.* Clinical Plasma Medicine **2018**, 10.1016/j.cpme.2018.04.001.

230. Bauer, G.; Graves, D.B.; Schuster, M.; Metelmann, H.-R. Side effect management. In *Comprehensive clinical plasma medicine*, Metelmann, H.-R.; von Woedtke, T.; Weltmann, K.-D., Eds. Springer International Publishing: Cham, 2018; pp 301–318.

231. Nguyen, T.N.S.; Jiunn Der, L.; Huynh, L.Q.; Ngo, H.T.M. *Stimulation of wound healing process through ros/rns signals indirectly generated by n2/ar micro-plasma—In vitro and in vivo studies.* Science and Technology Development Journal **2015**, 10.32508/stdj.v18i2.1069.

232. Shi, X.M.; Xu, G.M.; Zhang, G.J.; Liu, J.R.; Wu, Y.M.; Gao, L.G.; Yang, Y.; Chang, Z.S.; Yao, C.W. *Low-temperature plasma promotes fibroblast proliferation in wound healing by ros-activated nf-kappab signaling pathway.* Current Medical Science **2018**, 10.1007/s11596-018-1853-x.

233. Pan, S.; Xu, M.; Gan, L.; Zhang, S.; Chen, H.; Liu, D.; Li, Y.; Lu, X. *Plasma activated radix arnebiae oil as innovative antimicrobial and burn wound healing agent.* Journal of Physics D: Applied Physics **2019**, 10.1088/1361-6463/ab234c.

234. Xu, M.; Li, Y. *Infected wound healing using plasma activated oil.* IEEE Transactions on Plasma Science **2019**, 10.1109/tps.2019.2928590.

235. Xu, D.; Wang, S.; Li, B.; Qi, M.; Feng, R.; Li, Q.; Zhang, H.; Chen, H.; Kong, M.G. *Effects of plasma-activated water on skin wound healing in mice.* Microorganisms **2020**, 10.3390/microorganisms8071091.

236. Darmawati, S.; Nasruddin, N.; Putri, G.S.A.; Iswara, A.; Kurniasiwi, P.; Wahyuningtyas, E.S.; Nurani, L.H.; Hayati, D.N.; Ishijima, T.; Nakatani, T.; Sugama, J. *Accelerated healing of chronic wound under the combinatorial therapeutic regimen based on contact and non-contact styles of the cold atmospheric plasma jet.* Plasma Medicine **2021**, 10.1615/PlasmaMed.2021039083.

237. Lee, H.R.; Lee, H.-Y.; Heo, J.; Jang, J.Y.; Shin, Y.S.; Kim, C.-H. *Liquid-type nonthermal atmospheric plasma enhanced regenerative potential of silk—Fibrin composite gel in radiation-induced wound failure.* Materials Science and Engineering: C **2021**, 10.1016/j.msec.2021.112304.

238. Wang, S.; Xu, D.; Qi, M.; Li, B.; Peng, S.; Li, Q.; Zhang, H.; Liu, D. *Plasma-activated water promotes wound healing by regulating inflammatory responses.* Biophysica **2021**, 10.3390/biophysica1030022.

239. Choi, K.Y.; Sultan, M.T.; Ajiteru, O.; Hong, H.; Lee, Y.J.; Lee, J.S.; Lee, H.; Lee, O.J.; Kim, S.H.; Lee, J.S.; Park, S.J.; Eden, J.G.; Park, C.H. *Treatment of fungal-infected diabetic wounds with low temperature plasma.* Biomedicines **2021**, 10.3390/biomedicines10010027.

240. Bender, C.; Kramer, A. *Options for antiseptic wound treatment in veterinary practice with special consideration of tissue compatibility.* Kleintierpraxis **2017**, 10.2377/0023-2076-62-373.

241. Khan, A.; Malik, S.; Walia, J.; Fridman, G.; Fridman, A.; Friedman, P.C. *Tolerability of six months indirect cold (physical) plasma treatment of the scalp for hair loss.* Journal of Drugs Dermatology **2020**, 10.36849/JDD.2020.5186.

10 Plasma-Induced Immunogenic Cancer-Cell Death

Vandana Miller, Sander Bekeschus

10.1 INTRODUCTION TO PLASMA CANCER TREATMENT

While plasma treatment has become one of the standard treatments for wound care in many European countries [1], its use for treating cancers remains experimental. Since the early 2010s, several *in vivo* studies have provided evidence of tumor control via plasma treatment [2,3], laying the foundation for the field of plasma oncology. This effect results from direct debulking of tumors and immunomodulatory effects of a range of long and short-lived reactive species produced in gas plasmas, among other effectors [4]. The limitation of accessibility of gas plasmas to deeper tumors is overcome by using plasma-conditioned liquids that contain largely long-lived oxidants such as hydrogen peroxide, hypochlorous acid, nitrite, and nitrate [5]. A shortcoming of most studies in plasma oncology using plasma-conditioned liquids is that they are performed with liquids not approved for medical use, limiting the translatability of the results [6]. In general, anticancer effects have been demonstrated with different types of plasma sources and for several tumor types in murine models [7–13]. For an in-depth analysis of these studies, the interested reader is referred to recent reviews [14–17].

With significant advances in oncoimmunology, more efforts and resources are being devoted to developing newer, more effective, and less expensive immunotherapies for cancer. Current immunotherapy for cancer relies heavily on "manufactured" or "ready-made" antibodies, cytokines, and cell-based therapies, described as passive immunotherapy. Since there is no "engagement" of the patient's immune responses and no immune memory is generated, regular administration of these agents is required. While some patients achieve a "cure," these treatments are expensive and not effective in every case, and they can engender resistance. A more durable approach is to set up specific antitumor immune responses in patients to fight cancer [18]. This paradigm has been tested with plasma, demonstrating that exposure to plasma increases the tumors' immunogenicity. This enables secondary plasma effects in the form of stimulated antitumor immunity, which may aid tumor control by endogenous effectors. A single treatment during surgical removal of internal tumors, for example, could be sufficient to trigger new or engage existing immune responses against cancer cells. This concept is analogous to what had been proposed in the community of photodynamic therapy (PDT), where mainly singlet delta oxygen is generated locally in tumor tissue via photo-oxidation of a photo-sensitizer to induce immunogenic cell death (ICD) [19,20]. Based on our previous reviews of this topic [21–23] and focusing on cancer, we here describe the main features of ICD and the role of ICD and reactive species in plasma oncology.

10.2 IMMUNOGENIC CELL DEATH (ICD)

Immunogenic cell death (ICD) is a form of regulated cell death capable of provoking adaptive immune responses [24]. ICD is elicited in response to different types of stimuli, including viral infections, certain drugs, radiotherapy regimens, and PDT. Dead and dying cells alarm the innate immune cells by releasing or displaying danger molecules on their surface in so-called damage-associated molecular patterns (DAMPs). DAMPs can bind pattern recognition receptors (PRRs) of professional antigen-presenting cells (APCs) to activate them (Figure 10.1). Activated APCs have an increased capacity

DOI: 10.1201/9781003328056-12

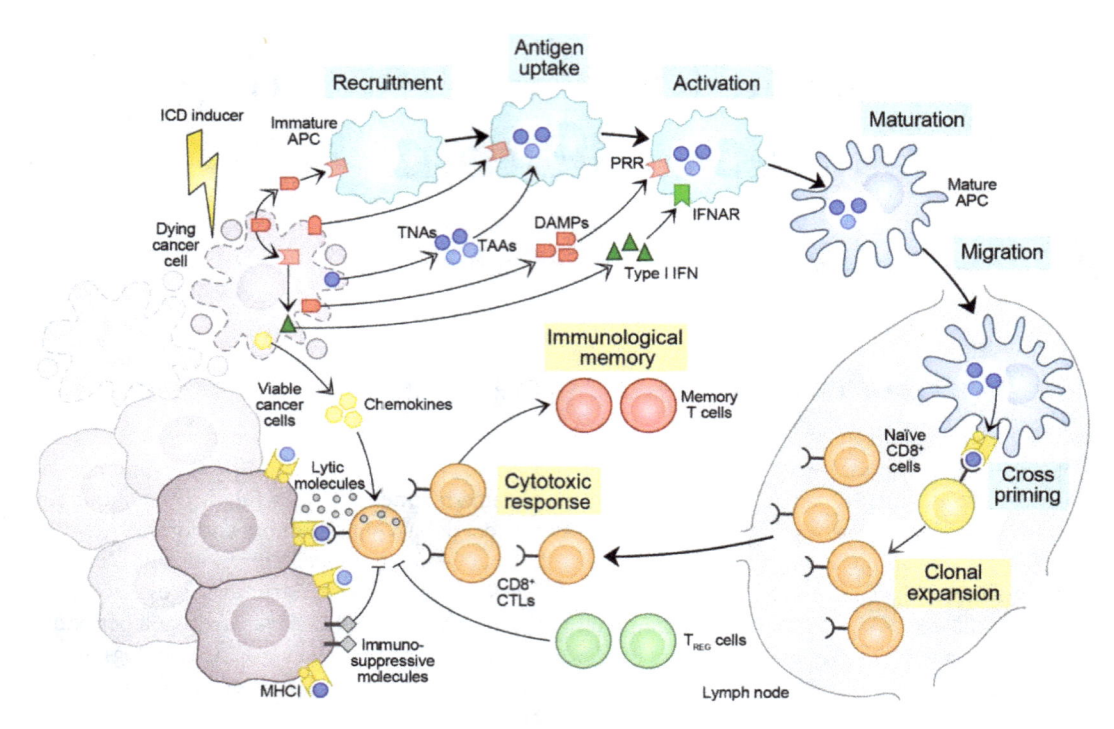

FIGURE 10.1   Major factors dictating the immunogenicity of cell death. Image is reproduced with permission from [79].

to migrate from the site of phagocytosis to nearby secondary lymphoid organs. There, they display material from previously engulfed dying or dead cells to cognate T cells as the cellular arm of adaptive immunity. This process is more efficient in the presence of DAMPs, which simultaneously provide further co-stimulatory signals to T cells for their activation and proliferation. This entire process is termed the "cancer-immunity cycle" and relies on the body's intrinsic ability to engage immune responses in afflicted tissues via inflammatory signaling [25,26]. Guido Kroemer and his team discovered the first elements of this paradigm in the early 2000s [27]. This group also highlighted the importance of calreticulin (CRT), a major DAMP, which plays decisive roles in determining the immunogenic potential of anticancer therapeutics, including drugs and physical treatment modalities [28–32].

In the body, billions of cells die daily as part of normal wear and tear and developmental processes [33]. Dead cell corpses are taken up by professional phagocytes and removed, and the components are recycled. This process is typically immunologically silent, because phagocyte activation during phagocytosis is suppressed by molecules such as phosphatidylserine expressed on the membrane of apoptotic cells [34]. This suppression, however, can be outweighed by sufficient activating signaling via DAMPs provided. DAMP expression and release, however, only occurs in response to ICD-inducers. The oxidative stress imposed during plasma-induced cell death induces the release and display of critical DAMPs, suggesting that plasma is an ICD inducer, although the exact mechanisms have yet to be established.

10.3   PLASMA-INDUCED ICD IN IMMUNO-ONCOLOGY

Two lines of evidence support the notion of plasma treatment as an ICD inducer in cancer cells. While indicators of ICD induction can be measured *in vitro*, valid proof can be obtained only by *in vivo* vaccination experiments [35]. To measure the hallmarks of ICD *in vitro*, several key DAMP molecules have been quantified (Figure 10.2). For instance, the translocation of CRT and

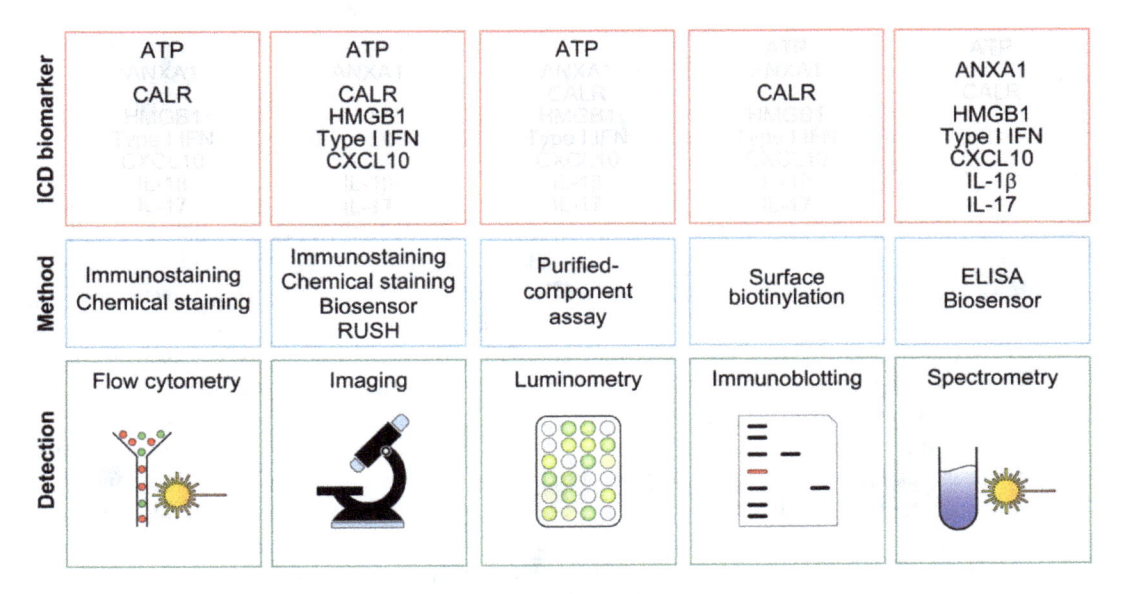

FIGURE 10.2   Main methodological approaches to measuring ICD biomarkers *in vitro*. Image is reproduced with permission from [37].

the heat-shock proteins 70 and 90 (HSP70, HSP90) on plasma-exposed cells has been demonstrated [36]. In addition, secretion of adenosine triphosphate (ATP), High mobility group box 1 protein (HMGB1), and ICD-associated chemokines, such as (C-X-C motif) ligand 1 (CXCL1) and CXCL10, has been evaluated [37]. The translocation of HMGB1 from the nucleus to the cytosol has also been assessed. On the molecular level, ICD is associated with stress of the endoplasmic reticulum (ER stress) accompanied by intracellular ROS and calcium increase, PERK activation, and phosphorylation and translation of the transcription factor eukaryotic translation initiation factor 2 subunit-α (eIF2α) [38]. However, not all ICD inducers trigger all hallmarks associated with ICD. For instance, radiotherapy elicits strong inflammasome activation and interleukin 1 beta (IL-1β) secretion, but PDT does not. As another example, only chemotherapy can trigger Annexin A1 release [37].

Plasma-triggered ICD hallmarks were observed in several cell lines *in vitro*. The broadest data are available for the atmospheric pressure argon plasma jet kINPen, a certified plasma source with a high degree of physico-chemical characterization [39]. Exposure to the kINPen plasma *in vitro* elicits CRT exposure in murine B16 melanoma cells [7,40,41] and murine CT26 colorectal cancer cells [42]. In the latter, increased ATP and HMGB1 release and phagocytosis by human primary monocyte-derived dendritic cells (moDCs) were also observed following treatment with kINPen but not with a dual neon plasma jet [43]. In human leukemia Jurkat and its derivative J-Lat and TK6 cells, kINPen treatment increased CRT and HSP90 and HSP70 translocation to the outer surface of the cell membrane [44,45]. In addition, kINPen helium plasma treatment augmented CRT, HSP70, and HSP90 display as well as ATP, HSP70, and interferon (IFN) α2 release in MCF-7 and MDA-MB-231 breast cancer cells. The same study also observed elevated CRT membrane translocation in plasma-treated 3D breast cancer spheroids [46]. In murine and human melanoma cells, direct kINPen argon plasma jet treatment also elevated the release of ATP, CXCL1, and CXCL10 [47,48]. The DBD plasma source with the most ICD-related data generated is a floating-electrode DBD. In A549 lung cancer cells exposed to this DBD *in vitro*, elevated CRT membrane translocation and ATP release were found [49]. Similar findings were made with CT26 cells *in vitro* [12] and CNE-1 cells for ATP release [36]. A DBD study comparing Jurkat responses to those found with the leukemia cell line THP-1 showed that CRT, HSP70, and HSP90 membrane translocation and phagocytosis by THP-1-derived macrophages were markedly elevated in Jurkat cells compared to THP-1 cells [50]. Using another DBD setup, Troitskaya and colleagues showed increased HMGB1 release in plasma-exposed CT26, A431, and MX-7 cells, and, in the latter cell type, elevated levels of ecto-CRT and ecto-HSP70 [51] as well as phagocytosis via DCs [52]. A recent DBD setup intended for transdermal needle application also elevated CRT membrane translocation in melanoma cells and showed moDC activation (increased CD80 and CD86 co-stimulatory ligand expression) profiles after these cells engulfed plasma-treated B16 melanoma cells [53]. Using another DBD setup, the same author reiterated the data in B16 melanoma and 4T1 breast cancer cells [54]. In addition, using a microwave argon plasma source, colorectal HCT-116 and thyroid BCPAP cells showed elevated HMGB1 expression and nuclear translocation following direct plasma treatment [55].

There are also several reports suggestive of ICD *in vivo* in syngeneic tumor models. A recent review on this topic provides a comprehensive overview [23]. Three main lines of evidence of *in vivo* ICD following plasma treatment exist. The first involves analysis of intratumoral hallmarks of ICD after tumor explantation using immunohistological analysis. This is a descriptive and relatively weak indicator for *in vivo* ICD. In plasma medicine, increased HMGB1 and CRT, along with elevated presence of immune cells, including DCs, were found in floating-electrode DBD-treated CT26 subcutaneous tumors *in vivo* [12]. In melanoma tumors *in vivo* treated with the same source, intratumoral T-cell levels were not affected, while DCs and pan-tissue CRT membrane translocation were found to be increased [56]. By contrast, microneedle plasma application in the same model elevated intratumoral T-cells and their proliferation and granzyme B expression and decreased regulatory T-cells [53]. The same author team reported similar results with another plasma source and in the same B16 model and a 4T1 breast cancer model [54]. Using a helium plasma jet, increased CD4$^+$ total and T_H17 and CD8$^+$ T-cells, as well as CRT membrane translocation, were found in directly treated breast cancers *in vivo*. At the same time, regulatory (FOXP3$^+$) T-cells (T_{reg}) were decreased [46]. With kINPen treatment, T-helper cells, cytotoxic T-cells, macrophages, and DCs were found in elevated numbers in plasma-treated melanomas [7]. In addition, that study provided the first evidence that such responses depend on not only the type of plasma source but also the mixture of reactive species it produces, as some plasma conditions outcompeted others with respect to increasing tumor-infiltrating lymphocytes (TIL) for the same type of jet. Moreover, using colorectal tumors inoculated on the chorioallantoic membrane of chicken embryos in the TUM-CAM model [57], kiNPen treatment elevated CRT, HSP70, and HSP90 membrane translocation or expression in tumors [58]. For another plasma jet operated with humidified oxygen admixture, intratumor CD8$^+$ T-cell levels were increased in plasma-treated melanomas [59].

The second line of evidence comes from the gold standard in ICD testing and is based on a vaccination protocol, allowing firm conclusions on ICD induction [35]. Briefly, cell lines are exposed to the agent in question, such as plasma, to induce DAMPs. As control conditions, cell killing through poorly immunogenic drugs (e.g., mitomycin C, MMC) or highly immunogenic drugs (e.g., mitoxantrone, MTX) is used. The tumor cells killed *in vitro* are subsequently injected into immunocompetent mice as a vaccine to allow the development of a protective immune response. One to two weeks later, animals are challenged with live tumor cells. The incidence and extent of tumor growth allow the investigators to determine the immunogenicity of the agent responsible for the *in vitro* killing of tumor cells. The gold standard test for plasma has been employed in several tumor models. DBD-killed CT26 colorectal cancer cells were more immunogenic than cisplatin-killed cells in a mouse model of the disease [12]. The same type of plasma source was used to create a B16 melanoma vaccine [12], which mediated superior survival compared to MTX-killed cells [60]. In comparison, a plasma jet-killed vaccine provided less tumor control than MTX but better tumor protection than MMC in the B16 melanoma model [7]. These results align with the 100-fold higher CRT levels induced *in vitro* by MTX compared to plasma conditions, which, in turn, were significantly higher than untreated controls. The differences in survival efficacy between the two different plasma source-created vaccines can be explained by the extent of CRT display on the cell surface, providing strong support for the role of plasma-induced ICD in generating immune responses against cancer cells.

The third line of evidence comes from studies investigating the so-called abscopal effect. Here, two spatially spread-out tumors are set in each mouse, one on the left and one on the right flank. Only one tumor is exposed to the local treatment modality, such as plasma, while the remote (opposite flank) tumor remains untreated. Tumor growth is monitored for both tumor sites. A size reduction, decelerated growth, or decline of the tumor on the untreated flank is a clear indicator of a systemic anticancer immune response. Using B16 melanoma cells, an early study attempted to test this model, but the overall results on plasma-induced abscopal effects were inconclusive [61]. By contrast, two other studies using B16 melanoma [53] and 4T1 breast cancer [46] provided clear evidence that DBD and plasma jet treatment are capable of eliciting therapeutically effective antitumor immunity in situ in the host-bearing tumors. In the latter report, extensive immune profiling of tumor tissue sections indicated that not only the directly plasma-treated tumor but also its untreated remote counterpart showed higher levels of apoptosis and TIL infiltration.

10.4 ROLES OF REACTIVE SPECIES IN PLASMA-DRIVEN ICD *IN VITRO*

The relationship between reactive oxygen and nitrogen species and plasma-mediated biological effects is well studied [62]. This is especially true in *in vitro* cell culture experiments characterized by excess cell culture or other buffered media. For plasma jets with relatively strong gas fluxes that disallow the direct *in vitro* exposure of cells to the plasma to prevent drying of cells, the effects of long-lived species on ICD have been well studied. In such plasma jet treatment schemes, the excess liquid needed to protect the cells buffers other plasma effects, e.g., UV radiation, and leads to a dominant effect of long-lived reactive species such as hydrogen peroxide because short-lived species quickly react with biomolecules and are restricted in terms of diffusion distances. The significant role of (long-lived) reactive species has been shown for plasma treatment and experiments using plasma-treated liquids with and without antioxidants or hydrogen-peroxide-removing enzymes such as catalase [6,63]. Likewise, hydrogen peroxide and nitrate provided at the same concentrations with plasma treatment in sodium chloride provoked similar expression of DAMP molecules in cancer cells as did direct treatment with kINPen [64]. In addition, it has been shown that adding antioxidants or pharmacologically inhibiting ROS-producing enzymes such as NADPH-oxidases reduces CRT membrane translocation in response to plasma [49]. Conversely, *in vitro* exposure of several gastrointestinal tumor cell lines to hydrogen peroxide, peroxynitrite, or hypochlorous acid induced hallmarks of ICD [65], albeit experimentally added species do not always fully recapitulate the extent of ICD hallmark expression as seen with plasma setups associated with prominent electromagnetic fields [49,60]. Finally, the evidence that reactive species produced by plasma are responsible for eliciting hallmarks of ICD comes from feed gas modulation studies to alter the composition of reactive species emitted from the plasma jet and its influence on the extent of CRT membrane translocation [41]. Previous work shows that agents promoting excess exogenous and endogenous ROS levels may serve as ICD triggers, including in response to PDT [19,66,67]. Therefore, it is no surprise that plasma treatment elicits hallmarks of ICD.

Nevertheless, the generation of DAMPs in response to reactive species should not be overestimated. Compared to anthracycline drugs such as mitoxantrone, plasma-induced CRT and HSP70 membrane display is up to a hundred-fold lower [7], indicating that the potency of pharmacological ER stressors that may be given systemically to patients outcompete plasmas in terms of ER-stress-related ICD by far. In addition, it should be noted that studies in the plasma medicine field have not investigated the redox state of released HMGB1. It is known that only reduced HMGB1 leads to immunogenic effects, while HMGB1 oxidized at the cysteine 106 leads to tolerance induction in DCs [68]. If plasma exposure contributes to oxidation of HMGB1, whose release, however, appears much later compared to the treatment time point, tolerogenic HMGB1 may dampen immunogenicity induced by other ICD hallmarks. In general, the timing of immunogenic signaling is critical as it needs additive stimuli to promote above-threshold activation and maturation signals in DCs to provoke adaptive immune responses [69].

Mitochondria are another source of ICD-relevant ROS enmeshed with ER stress [70]. Mitochondrial ROS are generated secondary to cytotoxic insults and have been observed in several *in vitro* studies of plasma-exposed cancer cells [36,71–73]. Accordingly, mitochondrial ROS production has been linked to the membrane translocation and expression of CRT, HSP70, and/or HSP90 [40,44,50,74]. A recent study explicitly targeting the mitochondrial membrane potential and eliciting mitochondrial dysfunction and mitochondrial-derived ROS, ER stress, and ICD has established a clear connection between these processes [75]. At the same time, it is clear that the plasma-induced mitochondrial ROS production is secondary to plasma-induced oxidative stress since the former can be repressed if the latter is abrogated by the use of, e.g., antioxidants [42,76].

10.5 CONCLUSION AND OUTLOOK

It is established that plasmas generate reactive species and that these may play an essential role in inducing ICD *in vitro*. However, additional comprehensive comparisons of plasma with other established ICD inducers, including drugs, PDT, and radiotherapy *in vitro*, are needed to understand the relative magnitude of plasma-induced ICD hallmarks. Despite attempts to link plasma components and reactive species profiles to ICD [7,41,49,60], the fundamental knowledge of which parameters maximize ICD induction is lacking, leading to vast gaps in our understanding of the mechanism of plasma-induced ICD. Likewise, it is unclear to what extent the observed plasma effects depend on the cell lines and tumor models investigated, since comparative studies are also few and far between. The differences among the numerous plasma sources in terms of their potencies and limitations are also unknown, except for one study that compared two different plasma jets in the same biological setting [43]. Although these questions seem somewhat academic, answering them would provide a mechanistic understanding of how to control and tune plasma cancer treatment for maximum immunogenicity to achieve better tumor immune control.

While the efficacy of plasma in inducing ICD *in vivo* is demonstrated, deciphering the exact functions of plasma effectors, including reactive species, is hampered by the technical challenges of singling out individual plasma components for each source to study their respective contributions to the effects observed. Furthermore, the depth of penetration of these species *in vivo* will influence the ability of plasma to induce ICD in deep-seated tumors. Current studies employ subcutaneous tumor models in mice where the stratum corneum is 4 μm compared to that of humans with 17 μm, and total murine skin thickness is 200 μm compared to 2400 μm thick human skin [77]. Because the plasma-derived short-lived reactive species are highly reactive, it is unlikely that they diffuse unaffected through 200 μm murine skin to reach tumors. It is conceivable that plasma antitumor immune effects result from the action of plasma-derived electric fields or indirect events such as cell-cell communication that take place in the skin above the tumor. Further studies are needed to shed light on the mechanisms of plasma-tissue-effects. In addition, more systematic studies providing direct evidence that plasma-induced immuno-stimulation promotes adaptive antitumor immune responses, such as T-cell depletion experiments *in vivo*, are needed.

Finally, the intra-surgical plasma treatment of tumor beds following surgical resection is a promising future treatment modality. In a recently published phase I study, the safety of plasma application was demonstrated in patients [78]. Prospectively, if the field continues translational plasma research in cancer immunology, ICD induction in such treatment settings could be a valuable strategy to improve patients' antitumor immunity, decrease tumor burden, and/or lower recurrences in the future.

REFERENCES

1. Bekeschus, S.; von Woedtke, T.; Emmert, S.; Schmidt, A. *Medical gas plasma-stimulated wound healing: Evidence and mechanisms.* Redox Biology **2021**, 10.1016/j.redox.2021.102116.
2. Keidar, M.; Walk, R.; Shashurin, A.; Srinivasan, P.; Sandler, A.; Dasgupta, S.; Ravi, R.; Guerrero-Preston, R.; Trink, B. *Cold plasma selectivity and the possibility of a paradigm shift in cancer therapy.* British Journal of Cancer **2011**, 10.1038/bjc.2011.386.

3. Brulle, L.; Vandamme, M.; Ries, D.; Martel, E.; Robert, E.; Lerondel, S.; Trichet, V.; Richard, S.; Pouvesle, J.M.; Le Pape, A. *Effects of a non thermal plasma treatment alone or in combination with gemcitabine in a mia paca2-luc orthotopic pancreatic carcinoma model.* PLoS One **2012**, 10.1371/journal.pone.0052653.

4. Laroussi, M.; Bekeschus, S.; Keidar, M.; Bogaerts, A.; Fridman, A.; Lu, X.; Ostrikov, K.; Hori, M.; Stapelmann, K.; Miller, V.; Reuter, S.; Laux, C.; Mesbah, A.; Walsh, J.; Jiang, C.; Thagard, S.M.; Tanaka, H.; Liu, D.; Yan, D.; Yusupov, M. *Low-temperature plasma for biology, hygiene, and medicine: Perspective and roadmap.* IEEE Transactions on Radiation and Plasma Medical Sciences **2022**, 10.1109/trpms.2021.3135118.

5. Wende, K.; von Woedtke, T.; Weltmann, K.D.; Bekeschus, S. *Chemistry and biochemistry of cold physical plasma derived reactive species in liquids.* Biological Chemistry **2018**, 10.1515/hsz-2018-0242.

6. Freund, E.; Bekeschus, S. *Gas plasma-oxidized liquids for cancer treatment: Preclinical relevance, immuno-oncology, and clinical obstacles.* IEEE Transactions on Radiation and Plasma Medical Sciences **2021**, 10.1109/trpms.2020.3029982.

7. Bekeschus, S.; Clemen, R.; Niessner, F.; Sagwal, S.K.; Freund, E.; Schmidt, A. *Medical gas plasma jet technology targets murine melanoma in an immunogenic fashion.* Advanced Science **2020**, 10.1002/advs.201903438.

8. Binenbaum, Y.; Ben-David, G.; Gil, Z.; Slutsker, Y.Z.; Ryzhkov, M.A.; Felsteiner, J.; Krasik, Y.E.; Cohen, J.T. *Cold atmospheric plasma, created at the tip of an elongated flexible capillary using low electric current, can slow the progression of melanoma.* PLoS One **2017**, 10.1371/journal.pone.0169457.

9. Chen, Z.; Simonyan, H.; Cheng, X.; Gjika, E.; Lin, L.; Canady, J.; Sherman, J.H.; Young, C.; Keidar, M. *A novel micro cold atmospheric plasma device for glioblastoma both in vitro and in vivo.* Cancers **2017**, 10.3390/cancers9060061.

10. Mashayekh, S.; Rajaee, H.; Akhlaghi, M.; Shokri, B.; Hassan, Z.M. *Atmospheric-pressure plasma jet characterization and applications on melanoma cancer treatment (b/16-f10).* Physics of Plasmas **2015**, 10.1063/1.4930536.

11. Mirpour, S.; Piroozmand, S.; Soleimani, N.; Jalali Faharani, N.; Ghomi, H.; Fotovat Eskandari, H.; Sharifi, A.M.; Mirpour, S.; Eftekhari, M.; Nikkhah, M. *Utilizing the micron sized non-thermal atmospheric pressure plasma inside the animal body for the tumor treatment application.* Scientific Reports **2016**, 10.1038/srep29048.

12. Lin, A.G.; Xiang, B.; Merlino, D.J.; Baybutt, T.R.; Sahu, J.; Fridman, A.; Snook, A.E.; Miller, V. *Non-thermal plasma induces immunogenic cell death in vivo in murine ct26 colorectal tumors.* Oncoimmunology **2018**, 10.1080/2162402X.2018.1484978.

13. Kordt, M.; Trautmann, I.; Schlie, C.; Lindner, T.; Stenzel, J.; Schildt, A.; Boeckmann, L.; Bekeschus, S.; Kurth, J.; Krause, B.J.; Vollmar, B.; Grambow, E. *Multimodal imaging techniques to evaluate the anticancer effect of cold atmospheric pressure plasma.* Cancers **2021**, 10.3390/cancers13102483.

14. Kumar Dubey, S.; Dabholkar, N.; Narayan Pal, U.; Singhvi, G.; Kumar Sharma, N.; Puri, A.; Kesharwani, P. *Emerging innovations in cold plasma therapy against cancer: A paradigm shift.* Drug Discovery Today **2022**, 10.1016/j.drudis.2022.05.014.

15. Bekeschus, S.; Saadati, F.; Emmert, S. *The potential of gas plasma technology for targeting breast cancer.* Clinical and Translational Medicine **2022**, 10.1002/ctm2.1022.

16. Berner, J.; Seebauer, C.; Sagwal, S.K.; Boeckmann, L.; Emmert, S.; Metelmann, H.-R.; Bekeschus, S. *Medical gas plasma treatment in head and neck cancer—challenges and opportunities.* Applied Sciences **2020**, 10.3390/app10061944.

17. Adhikari, M.; Adhikari, B.; Adhikari, A.; Yan, D.; Soni, V.; Sherman, J.; Keidar, M. *Cold atmospheric plasma as a novel therapeutic tool for the treatment of brain cancer.* Current Pharmaceutical Design **2020**, 10.2174/1381612826666200302105715.

18. Disis, M.L. *Immune regulation of cancer.* Journal of Clinical Oncology **2010**, 10.1200/JCO.2009.27.2146.

19. Krysko, D.V.; Garg, A.D.; Kaczmarek, A.; Krysko, O.; Agostinis, P.; Vandenabeele, P. *Immunogenic cell death and damps in cancer therapy.* Nature Reviews Cancer **2012**, 10.1038/nrc3380.

20. Galluzzi, L.; Kepp, O.; Kroemer, G. *Enlightening the impact of immunogenic cell death in photodynamic cancer therapy.* EMBO Journal **2012**, 10.1038/emboj.2012.2.

21. Bekeschus, S.; Clemen, R.; Metelmann, H.-R. *Potentiating anti-tumor immunity with physical plasma.* Clinical Plasma Medicine **2018**, 10.1016/j.cpme.2018.10.001.

22. Khalili, M.; Daniels, L.; Lin, A.; Krebs, F.C.; Snook, A.E.; Bekeschus, S.; Bowne, W.B.; Miller, V. *Non-thermal plasma-induced immunogenic cell death in cancer: A topical review.* Journal of Physics D-Applied Physics **2019**, 10.1088/1361-6463/ab31c1.

23. Bekeschus, S.; Clemen, R. *Plasma, cancer, immunity.* Journal of Physics D-Applied Physics **2022**, 10.1088/1361-6463/ac9398.

24. Galluzzi, L.; Vitale, I.; Aaronson, S.A.; Abrams, J.M.; Adam, D.; Agostinis, P.; Alnemri, E.S.; Altucci, L.; Amelio, I.; Andrews, D.W.; Annicchiarico-Petruzzelli, M.; Antonov, A.V.; Arama, E.; Baehrecke, E.H.; Barlev, N.A.; Bazan, N.G.; Bernassola, F.; Bertrand, M.J.M.; Bianchi, K.; Blagosklonny, M.V.; Blomgren, K.; Borner, C.; Boya, P.; Brenner, C.; Campanella, M.; Candi, E.; Carmona-Gutierrez, D.; Cecconi, F.; Chan, F.K.; Chandel, N.S.; Cheng, E.H.; Chipuk, J.E.; Cidlowski, J.A.; Ciechanover, A.; Cohen, G.M.; Conrad, M.; Cubillos-Ruiz, J.R.; Czabotar, P.E.; D'Angiolella, V.; Dawson, T.M.; Dawson, V.L.; De Laurenzi, V.; De Maria, R.; Debatin, K.M.; DeBerardinis, R.J.; Deshmukh, M.; Di Daniele, N.; Di Virgilio, F.; Dixit, V.M.; Dixon, S.J.; Duckett, C.S.; Dynlacht, B.D.; El-Deiry, W.S.; Elrod, J.W.; Fimia, G.M.; Fulda, S.; Garcia-Saez, A.J.; Garg, A.D.; Garrido, C.; Gavathiotis, E.; Golstein, P.; Gottlieb, E.; Green, D.R.; Greene, L.A.; Gronemeyer, H.; Gross, A.; Hajnoczky, G.; Hardwick, J.M.; Harris, I.S.; Hengartner, M.O.; Hetz, C.; Ichijo, H.; Jaattela, M.; Joseph, B.; Jost, P.J.; Juin, P.P.; Kaiser, W.J.; Karin, M.; Kaufmann, T.; Kepp, O.; Kimchi, A.; Kitsis, R.N.; Klionsky, D.J.; Knight, R.A.; Kumar, S.; Lee, S.W.; Lemasters, J.J.; Levine, B.; Linkermann, A.; Lipton, S.A.; Lockshin, R.A.; Lopez-Otin, C.; Lowe, S.W.; Luedde, T.; Lugli, E.; MacFarlane, M.; Madeo, F.; Malewicz, M.; Malorni, W.; Manic, G.; Marine, J.C.; Martin, S.J.; Martinou, J.C.; Medema, J.P.; Mehlen, P.; Meier, P.; Melino, S.; Miao, E.A.; Molkentin, J.D.; Moll, U.M.; Munoz-Pinedo, C.; Nagata, S.; Nunez, G.; Oberst, A.; Oren, M.; Overholtzer, M.; Pagano, M.; Panaretakis, T.; Pasparakis, M.; Penninger, J.M.; Pereira, D.M.; Pervaiz, S.; Peter, M.E.; Piacentini, M.; Pinton, P.; Prehn, J.H.M.; Puthalakath, H.; Rabinovich, G.A.; Rehm, M.; Rizzuto, R.; Rodrigues, C.M.P.; Rubinsztein, D.C.; Rudel, T.; Ryan, K.M.; Sayan, E.; Scorrano, L.; Shao, F.; Shi, Y.; Silke, J.; Simon, H.U.; Sistigu, A.; Stockwell, B.R.; Strasser, A.; Szabadkai, G.; Tait, S.W.G.; Tang, D.; Tavernarakis, N.; Thorburn, A.; Tsujimoto, Y.; Turk, B.; Vanden Berghe, T.; Vandenabeele, P.; Vander Heiden, M.G.; Villunger, A.; Virgin, H.W.; Vousden, K.H.; Vucic, D.; Wagner, E.F.; Walczak, H.; Wallach, D.; Wang, Y.; Wells, J.A.; Wood, W.; Yuan, J.; Zakeri, Z.; Zhivotovsky, B.; Zitvogel, L.; Melino, G.; Kroemer, G. *Molecular mechanisms of cell death: Recommendations of the nomenclature committee on cell death 2018.* Cell Death and Differentiation **2018**, 10.1038/s41418-017-0012-4.
25. Galluzzi, L.; Buque, A.; Kepp, O.; Zitvogel, L.; Kroemer, G. *Immunogenic cell death in cancer and infectious disease.* Nature Reviews Immunology **2017**, 10.1038/nri.2016.107.
26. Chen, D.S.; Mellman, I. *Elements of cancer immunity and the cancer-immune set point.* Nature **2017**, 10.1038/nature21349.
27. Casares, N.; Pequignot, M.O.; Tesniere, A.; Ghiringhelli, F.; Roux, S.; Chaput, N.; Schmitt, E.; Hamai, A.; Hervas-Stubbs, S.; Obeid, M.; Coutant, F.; Metivier, D.; Pichard, E.; Aucouturier, P.; Pierron, G.; Garrido, C.; Zitvogel, L.; Kroemer, G. *Caspase-dependent immunogenicity of doxorubicin-induced tumor cell death.* Journal of Experimental Medicine **2005**, 10.1084/jem.20050915.
28. Apetoh, L.; Obeid, M.; Tesniere, A.; Ghiringhelli, F.; Fimia, G.M.; Piacentini, M.; Kroemer, G.; Zitvogel, L. *Immunogenic chemotherapy: Discovery of a critical protein through proteomic analyses of tumor cells.* Cancer Genomics—Proteomics **2007**.
29. Obeid, M.; Panaretakis, T.; Joza, N.; Tufi, R.; Tesniere, A.; van Endert, P.; Zitvogel, L.; Kroemer, G. *Calreticulin exposure is required for the immunogenicity of gamma-irradiation and uvc light-induced apoptosis.* Cell Death and Differentiation **2007**, 10.1038/sj.cdd.4402201.
30. Obeid, M.; Panaretakis, T.; Tesniere, A.; Joza, N.; Tufi, R.; Apetoh, L.; Ghiringhelli, F.; Zitvogel, L.; Kroemer, G. *Leveraging the immune system during chemotherapy: Moving calreticulin to the cell surface converts apoptotic death from "silent" to immunogenic.* Cancer Research **2007**, 10.1158/0008-5472.CAN-07–1622.
31. Obeid, M.; Tesniere, A.; Ghiringhelli, F.; Fimia, G.M.; Apetoh, L.; Perfettini, J.L.; Castedo, M.; Mignot, G.; Panaretakis, T.; Casares, N.; Metivier, D.; Larochette, N.; van Endert, P.; Ciccosanti, F.; Piacentini, M.; Zitvogel, L.; Kroemer, G. *Calreticulin exposure dictates the immunogenicity of cancer cell death.* Nature Medicine **2007**, 10.1038/nm1523.
32. Obeid, M.; Tesniere, A.; Panaretakis, T.; Tufi, R.; Joza, N.; van Endert, P.; Ghiringhelli, F.; Apetoh, L.; Chaput, N.; Flament, C.; Ullrich, E.; de Botton, S.; Zitvogel, L.; Kroemer, G. *Ecto-calreticulin in immunogenic chemotherapy.* Immunological Reviews **2007**, 10.1111/j.1600-065X.2007.00567.x.
33. Kolb, J.P.; Oguin, T.H., 3rd; Oberst, A.; Martinez, J. *Programmed cell death and inflammation: Winter is coming.* Trends in Immunology **2017**, 10.1016/j.it.2017.06.009.
34. Kepp, O.; Tesniere, A.; Schlemmer, F.; Michaud, M.; Senovilla, L.; Zitvogel, L.; Kroemer, G. *Immunogenic cell death modalities and their impact on cancer treatment.* Apoptosis **2009**, 10.1007/s10495-008-0303-9.
35. Kroemer, G.; Galluzzi, L.; Kepp, O.; Zitvogel, L. *Immunogenic cell death in cancer therapy.* Annual Reviews in Immunology **2013**, 10.1146/annurev-immunol-032712–100008.
36. Lin, A.; Truong, B.; Pappas, A.; Kirifides, L.; Oubarri, A.; Chen, S.Y.; Lin, S.J.; Dobrynin, D.; Fridman, G.; Fridman, A.; Sang, N.; Miller, V. *Uniform nanosecond pulsed dielectric barrier discharge plasma enhances anti-tumor effects by induction of immunogenic cell death in tumors and stimulation of macrophages.* Plasma Processes and Polymers **2015**, 10.1002/ppap.201500139.

37. Fucikova, J.; Kepp, O.; Kasikova, L.; Petroni, G.; Yamazaki, T.; Liu, P.; Zhao, L.; Spisek, R.; Kroemer, G.; Galluzzi, L. *Detection of immunogenic cell death and its relevance for cancer therapy.* Cell Death and Discovery **2020**, 10.1038/s41419-020-03221-2.

38. Kepp, O.; Menger, L.; Vacchelli, E.; Locher, C.; Adjemian, S.; Yamazaki, T.; Martins, I.; Sukkurwala, A.Q.; Michaud, M.; Senovilla, L.; Galluzzi, L.; Kroemer, G.; Zitvogel, L. *Crosstalk between er stress and immunogenic cell death.* Cytokine Growth Factor Reviews **2013**, 10.1016/j.cytogfr.2013.05.001.

39. Reuter, S.; von Woedtke, T.; Weltmann, K.D. *The kinpen-a review on physics and chemistry of the atmospheric pressure plasma jet and its applications.* Journal of Physics D-Applied Physics **2018**, 10.1088/1361-6463/aab3ad.

40. Bekeschus, S.; Rodder, K.; Fregin, B.; Otto, O.; Lippert, M.; Weltmann, K.D.; Wende, K.; Schmidt, A.; Gandhirajan, R.K. *Toxicity and immunogenicity in murine melanoma following exposure to physical plasma-derived oxidants.* Oxidative Medicine and Cellular Longevity **2017**, 10.1155/2017/4396467.

41. Bekeschus, S.; Schmidt, A.; Niessner, F.; Gerling, T.; Weltmann, K.D.; Wende, K. *Basic research in plasma medicine—a throughput approach from liquids to cells.* Journal of Visualized Experiments **2017**, 10.3791/56331.

42. Bekeschus, S.; Mueller, A.; Miller, V.; Gaipl, U.; Weltmann, K.-D. *Physical plasma elicits immunogenic cancer cell death and mitochondrial singlet oxygen.* IEEE Transactions on Radiation and Plasma Medical Sciences **2018**, 10.1109/trpms.2017.2766027.

43. Miebach, L.; Freund, E.; Horn, S.; Niessner, F.; Sagwal, S.K.; von Woedtke, T.; Emmert, S.; Weltmann, K.D.; Clemen, R.; Schmidt, A.; Gerling, T.; Bekeschus, S. *Tumor cytotoxicity and immunogenicity of a novel v-jet neon plasma source compared to the kinpen.* Scientific Reports **2021**, 10.1038/s41598-020-80512-w.

44. Wolff, C.M.; Kolb, J.F.; Bekeschus, S. *Combined in vitro toxicity and immunogenicity of cold plasma and pulsed electric fields.* Biomedicines **2022**, 10.3390/biomedicines10123084.

45. Mohamed, H.; Clemen, R.; Freund, E.; Lackmann, J.W.; Wende, K.; Connors, J.; Haddad, E.K.; Dampier, W.; Wigdahl, B.; Miller, V.; Bekeschus, S.; Krebs, F.C. *Non-thermal plasma modulates cellular markers associated with immunogenicity in a model of latent hiv-1 infection.* PLoS One **2021**, 10.1371/journal.pone.0247125.

46. Mahdikia, H.; Saadati, F.; Freund, E.; Gaipl, U.S.; Majidzadeh, A.K.; Shokri, B.; Bekeschus, S. *Gas plasma irradiation of breast cancers promotes immunogenicity, tumor reduction, and an abscopal effect in vivo.* Oncoimmunology **2020**, 10.1080/2162402X.2020.1859731.

47. Rödder, K.; Moritz, J.; Miller, V.; Weltmann, K.-D.; Metelmann, H.-R.; Gandhirajan, R.; Bekeschus, S. *Activation of murine immune cells upon co-culture with plasma-treated b16f10 melanoma cells.* Applied Sciences **2019**, 10.3390/app9040660.

48. Sagwal, S.K.; Pasqual-Melo, G.; Bodnar, Y.; Gandhirajan, R.K.; Bekeschus, S. *Combination of chemotherapy and physical plasma elicits melanoma cell death via upregulation of slc22a16.* Cell Death and Diseases **2018**, 10.1038/s41419-018-1221-6.

49. Lin, A.; Truong, B.; Patel, S.; Kaushik, N.; Choi, E.H.; Fridman, G.; Fridman, A.; Miller, V. *Nanosecond-pulsed dbd plasma-generated reactive oxygen species trigger immunogenic cell death in a549 lung carcinoma cells through intracellular oxidative stress.* International Journal of Molecular Sciences **2017**, 10.3390/ijms18050966.

50. Mohamed, H.; Gebski, E.; Reyes, R.; Beane, S.; Wigdahl, B.; Krebs, F.C.; Stapelmann, K.; Miller, V. *Differential effect of non-thermal plasma rons on two human leukemic cell populations.* Cancers **2021**, 10.3390/cancers13102437.

51. Troitskaya, O.; Golubitskaya, E.; Biryukov, M.; Varlamov, M.; Gugin, P.; Milakhina, E.; Richter, V.; Schweigert, I.; Zakrevsky, D.; Koval, O. *Non-thermal plasma application in tumor-bearing mice induces increase of serum hmgb1.* International Journal of Molecular Sciences **2020**, 10.3390/ijms21145128.

52. Troitskaya, O.; Novak, D.; Varlamov, M.; Biryukov, M.; Nushtaeva, A.; Kochneva, G.; Zakrevsky, D.; Schweigert, I.; Richter, V.; Koval, O. *Immunological effects of cold atmospheric plasma-treated cells in comparison with those of cells treated with lactaptin-based anticancer drugs.* Biophysica **2022**, 10.3390/biophysica2030025.

53. Chen, G.; Chen, Z.; Wen, D.; Wang, Z.; Li, H.; Zeng, Y.; Dotti, G.; Wirz, R.E.; Gu, Z. *Transdermal cold atmospheric plasma-mediated immune checkpoint blockade therapy.* Proceedings of the National Academy of Sciences of the United States of America **2020**, 10.1073/pnas.1917891117.

54. Chen, G.; Chen, Z.; Wang, Z.; Obenchain, R.; Wen, D.; Li, H.; Wirz, R.E.; Gu, Z. *Portable air-fed cold atmospheric plasma device for postsurgical cancer treatment.* Science Advances **2021**, 10.1126/sciadv.abg5686.

55. Yoon, Y.; Ku, B.; Lee, K.; Jung, Y.J.; Baek, S.J. *Cold atmospheric plasma induces hmgb1 expression in cancer cells.* Anticancer Research **2019**, 10.21873/anticanres.13358.

56. Lin, A.; De Backer, J.; Quatannens, D.; Cuypers, B.; Verswyvel, H.; De La Hoz, E.C.; Ribbens, B.; Siozopoulou, V.; Van Audenaerde, J.; Marcq, E.; Lardon, F.; Laukens, K.; Vanlanduit, S.; Smits, E.; Bogaerts, A. *The effect of local non-thermal plasma therapy on the cancer-immunity cycle in a melanoma mouse model.* Bioengineering and Translational Medicine **2022**, 10.1002/btm2.10314.
57. Miebach, L.; Berner, J.; Bekeschus, S. *In ovo model in cancer research and tumor immunology.* Frontiers in Immunology **2022**, 10.3389/fimmu.2022.1006064.
58. Berner, J.; Miebach, L.; Herold, L.; Höft, H.; Gerling, T.; Mattern, P.; Bekeschus, S. *Gas flow shaping via novel modular nozzle system (monos) augments kinpen-mediated toxicity and immunogenicity in tumor organoids.* Cancers **2023**, 10.3390/cancers15041254
59. Mizuno, K.; Shirakawa, Y.; Sakamoto, T.; Ishizaki, H.; Nishijima, Y.; Ono, R. *Plasma-induced suppression of recurrent and reinoculated melanoma tumors in mice.* IEEE Transactions on Radiation and Plasma Medical Sciences **2018**, 10.1109/trpms.2018.2809673.
60. Lin, A.; Gorbanev, Y.; De Backer, J.; Van Loenhout, J.; Van Boxem, W.; Lemiere, F.; Cos, P.; Dewilde, S.; Smits, E.; Bogaerts, A. *Non-thermal plasma as a unique delivery system of short-lived reactive oxygen and nitrogen species for immunogenic cell death in melanoma cells.* Advanced Science **2019**, 10.1002/advs.201802062.
61. Mizuno, K.; Yonetamari, K.; Shirakawa, Y.; Akiyama, T.; Ono, R. *Anti-tumor immune response induced by nanosecond pulsed streamer discharge in mice.* Journal of Physics D-Applied Physics **2017**, 10.1088/1361-6463/aa5dbb.
62. Privat-Maldonado, A.; Schmidt, A.; Lin, A.; Weltmann, K.D.; Wende, K.; Bogaerts, A.; Bekeschus, S. *Ros from physical plasmas: Redox chemistry for biomedical therapy.* Oxidative Medicine and Cellular Longevity **2019**, 10.1155/2019/9062098.
63. Bekeschus, S.; Wende, K.; Hefny, M.M.; Rodder, K.; Jablonowski, H.; Schmidt, A.; Woedtke, T.V.; Weltmann, K.D.; Benedikt, J. *Oxygen atoms are critical in rendering thp-1 leukaemia cells susceptible to cold physical plasma-induced apoptosis.* Scientific Reports **2017**, 10.1038/s41598-017-03131-y.
64. Miebach, L.; Freund, E.; Clemen, R.; Kersting, S.; Partecke, L.I.; Bekeschus, S. *Gas plasma-oxidized sodium chloride acts via hydrogen peroxide in a model of peritoneal carcinomatosis.* Proceedings of the National Academy of Sciences of the United States of America (PNAS) **2022**, 10.1073/pnas.2200708119.
65. Freund, E.; Miebach, L.; Stope, M.B.; Bekeschus, S. *Hypochlorous acid selectively promotes toxicity and the expression of danger signals in human abdominal cancer cells.* Oncology Reports **2021**, 10.3892/or.2021.8022.
66. Adkins, I.; Fucikova, J.; Garg, A.D.; Agostinis, P.; Spisek, R. *Physical modalities inducing immunogenic tumor cell death for cancer immunotherapy.* Oncoimmunology **2014**, 10.4161/21624011.2014.968434.
67. Tanaka, M.; Kataoka, H.; Yano, S.; Sawada, T.; Akashi, H.; Inoue, M.; Suzuki, S.; Inagaki, Y.; Hayashi, N.; Nishie, H.; Shimura, T.; Mizoshita, T.; Mori, Y.; Kubota, E.; Tanida, S.; Takahashi, S.; Joh, T. *Immunogenic cell death due to a new photodynamic therapy (pdt) with glycoconjugated chlorin (g-chlorin).* Oncotarget **2016**, 10.18632/oncotarget.9725.
68. Kazama, H.; Ricci, J.E.; Herndon, J.M.; Hoppe, G.; Green, D.R.; Ferguson, T.A. *Induction of immunological tolerance by apoptotic cells requires caspase-dependent oxidation of high-mobility group box-1 protein.* Immunity **2008**, 10.1016/j.immuni.2008.05.013.
69. Skoberne, M.; Beignon, A.S.; Bhardwaj, N. *Danger signals: A time and space continuum.* Trends in Molecular Medicine **2004**, 10.1016/j.molmed.2004.04.001.
70. Feng, X.; Lin, T.; Chen, D.; Li, Z.; Yang, Q.; Tian, H.; Xiao, Y.; Lin, M.; Liang, M.; Guo, W.; Zhao, P.; Guo, Z. *Mitochondria-associated er stress evokes immunogenic cell death through the ros-perk-eif2alpha pathway under ptt/cdt combined therapy.* Acta Biomaterialia **2023**, 10.1016/j.actbio.2023.02.011.
71. Ahn, H.J.; Kim, K.I.; Hoan, N.N.; Kim, C.H.; Moon, E.; Choi, K.S.; Yang, S.S.; Lee, J.S. *Targeting cancer cells with reactive oxygen and nitrogen species generated by atmospheric-pressure air plasma.* PLoS One **2014**, 10.1371/journal.pone.0086173.
72. Choi, J.S.; Kim, J.; Hong, Y.J.; Bae, W.Y.; Choi, E.H.; Jeong, J.W.; Park, H.K. *Evaluation of non-thermal plasma-induced anticancer effects on human colon cancer cells.* Biomedical Optical Express **2017**, 10.1364/BOE.8.002649.
73. Conway, G.E.; Casey, A.; Milosavljevic, V.; Liu, Y.; Howe, O.; Cullen, P.J.; Curtin, J.F. *Non-thermal atmospheric plasma induces ros-independent cell death in u373mg glioma cells and augments the cytotoxicity of temozolomide.* British Journal of Cancer **2016**, 10.1038/bjc.2016.12.

74. Miebach, L.; Mohamed, H.; Wende, K.; Miller, V.; Bekeschus, S. *Pancreatic cancer cells undergo immunogenic cell death upon exposure to gas plasma-oxidized ringers lactate.* Cancers **2023**, 10.3390/cancers15010319.

75. Jeong, S.D.; Jung, B.K.; Ahn, H.M.; Lee, D.; Ha, J.; Noh, I.; Yun, C.O.; Kim, Y.C. *Immunogenic cell death inducing fluorinated mitochondria-disrupting helical polypeptide synergizes with pd-l1 immune checkpoint blockade.* Advanced Science **2021**, 10.1002/advs.202001308.

76. Oh, C.; Won, H.R.; Kang, W.S.; Kim, D.W.; Jung, S.N.; Im, M.A.; Liu, L.; Jin, Y.L.; Piao, Y.; Kim, H.J.; Kang, Y.E.; Lee, M.J.; Heo, J.Y.; Jun, S.; Sim, N.S.; Lee, J.H.; Song, K.; Kim, Y.I.; Chang, J.W.; Koo, B.S. *Head and neck cancer cell death due to mitochondrial damage induced by reactive oxygen species from nonthermal plasma-activated media: Based on transcriptomic analysis.* Oxidative Medicine and Cellular Longevity **2021**, 10.1155/2021/9951712.

77. Wei, J.C.J.; Edwards, G.A.; Martin, D.J.; Huang, H.; Crichton, M.L.; Kendall, M.A.F. *Allometric scaling of skin thickness, elasticity, viscoelasticity to mass for micro-medical device translation: From mice, rats, rabbits, pigs to humans.* Scientific Reports **2017**, 10.1038/s41598-017-15830-7.

78. Canady, J.; Cheng, X.; Zhuang, T.; Murthy, S.; Nissan, A.; Gitelis, S.; Ly, L.; Jones, O.; Adele, M.; Blank, A.T.; O'Donoghue, C.; Stenson, K.; O'Hara, K.; Keidar, M.; Basadonna, G. *The first cold atmospheric plasma phase i clinical trial for the treatment of advanced solid tumors: A novel 4th treatment arm for cancer.* Journal of Clinical Oncology **2023**, 10.1200/JCO.2023.41.16_suppl.e14596.

79. Galluzzi, L.; Vitale, I.; Warren, S.; Adjemian, S.; Agostinis, P.; Martinez, A.B.; Chan, T.A.; Coukos, G.; Demaria, S.; Deutsch, E.; Draganov, D.; Edelson, R.L.; Formenti, S.C.; Fucikova, J.; Gabriele, L.; Gaipl, U.S.; Gameiro, S.R.; Garg, A.D.; Golden, E.; Han, J.; Harrington, K.J.; Hemminki, A.; Hodge, J.W.; Hossain, D.M.S.; Illidge, T.; Karin, M.; Kaufman, H.L.; Kepp, O.; Kroemer, G.; Lasarte, J.J.; Loi, S.; Lotze, M.T.; Manic, G.; Merghoub, T.; Melcher, A.A.; Mossman, K.L.; Prosper, F.; Rekdal, O.; Rescigno, M.; Riganti, C.; Sistigu, A.; Smyth, M.J.; Spisek, R.; Stagg, J.; Strauss, B.E.; Tang, D.; Tatsuno, K.; van Gool, S.W.; Vandenabeele, P.; Yamazaki, T.; Zamarin, D.; Zitvogel, L.; Cesano, A.; Marincola, F.M. *Consensus guidelines for the definition, detection and interpretation of immunogenic cell death.* Journal of Immunotherapy in Cancer **2020**, 10.1136/jitc-2019-000337.

11 Anticancer Vaccination Approaches Using Plasma

Ramona Clemen, Sander Bekeschus

11.1 INTRODUCTION

The COVID-19 pandemic made us remember the importance of vaccination in promoting human health. During this time, societies were voluntarily or involuntarily engaged with the principal actions and different types of vaccines. During the lockdowns in many countries, it became clear that there is a great need for optimizing vaccination approaches to prepare for present and future threats. In addition to the now popular mRNA or vector vaccines for protection against COVID-19 infection, there are vaccine preparations from living or dead pathogens or their fragments, such as proteins. They all have the same goal: stimulating immune cells to induce an adaptive immune response. Various agents, such as adjuvants, enhance immune cell activation, which is crucial to generating an immune response and successful vaccination-induced protection based on trained immunity. Examples of adjuvants are bacterial proteins or oligonucleotides. The adjuvant dissolvent (e.g., oil), adding non-organic structures (e.g., aluminum hydroxide), or encapsulation of the vaccine (e.g., in lipid nanoparticles) can further increase the vaccine's immunogenicity [1]. The vaccines mentioned so far make the immune system aware of new structures and create cells with memory functions for recognizing the structure to fight the pathogen. In infection biology, one also speaks of immunization, through which an organism is protected against developing a severe course of a disease to be exposed to in the future.

Immunization is rarely used in the context of cancer prevention. This is because tumors are seldom characterized by one structure to which the vaccine can be generated but instead by DNA mutational heterogeneity or mutations that do not directly translate into immunogenic peptides. Exceptions are malignancies induced by a virus (e.g., human papillomavirus, responsible for the appearance of cervical and head and neck cancers) [2]. There are vaccines directed against "characteristic" tumor-associated antigens (TAA), neoantigens, or tumor-specific antigens (TSA)—proteins that are associated with certain tumors and are likely to be expressed [3]. In this context, therapeutic vaccination (i.e., vaccination performed when the disease is already present) can induce an antitumor immune response by stimulating and (re-)activating immune cells. Hence, this type of vaccine is given after cancer diagnosis and aims to promote immune cells that target tumor cells expressing specific peptides. To this end, a biopsy is taken to identify TAA or TSA for generating a personalized vaccine. Besides the vaccine preparations mentioned above, vaccines can also be tailored from cells and administered in a process called adoptive cell transfer therapy—for instance, genetically modified T-cells (e.g., CAR T-cells [4]) with specific receptors to recognize and attack cancer cells—,in which immune cells are isolated from the patient, modified (stimulated or transgenically modified) under sterile conditions, and injected back into the patient. In addition, peptide-loaded dendritic-cell-based (DC) vaccines generate an antitumor immune response [5]. However, this is a complex process requiring use of tools from molecular biology and bioinformatics, production of peptides in significant amounts, and feeding of peptides to patient-derived immune cells generated in large quantities in specific, pathogen-free laboratory environments before being injected back into the patient [6,7]. This is a costly process. Injecting peptides alone into patients without routing them through immune cells has been less successful [8].

DOI: 10.1201/9781003328056-13

DCs and other antigen-presenting cells (APCs) are essential in eliciting immune responses. Their function, in brief, is protein uptake and degradation to present antigens (usually peptides of the ingested protein) on the cell surface to adaptive immune cells, such as T-cells. If the T-cells express the T-cell receptor (TCR) specifically targeting this antigenic epitope, T-cell activation and differentiation occur to induce an immune response. Optimizations for all vaccine approaches are possible by promoting specific sets of immune cells and their pre-stimulation or by modifying the vaccine. We propose gas plasma technology to upgrade vaccines by increasing adjuvanticity and antigenicity due to oxidation.

11.2 PHYSICAL PLASMA FOR ENHANCING TUMOR VACCINATION RESPONSES

Cold physical plasma is known to oxidatively modify biomolecules, such as proteins and lipids, by simultaneously producing many reactive oxygen and nitrogen species (ROS/RNS). Along these lines, plasma treatment of cells generates oxidative modification as primary plasma effects followed by secondary responses on the cellular level by affecting their function and signaling [9]. As such, plasma treatment is a process within applied redox biology and—at moderate doses—within oxidative eustress [10]. Clinically, this is being taken advantage of in treating chronic wounds using medically certified plasma devices [11]. The exact mechanisms are unclear, but it seems plausible to assume that enhanced healing responses in plasma-treated wounds may be related to oxidatively modified biomolecules in the wound bed. Even without plasma treatment, an abundance of such modifications on hundreds of proteins are seen during acute surgical wound healing [12]. It is known that oxidative post-translational modifications (oxPTMs) are associated with autoimmunity following chronic inflammation [13,14], i.e., in the event of continuous inflammation-associated ROS/RNS production. Naturally, the question arises of whether this phenomenon might be useful in oncological therapy. One hurdle concerns the ability to produce several types of inflammation-related ROS/RNS simultaneously, as seen in the body. At this point, plasma technology fills the gap to employ multi-ROS production for generating vastly oxidized oxPTMs in tumor materials.

Based on this strategy, novel research lines in plasma vaccination were recently encouraged [15,16]. The idea is to use ROS/RNS as an adjuvant to increase biological structures' (proteins') immunogenicity and modulate immune cell activity. In 2016, Wang and colleagues used plasma technology to oxidize virus-vaccine for optimization [17]. Indeed, plasma treatment of Newcastle disease virus (NDV) led to deformation, and vaccinated chicken showed higher antibody titer than chicken vaccinated with a standard vaccine. However, additional evidence of successful protective anti-viral vaccination *in vivo* is lacking. Plasma application for optimizing vaccines has been envisioned as using plasma-treated proteins or tumor cells either as vaccine directly or given to dendritic cells for the cell-based vaccines. Another cell-based vaccine approach is plasma-treating dendritic cells before stimulation with antigens to increase their activity. We have recently found oxidative stress-induced human DCs activation *in vitro* to be profound to modest, depending on the oxidizing agent and cell surface molecule or release factor to be investigated [18]. The primary strategy, however, remains the induction of multiple oxPTMs directly via plasma processes of tumor material to increase its immunogenicity and, thereby, the alertness of DCs in the tissue to which the vaccine is given. Given the importance of tumor mutations dynamically changing throughout the disease course [19,20], it is plausible to apply such a vaccination approach therapeutically (i.e., during manifested disease) rather than prophylactically (i.e., before disease sets in).

There is only limited evidence of increased therapeutic vaccine efficiency after cold physical plasma treatment. In a study first presented in 2018 [21], we made a vaccine from plasma-treated murine melanoma cells that were killed through this plasma process or via immunogenic or non-immunogenic chemicals (positive controls) and subsequently injected subcutaneously into mice as a vaccine. After repeating this procedure, the mice received injections with live tumor cells. The higher the immunogenicity of the *in vitro* killing procedure, the better the vaccine elicited antitumor immunity. Plasma treatment was more protective than non-immunogenic cancer drugs but less immunogenic than a pro-immunogenic drug, suggesting that the oxidatively-induced cell death is attributable to a certain degree of immunogenicity [22]. Hence, oxidized tumor cell lysates can mediate protective effects, underlining previous reports in murine syngeneic mouse models of melanoma [23] and colorectal cancer [24]. Several lessons can be learned from these studies. First, applying the plasma-oxidized whole-tumor lysate vaccine enables immunoprotection from tumor growth. Second, the process suggests that using whole tumor cells or tissues and their lysates is more convenient and economical than identifying and using dedicated tumor-specific proteins or peptides. Third, provided that the immune systems successfully filters the immunogenic antigens from all antigens in such complex cell-derived protein lysates, it is feasible to assume that a larger pool of potential tumor antigens may simultaneously serve to provide greater immuno-stimulation than can a single or a few cancer-specific peptides, as administered in today's vaccine schemes in oncology. Critically, the plasma process provides oxidatively modified proteins to enhance antigen-presenting cell alertness for sufficient co-stimulation during antigen presentation for above-threshold T-cell activation signaling. However, compared to the *in vivo* studies described above, cell culture lysates differ from three-dimensional tumors in an organism due to the tumor cells' genetic and phenotyping heterogeneity and the tumor microenvironment's complexity. To this end, proof-of-concept studies demonstrating plasma-enhanced whole-tumor lysate immunogenicity in more complex model systems are needed.

Concerning the translational potential of this approach, plasma-assisted whole-tumor lysate oxidation and vaccine preparation potentially enable personalized precision medicine in clinical oncology. Specifically, a tumor biopsy would be taken from the patient and used to develop a plasma-optimized vaccine *in vitro*, which would be administered to the patient as a therapeutic vaccine (Figure 11.1) [16]. This tumor biopsy would allow alternative vaccination schemes in place of direct vaccine injection. For example, plasma-treated whole-tumor lysate can be used to pulse DCs. In the second step, these cells can be administered as cell-based vaccines. Controlling DC antigen uptake and expression of co-stimulatory molecules in the laboratory before the cells are injected back into the patient ensures their proper priming to elicit antitumor T-cell immunity. When directly injecting vaccines into the patient, it is not always clear whether DCs would be sufficiently activated above threshold to induce anticancer effects *in vivo*. However, plasma-treated whole-tumor lysate may overcome this threshold by different mechanisms. First, there is evidence that DCs prefer the uptake of structures that have been oxidized with chemically produced single-agent ROS/RNS (e.g., hypochlorous acid/HOCl) [25], and of plasma-oxidized proteins [26]. Recently, we studied the response of DCs on different plasma-oxidized proteins, where more than 85% of the proteins

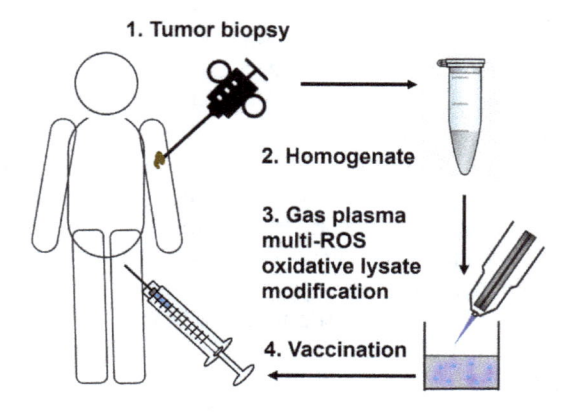

FIGURE 11.1   Simplified scheme of plasma-mediated multi-ROS-driven improvement of autologous whole tumor-lysate vaccines. Adapted from [16].

enhanced activation and maturation [27]. Second, DC vaccines based on cells pulsed with HOCl-oxidized tumor lysate are more efficient than DCs pulsed with non-oxidized cells and lead to reduced tumor burden [28–31]. Similar effects are also expected in DC vaccines after pulsing with plasma oxidized lysate, as DCs increase their activation and maturation after co-culture with plasma-killed tumor cells [32,33]. The activation and maturation of DCs are vital to activate lymphocytes, and it is a groundbreaking finding that plasma promotes this process. However, evidence of successful vaccination *in vivo* is lacking, and mechanisms behind increased immunogenicity of plasma-treated cells or proteins must be further deciphered along with optimal plasma treatment times, setups, gas phase ROS chemistries, and other parameters. In summary, such a vaccine could be based on two mechanisms that go hand in hand. On the one hand, plasma treatment induces cellular stress that makes a cancer cell present alarming signals on the cell surface—and more immunogenic (see next paragraph). On the other hand, plasma-generated ROS/RNS modify proteins and amino acids, which may act as alert signaling to APCs.

11.3 (PLASMA) OXIDIZED BIOMOLECULES ACTIVATE IMMUNE CELLS

When producing plasma-oxidized vaccines, the first question to consider is whether proteins shall be oxidized or cells modified to activate immune cells. This does not necessarily imply that plasma oxidatively modifies all biological structures, turning them into pathological ones. On the one hand, in viable cells, insoluble biological structures are anchored, and cells have antioxidants and repair mechanisms to ward off manageable oxidative stress [34]. On the other hand, oxPTMs are not alien, as immune cells produce ROS/RNS during any natural inflammatory response to kill pathogens, and oxPTMs on bacterial proteins alert immune cells. Furthermore, diseases characterized by chronic inflammation, such as diabetes, rheumatoid arthritis, chronic hyperglycemia, and atherosclerosis, are associated with oxPTM-bearing self-proteins. Indeed, auto-antibodies and auto-reactive T-cells against oxidized targets (such as insulin, collagen, or low-density lipoprotein/LDL) were observed in patients with autoimmune disorders, partly in a neoepitope-like fashion [35,36]. However, not all of the body's oxidized antigens are recognized as pathogenic structures since the site of initial recognition plays a role, and the induction of an immune response requires a complex interaction of different immune cells. For an autoimmune response, the corresponding antigen must be presented on cells in addition to randomly misprogrammed T-cells, drastically reducing the probability of the disease breaking out. Researchers have identified a pivotal role of ROS/RNS and oxPTMs in auto-immunity for decades, but current vaccine preparation strategies almost unanimously neglect this ancient evolutionary part of antigen modification.

When producing a whole tumor lysate as a vaccine, the time factor should be considered for activating signaling pathways that increase the expression of immunostimulatory molecules. Some anticancer agents trigger signaling cascades that ensure the presentation and secretion of immuno-stimulatory molecules, such as damage-associated molecular patterns (DAMP) [37]. These markers include, for example, calreticulin (CRT), high mobility group box 1 protein (HMGB1), heat shock proteins (HSP), and adenosine triphosphate (ATP), the energy currency of cells. In different tumor cells, plasma treatment augments the expression and secretion of DAMP molecules CRT and HMGB1 [38,39]. These DAMP molecules are characteristic of immunogenic cell death, and immune cells induce an immune response upon recognition. Plasma-treated tumors *in vivo* show increased infiltration of immune cells and activation of cytotoxic T-cells after direct exposure [22] or exposure to plasma-treated liquids [40]. Another immunogenic receptor raised in plasma-treated tumor cells is the major histocompatibility complex (MHC), which presents peptides of tumor anti-gens on the cell surface to activate T-cells. Interestingly, the spectra of peptides found on MHC were altered after plasma treatment [41], suggesting additional stimulation of interacting T-cells. T-cells' recognition of peptides on the MHC complex for activation proceeds briefly through a complex interaction between antigen-specific receptor (TCR) and the peptide-loaded MHC complex. In addition to the number of interacting receptors in a cell, the affinities between peptide-loaded MHC and TCR and the affinity of the peptide in the MHC binding groove also play important roles in crossing the threshold for T-cell activation. T-cells bearing high-affinity TCR directed against some TAAs are developmentally subjected to negative selection in the thymus and peripheral lymphoid organs and clonally deleted to avoid auto-immunity. In contrast, alternative antigens on tumor cells after plasma treatment or vaccine-generated T-cells with different TCR affinities may promote anti-tumor immune responses since the peptide-HLA binding form differs due to oxPTMs, as shown before with amino acid exchange and amino acid modification [42]. Still, there might be a risk of developing autoimmunity. In the context of cancer, vaccination with oxidized melanoma-associated antigens may induce vitiligo, but other immunoregulatory anticancer agents also turn an immune response upside down. Furthermore, vaccination with oxidized whole tumor lysate may contain oxidized self or near-to-self antigens that could lead to autoimmunity but potentially also TSA/TAAs that help reduce tumor burden. TSA and TAA trigger an immune response when recognized properly, and vaccination with an oxidized form ensures the activation of existing antigen-specific T-cells or generates new ones. The increased immunogenicity includes the response to randomly appearing TSA in oxidized whole tumor lysate vaccination (as described above) and oxPTM protein vaccine. Indeed, promising results were shown in a 15-year follow-up study of patients with breast cancer who received vaccination with oxidized breast cancer TSA (mannan-MUC1) and showed significantly fewer recurrences [43]. However, oxPTM on MUC1 was not investigated, although different amino acids are preferably oxidized and may be differentially recognized by immune cells. This gap is surprising because specific preferences for oxPTM have been investigated in detail in other research fields [44]. For instance, cysteine is affected by free oxygen radicals and can easily be converted into cystine. Furthermore, the oxidation of thiols to disulfides leads to disulfide bridges, resulting in structural changes. Seitz and colleagues used oxidation-induced intermolecular disulfide (S-S) bond formation to multimerize bacterial thioredoxin scaffold with an epitope of human papillomavirus (HPV) (TrxL2) to form a highly immunogenic aggregate. They showed that the reduced form led to a lower antibody titer in vaccinated animals [45]. Cold plasma-derived ROS/RNS also induce nitrosylation and oxidation of cysteine in peptides and proteins. Lackmann and colleagues tested the effect on cysteine after treatment with kINPen or COST-jet and found that plasma tuning not only by gas admixtures alone but by adjusting the surrounding atmosphere affected oxPTM [46]. Concerning vaccines, classifying amino acids' and proteins' potential oxidation sites helps identify which structures can provide a strong immune response. Recently, a study on the model protein ovalbumin served as a proof of concept study of the effectiveness of plasma-treated protein vaccine [26]. In brief, the protein was treated with two different gas compositions and injected into animals as a protective vaccine before they were challenged with viable tumor cells that express ovalbumin (Figures 11.2a-c). Animals that received the oxidized protein as vaccines showed less tumor growth than animals that received the native protein. Ovalbumin served as a model for a tumor-associated antigen but also showed that plasma can oxidatively modify proteins to increase their immunogenicity. After treatment with the different plasmas, oxidation, di- and trioxidation (on cysteines), and other modifications were found on ovalbumin (Figure 11.2d). Interestingly, one of the gas conditions (helium-oxygen) not only oxidized the protein but also led to chlorination and resulted in preferred uptake by DCs, elevated T-cell activation, and more effective immunization. Other studies confirm plasma's effect on protein activity, structure, and amino acids [47,48], but little on increased immunogenicity. Some studies have described altered binding affinity of antibodies to plasma-oxidzied proteins when performing enzyme-linked immunosorbent assay (ELISA) [49,50] and increased activity of antigen-presenting cells when incubated with oxidized proteins [27,51], but *in vivo* experiments are lacking. There are some challenges to consider when using plasma-generated ROS/RNS to induce oxPTM. First, the production, reaction kinetics, and specificity differ when using different plasma sources and gas compositions. Second, the measurement, identification, and analysis of oxPTMs require sophisticated infrastructure and bioinformatics. The complexity of the different spatial-temporal concentrations of each species along a plasma jet's axis also allows identification of chemical reactions. Subsequently, novel gas compositions lead to an enrichment of some types of ROS/RNS and a partial depletion of others to treat proteins

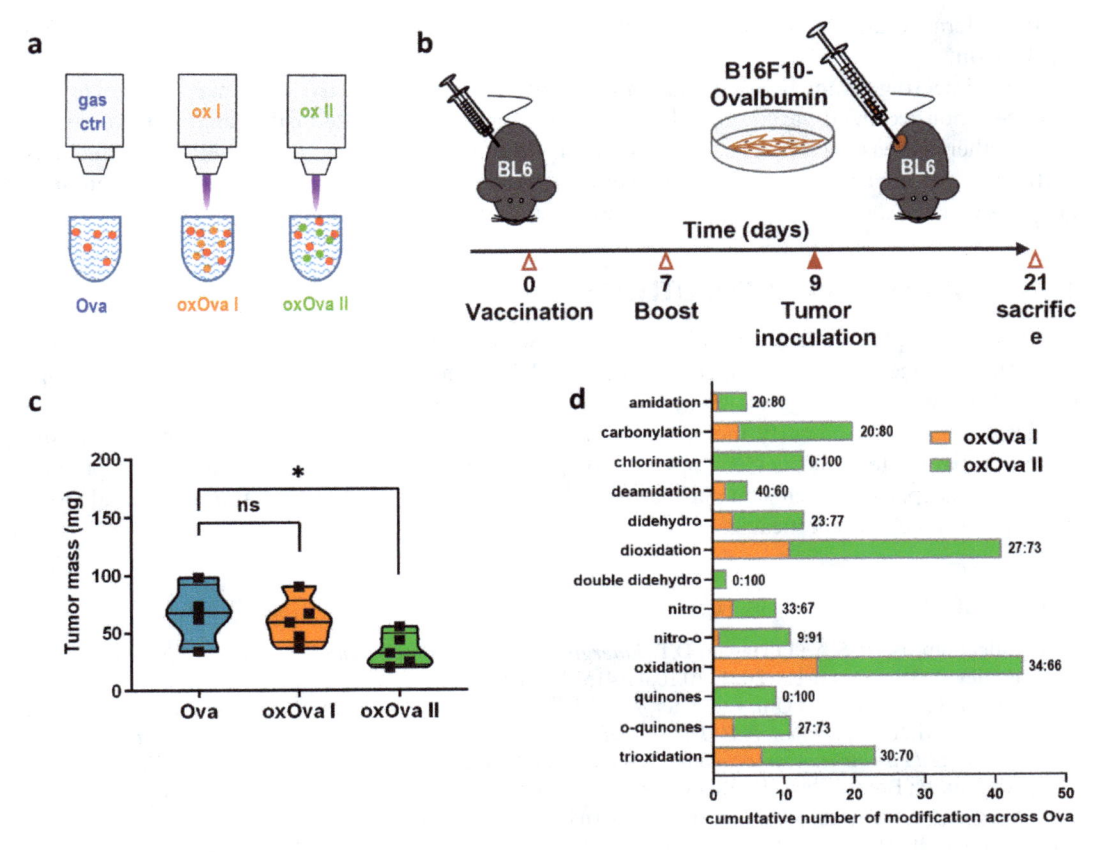

FIGURE 11.2 Argon plasma (ox I) and helium-oxygen plasma (ox II) treatment of chicken ovalbumin (Ova) lead to its oxidation (oxOva), which increases its immunogenicity. Adapted from [26].

dependent on their amino acid sequence to gain the most changes. Additional studies are needed to learn more about which plasma conditions lead to which oxPTMs and which may correlate to different extent of immunogenicity to optimize vaccination process.

11.4 AUTOIMMUNITY IN THE CONCEPT OF PLASMA-OXIDIZED BIOMOLECULES IN WHOLE TUMOR LYSATES

In contrast to the previously discussed immune response to separated oxidized proteins and peptides bound to MHC, the effect of oxPTM in the whole tumor lysate must be addressed. The human immune system recognizes endogenous substances as such and distinguishes them from those foreign to the body. This self-tolerance allows antigens to be identified as endogenous and tolerated by the immune cells, while oxidized proteins can break this tolerance. Once the whole tumor lysate is injected into the tissue, peripheral immune cells recognize the structures similarly to an injury. After APCs engulf extracellular proteins of tumor lysate, they re-present both native and oxidized structures for T-cells in the peripheral lymphoid organs. From this moment on, the T-cell's task is to classify the native and oxidized antigens. The diverse pool of T- and B-cells with an extensive receptor repertoire enables the classification of various antigens. Based on peripheral tolerance for self-antigens, T-cells decrease the immune response when recognizing self-antigens or get activated by altered binding affinity when identifying a foreign structure. By selection in the thymus, T-cells bind strongly to endogenous structures and are not released into the periphery to achieve central tolerance and avoid autoimmunity. However, autoreactive T-cells sometimes escape this mechanism.

When injecting tumor lysate, it cannot be ruled out that autoreactive cells may recognize and activate a non-oxidized variant of self-antigens throughout the lysate or that an oxidized variant may further increase activation. On the contrary, the immune system deals with many antigens daily, and the central and peripheral tolerance works adequately to distinguish between self-antigens and TSA.

11.5 CONCLUSION AND OUTLOOK

The concept of plasma-driven multi-oxidation of whole-tumor lysates for generating cancer vaccines that engage and alert antigen-presenting cells for improved mounting of anticancer adaptive immune responses suggests many possibilities that have been scarcely explored thus far. More studies are needed to identify the nature of the most immunogenic oxPTMs, to elucidate plasma treatment parameters and devices optimal for producing effective modifications, and to provide proof-of-concept experiments on plasma-treated cancer lysates from complex tumor material along with *in vivo* vaccination efficacies.

REFERENCES

[1] Pulendran, B.; P, S.A.; O'Hagan, D.T. *Emerging concepts in the science of vaccine adjuvants.* Nature Reviews: Drug Discovery **2021**, 10.1038/s41573-021-00163-y.

[2] Kruse, S.; Buchler, M.; Uhl, P.; Sauter, M.; Scherer, P.; Lan, T.C.T.; Zottnick, S.; Klevenz, A.; Yang, R.; Rosl, F.; Mier, W.; Riemer, A.B. *Therapeutic vaccination using minimal hpv16 epitopes in a novel mhc-humanized murine hpv tumor model.* Oncoimmunology **2019**, 10.1080/2162402X.2018.1524694.

[3] Alcazer, V.; Bonaventura, P.; Tonon, L.; Wittmann, S.; Caux, C.; Depil, S. *Neoepitopes-based vaccines: Challenges and perspectives.* European Journal of Cancer **2019**, 10.1016/j.ejca.2018.12.011.

[4] Bahmanyar, M.; Vakil, M.K.; Al-Awsi, G.R.L.; Kouhpayeh, S.A.; Mansoori, H.; Mansoori, Y.; Salahi, A.; Nikfar, G.; Tavassoli, A.; Behmard, E.; Moravej, A.; Ghasemian, A. *Opportunities and obstacles for the melanoma immunotherapy using t cell and chimeric antigen receptor t (car-t) applications: A literature review.* Molecular Biology Reports **2022**, 10.1007/s11033-022-07633-5.

[5] Mastelic-Gavillet, B.; Balint, K.; Boudousquie, C.; Gannon, P.O.; Kandalaft, L.E. *Personalized dendritic cell vaccines-recent breakthroughs and encouraging clinical results.* Frontiers in Immunology **2019**, 10.3389/fimmu.2019.00766.

[6] Butterfield, L.H. *Cancer vaccines.* BMJ **2015**, 10.1136/bmj.h988.

[7] Capietto, A.H.; Jhunjhunwala, S.; Delamarre, L. *Characterizing neoantigens for personalized cancer immunotherapy.* Current Opinion in Immunology **2017**, 10.1016/j.coi.2017.04.007.

[8] Calvo Tardon, M.; Allard, M.; Dutoit, V.; Dietrich, P.Y.; Walker, P.R. *Peptides as cancer vaccines.* Current Opinion in Pharmacology **2019**, 10.1016/j.coph.2019.01.007.

[9] Privat-Maldonado, A.; Schmidt, A.; Lin, A.; Weltmann, K.D.; Wende, K.; Bogaerts, A.; Bekeschus, S. *Ros from physical plasmas: Redox chemistry for biomedical therapy.* Oxidative Medicine and Cellular Longevity **2019**, 10.1155/2019/9062098.

[10] Sies, H.; Jones, D.P. *Reactive oxygen species (ros) as pleiotropic physiological signalling agents.* Nature Reviews: Molecular Cell Biology **2020**, 10.1038/s41580-020-0230-3.

[11] Bekeschus, S.; von Woedtke, T.; Emmert, S.; Schmidt, A. *Medical gas plasma-stimulated wound healing: Evidence and mechanisms.* Redox Biology **2021**, 10.1016/j.redox.2021.102116.

[12] Bekeschus, S.; Lackmann, J.W.; Gumbel, D.; Napp, M.; Schmidt, A.; Wende, K. *A neutrophil proteomic signature in surgical trauma wounds.* International Journal of Molecular Sciences **2018**, 10.3390/ijms19030761.

[13] Strollo, R.; Vinci, C.; Arshad, M.H.; Perrett, D.; Tiberti, C.; Chiarelli, F.; Napoli, N.; Pozzilli, P.; Nissim, A. *Antibodies to post-translationally modified insulin in type 1 diabetes.* Diabetologia **2015**, 10.1007/s00125-015-3746-x.

[14] Smallwood, M.J.; Nissim, A.; Knight, A.R.; Whiteman, M.; Haigh, R.; Winyard, P.G. *Oxidative stress in autoimmune rheumatic diseases.* Free Radical Biology and Medicine **2018**, 10.1016/j.freeradbiomed.2018.05.086.

[15] Mohamed, H.; Esposito, R.A.; Kutzler, M.A.; Wigdahl, B.; Krebs, F.C.; Miller, V. *Nonthermal plasma as part of a novel strategy for vaccination*. Plasma Process Polym **2020**, 10.1002/ppap.202000051.

[16] Clemen, R.; Bekeschus, S. *Ros cocktails as an adjuvant for personalized antitumor vaccination?* Vaccines **2021**, 10.3390/vaccines9050527.

[17] Wang, G.; Zhu, R.; Yang, L.; Wang, K.; Zhang, Q.; Su, X.; Yang, B.; Zhang, J.; Fang, J. *Non-thermal plasma for inactivated-vaccine preparation*. Vaccine **2016**, 10.1016/j.vaccine.2015.10.099.

[18] Bekeschus, S.; Meyer, D.; Arlt, K.; von Woedtke, T.; Miebach, L.; Freund, E.; Clemen, R. *Argon plasma exposure augments costimulatory ligands and cytokine release in human monocyte-derived dendritic cells*. International Journal of Molecular Sciences **2021**, 10.3390/ijms22073790.

[19] Jia, Q.; Chiu, L.; Wu, S.; Bai, J.; Peng, L.; Zheng, L.; Zang, R.; Li, X.; Yuan, B.; Gao, Y.; Wu, D.; Li, X.; Wu, L.; Sun, J.; He, J.; Robinson, B.W.S.; Zhu, B. *Tracking neoantigens by personalized circulating tumor DNA sequencing during checkpoint blockade immunotherapy in non-small cell lung cancer*. Adv Sci (Weinh) **2020**, 10.1002/advs.201903410.

[20] Montellier, E.; Gaucher, J. *Targeting the interplay between metabolism and epigenetics in cancer*. Current Opinion in Oncology **2019**, 10.1097/CCO.0000000000000501.

[21] Bekeschus, S.; von Woedtke, T.; Weltmann, K.-D.; Metelmann, H.-R. *Plasma, cancer, immunity*. Clinical Plasma Medicine **2018**.

[22] Bekeschus, S.; Clemen, R.; Niessner, F.; Sagwal, S.K.; Freund, E.; Schmidt, A. *Medical gas plasma jet technology targets murine melanoma in an immunogenic fashion*. Advanced Science **2020**, 10.1002/advs.201903438.

[23] Lin, A.; Gorbanev, Y.; De Backer, J.; Van Loenhout, J.; Van Boxem, W.; Lemiere, F.; Cos, P.; Dewilde, S.; Smits, E.; Bogaerts, A. *Non-thermal plasma as a unique delivery system of short-lived reactive oxygen and nitrogen species for immunogenic cell death in melanoma cells*. Adv Sci (Weinh) **2019**, 10.1002/advs.201802062.

[24] Lin, A.G.; Xiang, B.; Merlino, D.J.; Baybutt, T.R.; Sahu, J.; Fridman, A.; Snook, A.E.; Miller, V. *Non-thermal plasma induces immunogenic cell death in vivo in murine ct26 colorectal tumors*. Oncoimmunology **2018**, 10.1080/2162402X.2018.1484978.

[25] Prokopowicz, Z.M.; Arce, F.; Biedron, R.; Chiang, C.L.; Ciszek, M.; Katz, D.R.; Nowakowska, M.; Zapotoczny, S.; Marcinkiewicz, J.; Chain, B.M. *Hypochlorous acid: A natural adjuvant that facilitates antigen processing, cross-priming, and the induction of adaptive immunity*. Journal of Immunology **2010**, 10.4049/jimmunol.0902606.

[26] Clemen, R.; Freund, E.; Mrochen, D.; Miebach, L.; Schmidt, A.; Rauch, B.H.; Lackmann, J.W.; Martens, U.; Wende, K.; Lalk, M.; Delcea, M.; Broker, B.M.; Bekeschus, S. *Gas plasma technology augments ovalbumin immunogenicity and ot-ii t cell activation conferring tumor protection in mice*. Advanced Science **2021**, 10.1002/advs.202003395.

[27] Clemen, R.; Arlt, K.; Miebach, L.; von Woedtke, T.; Bekeschus, S. *Oxidized proteins differentially affect maturation and activation of human monocyte-derived cells*. Cells **2022**, 10.3390/cells11223659.

[28] Sarivalasis, A.; Boudousquie, C.; Balint, K.; Stevenson, B.J.; Gannon, P.O.; Iancu, E.M.; Rossier, L.; Martin Lluesma, S.; Mathevet, P.; Sempoux, C.; Coukos, G.; Dafni, U.; Harari, A.; Bassani-Sternberg, M.; Kandalaft, L.E. *A phase i/ii trial comparing autologous dendritic cell vaccine pulsed either with personalized peptides (pep-dc) or with tumor lysate (oc-dc) in patients with advanced high-grade ovarian serous carcinoma*. Journal of Translational Medicine **2019**, 10.1186/s12967-019-02133-w.

[29] Tanyi, J.L.; Bobisse, S.; Ophir, E.; Tuyaerts, S.; Roberti, A.; Genolet, R.; Baumgartner, P.; Stevenson, B.J.; Iseli, C.; Dangaj, D.; Czerniecki, B.; Semilietof, A.; Racle, J.; Michel, A.; Xenarios, I.; Chiang, C.; Monos, D.S.; Torigian, D.A.; Nisenbaum, H.L.; Michielin, O.; June, C.H.; Levine, B.L.; Powell, D.J., Jr.; Gfeller, D.; Mick, R.; Dafni, U.; Zoete, V.; Harari, A.; Coukos, G.; Kandalaft, L.E. *Personalized cancer vaccine effectively mobilizes antitumor t cell immunity in ovarian cancer*. Science Translational Medicine **2018**, 10.1126/scitranslmed.aao5931.

[30] Zhou, R.; Huang, W.J.; Ma, C.; Zhou, Y.; Yao, Y.Q.; Wang, Y.X.; Gou, L.T.; Yi, C.; Yang, J.L. *Hocl oxidation-modified ct26 cell vaccine inhibits colon tumor growth in a mouse model*. Asian Pacific Journal of Cancer Prevention **2012**, 10.7314/apjcp.2012.13.8.4037.

[31] Chiang, C.L.; Kandalaft, L.E.; Tanyi, J.; Hagemann, A.R.; Motz, G.T.; Svoronos, N.; Montone, K.; Mantia-Smaldone, G.M.; Smith, L.; Nisenbaum, H.L.; Levine, B.L.; Kalos, M.; Czerniecki, B.J.; Torigian, D.A.; Powell, D.J., Jr.; Mick, R.; Coukos, G. *A dendritic cell vaccine pulsed with autologous hypochlorous acid-oxidized ovarian cancer lysate primes effective broad antitumor immunity: From bench to bedside*. Clin Cancer Res **2013**, 10.1158/1078-0432.CCR-13-1185.

[32] Chen, G.; Chen, Z.; Wang, Z.; Obenchain, R.; Wen, D.; Li, H.; Wirz, R.E.; Gu, Z. *Portable air-fed cold atmospheric plasma device for postsurgical cancer treatment.* Science Advances **2021**, 10.1126/sciadv. abg5686.

[33] Tomic, S.; Petrovic, A.; Puac, N.; Skoro, N.; Bekic, M.; Petrovic, Z.L.; Colic, M. *Plasma-activated medium potentiates the immunogenicity of tumor cell lysates for dendritic cell-based cancer vaccines.* Cancers **2021**, 10.3390/cancers13071626.

[34] Aquilano, K.; Baldelli, S.; Ciriolo, M.R. *Glutathione: New roles in redox signaling for an old antioxidant.* Frontiers in Pharmacology **2014**, 10.3389/fphar.2014.00196.

[35] Burska, A.N.; Hunt, L.; Boissinot, M.; Strollo, R.; Ryan, B.J.; Vital, E.; Nissim, A.; Winyard, P.G.; Emery, P.; Ponchel, F. *Autoantibodies to posttranslational modifications in rheumatoid arthritis.* Mediators of Inflammation **2014**, 10.1155/2014/492873.

[36] Strollo, R.; Vinci, C.; Napoli, N.; Fioriti, E.; Maddaloni, E.; Akerman, L.; Casas, R.; Pozzilli, P.; Ludvigsson, J.; Nissim, A. *Antibodies to oxidized insulin improve prediction of type 1 diabetes in children with positive standard islet autoantibodies.* Diabetes/Metabolism Research and Reviews **2019**, 10.1002/dmrr.3132.

[37] Garg, A.D.; Galluzzi, L.; Apetoh, L.; Baert, T.; Birge, R.B.; Bravo-San Pedro, J.M.; Breckpot, K.; Brough, D.; Chaurio, R.; Cirone, M.; Coosemans, A.; Coulie, P.G.; De Ruysscher, D.; Dini, L.; de Witte, P.; Dudek-Peric, A.M.; Faggioni, A.; Fucikova, J.; Gaipl, U.S.; Golab, J.; Gougeon, M.L.; Hamblin, M.R.; Hemminki, A.; Herrmann, M.; Hodge, J.W.; Kepp, O.; Kroemer, G.; Krysko, D.V.; Land, W.G.; Madeo, F.; Manfredi, A.A.; Mattarollo, S.R.; Maueroder, C.; Merendino, N.; Multhoff, G.; Pabst, T.; Ricci, J.E.; Riganti, C.; Romano, E.; Rufo, N.; Smyth, M.J.; Sonnemann, J.; Spisek, R.; Stagg, J.; Vacchelli, E.; Vandenabeele, P.; Vandenberk, L.; Van den Eynde, B.J.; Van Gool, S.; Velotti, F.; Zitvogel, L.; Agostinis, P. *Molecular and translational classifications of damps in immunogenic cell death.* Frontiers in Immunology **2015**, 10.3389/fimmu.2015.00588.

[38] Van Loenhout, J.; Flieswasser, T.; Freire Boullosa, L.; De Waele, J.; Van Audenaerde, J.; Marcq, E.; Jacobs, J.; Lin, A.; Lion, E.; Dewitte, H.; Peeters, M.; Dewilde, S.; Lardon, F.; Bogaerts, A.; Deben, C.; Smits, E. *Cold atmospheric plasma-treated pbs eliminates immunosuppressive pancreatic stellate cells and induces immunogenic cell death of pancreatic cancer cells.* Cancers **2019**, 10.3390/cancers11101597.

[39] Mahdikia, H.; Saadati, F.; Freund, E.; Gaipl, U.S.; Majidzadeh, A.K.; Shokri, B.; Bekeschus, S. *Gas plasma irradiation of breast cancers promotes immunogenicity, tumor reduction, and an abscopal effect in vivo.* Oncoimmunology **2020**, 10.1080/2162402X.2020.1859731.

[40] Freund, E.; Liedtke, K.R.; van der Linde, J.; Metelmann, H.R.; Heidecke, C.D.; Partecke, L.I.; Bekeschus, S. *Physical plasma-treated saline promotes an immunogenic phenotype in ct26 colon cancer cells in vitro and in vivo.* Scientific Reports **2019**, 10.1038/s41598-018-37169-3.

[41] Mohamed, H.; Clemen, R.; Freund, E.; Lackmann, J.W.; Wende, K.; Connors, J.; Haddad, E.K.; Dampier, W.; Wigdahl, B.; Miller, V.; Bekeschus, S.; Krebs, F.C. *Non-thermal plasma modulates cellular markers associated with immunogenicity in a model of latent hiv-1 infection.* PLoS One **2021**, 10.1371/journal. pone.0247125.

[42] Webb, A.I.; Dunstone, M.A.; Chen, W.; Aguilar, M.I.; Chen, Q.; Jackson, H.; Chang, L.; Kjer-Nielsen, L.; Beddoe, T.; McCluskey, J.; Rossjohn, J.; Purcell, A.W. *Functional and structural characteristics of ny-eso-1-related hla a2-restricted epitopes and the design of a novel immunogenic analogue.* Journal of Biological Chemistry **2004**, 10.1074/jbc.M314066200.

[43] Vassilaros, S.; Tsibanis, A.; Tsikkinis, A.; Pietersz, G.A.; McKenzie, I.F.; Apostolopoulos, V. *Up to 15-year clinical follow-up of a pilot phase iii immunotherapy study in stage ii breast cancer patients using oxidized mannan-muc1.* Immunotherapy **2013**, 10.2217/imt.13.126.

[44] Davies, M.J. *Protein oxidation and peroxidation.* Biochemical Journal **2016**, 10.1042/BJ20151227.

[45] Seitz, H.; Dantheny, T.; Burkart, F.; Ottonello, S.; Muller, M. *Influence of oxidation and multimerization on the immunogenicity of a thioredoxin-l2 prophylactic papillomavirus vaccine.* Clinical and Vaccine Immunology **2013**, 10.1128/CVI.00195-13.

[46] Lackmann, J.W.; Bruno, G.; Jablonowski, H.; Kogelheide, F.; Offerhaus, B.; Held, J.; Schulz-von der Gathen, V.; Stapelmann, K.; von Woedtke, T.; Wende, K. *Nitrosylation vs. Oxidation—how to modulate cold physical plasmas for biological applications.* PLoS One **2019**, 10.1371/journal.pone.0216606.

[47] Krewing, M.; Stepanek, J.J.; Cremers, C.; Lackmann, J.W.; Schubert, B.; Muller, A.; Awakowicz, P.; Leichert, L.I.O.; Jakob, U.; Bandow, J.E. *The molecular chaperone hsp33 is activated by atmospheric-pressure plasma protecting proteins from aggregation.* Journal of The Royal Society Interface **2019**, 10.1098/rsif.2018.0966.

[48] Han, Y.X.; Cheng, J.H.; Sun, D.W. *Changes in activity, structure and morphology of horseradish peroxidase induced by cold plasma.* Food Chemistry **2019**, 10.1016/j.foodchem.2019.125240.

[49] Ng, S.W.; Lu, P.; Rulikowska, A.; Boehm, D.; O'Neill, G.; Bourke, P. *The effect of atmospheric cold plasma treatment on the antigenic properties of bovine milk casein and whey proteins.* Food Chemistry **2021**, 10.1016/j.foodchem.2020.128283.

[50] Zhang, Q.; Cheng, Z.; Zhang, J.; Nasiru, M.M.; Wang, Y.; Fu, L. *Atmospheric cold plasma treatment of soybean protein isolate: Insights into the structural, physicochemical, and allergenic characteristics.* Journal of Food Science **2021**, 10.1111/1750-3841.15556.

[51] Clemen, R.; Arlt, K.; von Woedtke, T.; Bekeschus, S. *Gas plasma protein oxidation increases immunogenicity and human antigen-presenting cell maturation and activation.* Vaccines (Basel) **2022**, 10.3390/vaccines10111814.

12 Plasma Cancer Treatment in Combination With Other Physical Modalities

Audrey Glory, Rodin Chermat

12.1 RADIOTHERAPY

Radiotherapy is the use of ionizing radiation to treat cancer. Briefly, ionizing radiation can be broken down into gamma radiation (photons), beta radiation (neutrons), and alpha radiation (helium atoms), with photons being the most used in clinical radiotherapy [4]. Additionally, particle radiotherapy (i.e., the use of beams of ionized atoms such as carbon) is an area of active investigation [5]. Although "radiotherapy" is often employed as a generic term, clinical RT sources can be divided into two broad categories: external beam radiotherapy (EBRT) and brachytherapy (BT). External beam radiotherapy is the most common form of clinical RT, in which patients are placed in or near the RT applicator, and their tumor is subjected to a radiation beam. Current trends in EBRT revolve around increased stereotactic targeting of tumor tissue and reliance on image guidance to ensure precise dose distribution across the geometry of the tumor [4,6–9]. In contrast to EBRT, BT consists of irradiating the tumor from the inside and shares with stereotactic radiotherapy the ability to impose a precise focal point and to fine-tune dose distribution to a tumor [10]. Both types of radiotherapy employ a variety of sources, dose rates, and energies, all of which find their application in cancer treatment depending on the tumor type and localization, but also on the adopted treatment strategy [10–12]. Recently, new modalities and techniques for both categories of RT have also been making their way into pre-clinical and clinical settings [13–16]. Finally, radiopharmaceuticals are a special class of radioisotope-carrying drugs that can be used for image-guidance and/or treatment of a tumor [17]. Therapeutic radiopharmaceuticals can take many forms (proteins, antibodies, peptides, small molecules) and carry radioisotopes that release high-energy alpha or beta particles. They are designed to target cancer cells while sparing normal tissue and to rapidly clear from blood and tissue [17,18].

Regardless of the type of radiation (electromagnetic or particle), the damaging effects of ionizing radiation are caused by the ejection of electrons from molecules within the cell, i.e., their ionization [4]. Although the radiation has sufficient energy to induce "direct" damage by ionization, most molecular damage is "indirect" and caused by a cascade of ejected electrons ionizing molecules along their path. Collisions and subsequent ionizations become more frequent as the speed of the electrons decreases, which leads to the formation of clusters of ionizations [4]. Most of the energy deposited by RT is initially absorbed by water, as water molecules account for around 80% of the composition of the cell [4,19]. Ionization of water leads to the rapid production of reactive radical intermediates that, in turn, interact with other molecules such as DNA, causing the aforementioned "indirect effects". DNA, the largest molecule within the cell, is a critical target for RT. Unlike other cellular molecules, DNA exists only in two copies with a limited turnover and is essential for cellular functions. Thus, when clusters of ionizations are formed in DNA, breaking chemical bonds and disrupting its structure and function, they eventually lead to cell death [4]. Of the reactive radical species, the most damaging is the hydroxyl radical ($\cdot$OH), which can produce a radical on DNA (DNA$\cdot$) when in range, potentially leading to sustained DNA damage [19,20]. Indeed, DNA$\cdot$ immediately enters a competition for reduction or oxidation. Its reduction by glutathione or cysteine (thiol -SH containing

 DOI: 10.1201/9781003328056-14

molecules) restores the DNA to its original form, while its oxidation by oxygen permanently fixes the damage [4,19,20]. Biological effects of RT vary depending on key characteristics of the cell but also of the radiation itself. For example, radiation types can differ in their linear energy transfer (LET), the average energy per unit of distance (usually keV/μm) deposited on their trajectory. Photons are described as low-LET, with a LET of around 0.3 keV/μm, while α-particles and their 100 keV/μm LET are a good example of "high-LET" radiation. Higher LET directly implies higher density of displaced electrons across the trajectory of the particle and, thus, higher numbers and larger clusters of DNA damage within the cell. Additionally, direct interactions are more probable than indirect in high-LET radiation, leading to less dependency on oxygenation to cause lasting DNA damage [4].

In mammalian cells, DNA damage triggers a complex system of pathways known as the DNA damage response (DDR), which directly determines the cellular outcome after irradiation. Briefly, DDR consists of DNA damage sensing and subsequent transducing of signals to engage one of three effector pathways: the repair of proteins, cell death mechanisms, or cell-cycle checkpoints [21]. Overall, radiation-induced DNA damage will result either in direct cell killing (apoptosis, autophagy, or necrosis) or, more often, in indirect cell killing, either by halting cell division (cellular senescence) or through mitotic catastrophe [4,22]. Although senescence (the state of a cell that has permanently lost its ability to divide) technically qualifies as cell-cycle arrest, it is listed as a form of cell death in the context of irradiation due to the effective absence of cell division, equivalent to clonogenic cell death [23,24]. As previously mentioned, outcomes of irradiation depend not only on radiation parameters but also on cellular context. As such, all known forms of cell death have been observed after irradiation and are most often classified as either "early" cell death, meaning pre-mitosis, or "late" cell death, meaning post-mitosis. Pre-mitotic cell death occurs rapidly after irradiation and in general is rarely observed in solid cancers but rather is confined to specific cell types and their associated cancers (e.g., lymphomas). In most cases, however, irradiated cells will attempt mitosis despite suffering from extensive DNA damage, progressively leading to mitotic catastrophe [4]. Mitotic catastrophe is characterized by the presence of micronuclei and chromosomal aberrations within the cells, which can trigger the previously described cell deaths after a few rounds of replication [22,25,26]. Additionally, a form of cell death called "bystander" death has been described and consists of cell death due to the irradiation (and death) of a neighboring cell. However, the exact mechanism and importance of this bystander death have yet to be determined [4,27,28].

Like any other anticancer therapy, RT is used mostly in combination with other treatments. Besides chemotherapy [29–31], RT can be applied in combination with immunotherapy: low doses of RT can modify the tumor microenvironment to render it pro-immunogenic and, therefore, potentiate the immunotherapy treatment, and RT is known to induce immunogenic cell death in cancer cells [32,33]. Another example is the application of RT during surgery, called intraoperative RT, which has been proven to help reduce the risk of local recurrences [34–37].

Only a few studies have researched the combination of RT and CAP. Lin et al. have tested a combination of CAP and RT (2 Gy) on lung, cervical, and liver cancer cells and have found that CAP sensitizes the cells to the subsequent RT treatment, with the combination leading to more apoptosis, the production of more RONS and the accumulation of cells in the G2/M phase of the cell cycle, as well as an increase in DNA double-strand breaks (DSB) [38]. They have tested the combination of CAP and RT on liver tumors *in vivo* and have found it to have an additive effect on the reduction of tumor growth [38]. The radio-sensitization of cells by CAP was also proven by Lafontaine et al. on a panel of 7 breast cancer cell lines [39]. As with the individual treatments, the cell lines did not exhibit similar sensitivity to the combination of CAP and RT (4 Gy), independently of their characteristics (such as receptor status). Interestingly, the sensitivity of a cell line to RT correlated with its sensitivity to CAP, which provides a useful tool for CAP researchers to select cell lines that could be sensitive or resistant to CAP treatment. The combination of both CAP and RT with the Poly(ADP-ribose) polymerase (PARP) inhibitor Olaparib showed that their shared ability to produce DNA damage could lead to the clinical use of either CAP, RT, or a combination of both that would allow for the reduction of RT dose without compromising efficacy [39]. Moniruzzaman et al. have also shown the combination of CAP and RT (5 Gy) to be additively cytotoxic in monocytes [40]. Finally, Pasqual-Melo et al. have used a combination of CAP and RT as a single treatment of 2 or 8 Gy or as three treatments at these doses, 24h apart. They have found that the association of CAP with multiple low doses (2 Gy) of RT had the best additive effect and led to increased toxicity in a murine melanoma cell line and increased immunostimulatory potential [41]. In these four studies, CAP was used as a sensitizer to RT and was applied 24h or immediately before RT treatment [38–41]. In one case, the combination of CAP and RT did not improve their individual cytotoxic effects, but the dose of RT used was very low (1 Gy) [42]. As demonstrated in these studies, the rationale for the combination of CAP and RT is strong. The main cause of cytotoxicity from CAP treatment is the RONS that are produced in the CAP and will accumulate in the cells [43,44]. This accumulation will result in an oxidative environment that contributes to the fixation of DNA• and, therefore, of the damage that RT causes to the DNA. CAP will also activate apoptotic pathways in the cells that will lead to additional DNA damage [45] (Figure 12.1). Collet et al. have found that CAP is capable of increasing blood flow and tissue oxygenation *in vivo*, a mechanism that could sensitize cells to RT, fix RT-induced DNA damage, and help eliminate radio-resistant cells in hypoxic regions [46].

12.2 HYPERTHERMIA

Hyperthermia (HT) is a non-invasive technique that uses heat (39°C-45°C) to kill cancer cells while having a minimal impact on normal tissues [47]. Indeed, cancer cells are more sensitive to changes in temperature than normal cells, in part because heat shock proteins (Hsps), chaperones that help in both the refolding of proteins and resistance to other heat-related damage, are less present in cancer cells [48]. The irregular vasculature of a tumor also explains the sensitivity of cancer cells to heat, as inadequate vascularization means that the excess heat cannot be dissipated via the blood flow [49]. Depending on the dose used, HT can also cause the collapse of the tumor's vasculature, leading to

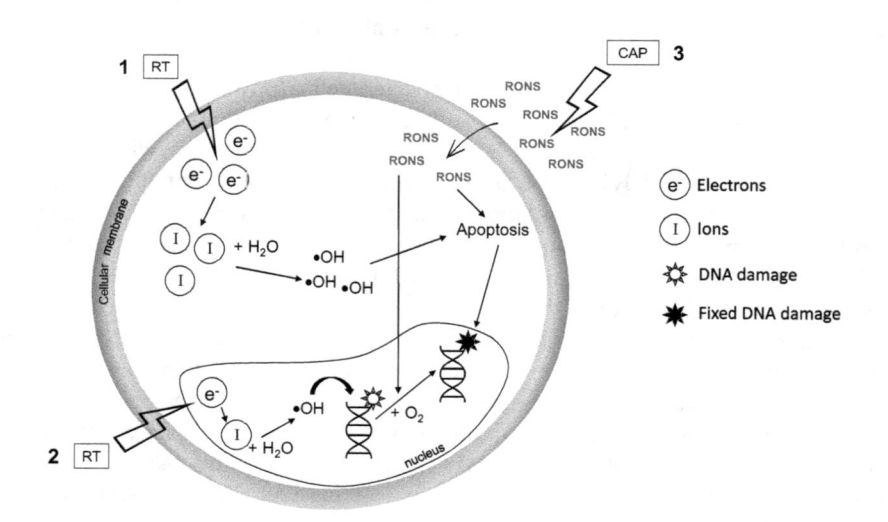

FIGURE 12.1   Simplified schematic representation of the combination of CAP and RT at the cellular level. 1. The ionizing radiation from RT induces a cascade of electrons (e-) and ions (I) that will react with water (H₂O) to produce the hydroxyl radical (·OH). ·OH will oxidize diverse cell components, possibly leading to apoptosis. 2. This cascade can also happen in the nucleus and cause damage to the DNA. In the presence of oxygen (O₂), such damage will become permanent, and the accumulation of DNA damage could lead to cell death. 3. CAP produces RONS that penetrate the cells. These RONS can be a source of oxygen to help fix the DNA damage caused by RT. They can also activate the pathways to apoptosis, resulting in more DNA damage and precipitating the cell to its death. The combination of these treatments allows a decrease in the dose required for therapeutic response and, therefore, in the risks of side effects. RT: Radiotherapy; RONS: Reactive oxygen and nitrogen species; CAP: Cold atmospheric plasma.

a decrease in oxygen and nutrients and an increase in pH that will render the neighboring cells even more sensitive to heat shocks [49]. The use of hyperthermia in patients presents very few side effects, mainly mild burns, fever, and vomiting [50,51]. HT causes the death of cancer cells by the denaturation and aggregation of proteins, as well as the inactivation of the synthesis and repair mechanisms of DNA. These changes are accompanied by an increase in the cells' metabolism, leading to an accumulation of RONS, damage to proteins and lipids, and eventually cell death [52,53]. HT can activate the three main pathways to apoptosis: the death receptors pathway [54], the endoplasmic reticulum (ER) pathway [55], and the mitochondrial pathway [56]. HT can also lead to cell death by mitotic catastrophe, which primarily affects cells with a high rate of division, such as cancer cells[57].

Just like plasma, heat can be used for ablation or at therapeutic doses. The two most frequent sources of heat used in the clinic are radiofrequencies and ultrasound, which can be used for both purposes. High-intensity focalized ultrasound (HIFU) is used to precisely heat a tumor but can also be used to permeabilize tumor cells, thanks to the creation of gas bubbles that oscillate and disrupt the cellular membranes [58–60].

Depending on the localization of the tumor, different types of HT can be used to ensure maximal tumor coverage. Local HT (41–45°C) is focused on the tumor, thanks to imaging techniques, and uses heat sources such as microwaves and ultrasound. Probes can also be used for tumors located in body cavities. Regional HT (41–45°C) is applied to an entire organ/tissue, with heat sources disposed all around the target. Liquids can also be treated: blood can be heated and combined with a cytotoxic agent before being reinjected in the patient, or chemotherapeutic-containing liquids can be heated and used to immerse the peritoneal cavity during hyperthermic intraperitoneal chemotherapy (HIPEC). Global HT (<42°C) is the heating of the entire body to treat metastatic cancers. Water baths, infrared lights, and heating pads can be used. This technique often requires the sedation of the patient and leads to more side effects as the treatment is not localized to the tumor [50,51,61]. Intracellular HT has been developed using different techniques: magnetic nanoparticles can be locally exposed to an oscillating magnetic field to create heat, or nanoparticles with photothermal agents (PTA) can be released and locally excited by light (usually near-infrared) to create heat [62]. These treatments can be combined with heat-sensitive nanoparticles for the local release of drugs [63,64].

HT can be combined with chemotherapy [65,66] and synergize with chemotherapeutic agents to treat resistant cells [67]. Depending on the dose used, HT can increase the blood flow inside the tumor to facilitate the transport of drugs and increase the presence of oxygen. HT is therefore considered a radio-sensitizer, as a higher presence of oxygen will create an oxidative environment in the cells and fix the damage caused by RT[68]. HT is capable of eliminating cancer cells in the center of the tumor in hypoxic and acidic conditions, where the cells are usually radio-resistant [48]. The combination of RT and HT has been shown to increase the rates of local control and complete remission in recurrent breast cancer patients [69], and HT has been found to inhibit radiation damage recovery [70]. HT can also be used in combination with immunotherapy, as only heated cancer cells (and not normal cells) will present Hsps on their surface, thereby becoming a target for monoclonal antibodies or dendritic cell therapies [71].

The combination of HT (42°C) with direct CAP treatment has been tested on monocytes and was found to be synergistic: the oxidation of lipids by HT leads to the disruption of the cellular membrane, allowing the RONS produced by CAP to massively enter the cells [40]. HT (42°C) has also been combined with CAP-modified acetated Ringer's lactate on lung-cancer cells, leading to a synergistic increase in cell death and DNA damage [72]. The combination of CAP and iron nanoparticles has been found to be synergistic *in vitro* due to the iron nanoparticles' RONS-inducing capacity. However, another combination option would be to use these particles for intracellular HT by exciting them using an alternating magnetic field [73]. The same principle can be applied to CAP and gold nanoparticles, which can also be used for photothermal therapy [74]. Qin et al. have tested this strategy *in vitro* and *in vivo*, combining CAP, gold nanoparticles, and laser excitation. They have found that adding CAP to photothermal therapy allows the use of lower energy lasers to obtain the same efficacy but with fewer side effects to neighboring normal cells [75]. The rationale for the combination of CAP and HT rests mainly on the fact that both modalities lead to the production of RONS, so their combination will overwhelm the already redox-challenged cancer cells (Figure 12.2). The fact that HT can also disrupt the cell membrane will also help CAP-induced RONS penetrate the cells and, therefore, will potentiate the CAP treatment [40].

12.3 PHOTOTHERAPY

Phototherapy is a general term for using light rays as a therapeutic tool. This section will mainly focus on photodynamic therapy (PDT), although the definition of phototherapy technically encompasses photon-based radiotherapy and photothermal therapy. In PDT, a light-activated molecule called photosensitizer induces therapeutic effects. Non-toxic photosensitizers exposed to light sources of a corresponding wavelength (most often in the visible range) will go from a ground state to an excited state. This will rapidly produce cytotoxic RONS upon interaction of the excited photosensitizer with oxygen, thereby killing malignant cells via apoptosis, necrosis, or autophagy [76–80]. In addition to direct cytotoxicity in tumor cells, PDT has been demonstrated to damage the tumor microvasculature and induce a potent local inflammatory reaction. These three interlinked mechanisms all contribute to PDT efficacy, depending on treatment parameters such as the selected compound, its dosage, and the scheduling of light exposure [77]. Photofrin (a commercially available hematoporphyrin derivative) was the first compound to receive regulatory approval for PDT, in 1993, to treat bladder cancer [76]. After this breakthrough, Photofrin was approved to treat patients with skin, lung, digestive, and genito-urinary tract cancers worldwide and rapidly became the most used clinically [76,81]. However, inherent limitations of its molecular structure necessitate the use of high doses of both compound and light [76]. As a result, developing new photosensitizers (especially ones with higher tumor specificity) has been a major focus of PDT research, with variable success in clinical trials [76,81]. In a clinical setting, photodynamic therapy is a minimally invasive procedure scheduled in a two-step process. Depending on tumor location, patients are first administered the photosensitizer orally, topically, or intravenously. This leads to a local accumulation of the compound in the tumor, which is then exposed to the light source (typically lasers or light emitting diodes (LEDs)) 1 to 3 days later. As the procedure is very short, it is most often conducted in an outpatient setting and can be safely and effectively combined with other cancer treatments [77]. For example, PDT has been shown to positively interact with immunotherapy, as PDT-induced cell death stimulates the immune system through acute tumor site inflammation and rapid cytokine release [78,81]. Photoimmunotherapy has now been developed with the direct combination of a photosensitizer and an antigen-specific

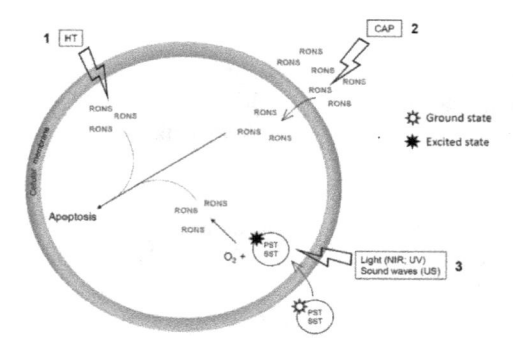

FIGURE 12.2 Simplified schematic representation of the combination of CAP and HT, PDT, and SDT at the cellular level. 1. The main mechanism of HT cytotoxicity is the production of intracellular RONS. 2. CAP produces RONS that penetrate the cells. 3. A PST/SST in the inactive ground state is administered to the cells. Upon contact with a light/sound source, the PST/SST is excited and can now interact with oxygen to produce RONS. The accumulation of RONS from these different treatment methods leads to apoptosis. Combining two or more of these treatments allows for a decrease in the dose used and, therefore, the risks of side effects. HT: Hyperthermia; RONS: Reactive oxygen and nitrogen species; CAP: Cold atmospheric plasma; PDT/SDT: Photo-/ Sonodynamic therapy; PST/SST: Photo-/Sonosensitizer; NIR: Near infrared; UV: Ultraviolet; US: Ultrasound.

recognition system that can lead to immunogenic cell death [82]. PDT can also be associated with RT and with nanoparticles [80,83,84]. As light does not penetrate deep into tissues, PDT is mostly used for superficial or natural cavity-located tumors. The use of fiberscopes now allows the application of light to tumors located deeper inside the body, and the development of new photosensitizers that are organ-specific or that can bind to specific receptors at the surface of cancer cells increases the specificity of the treatment and reduce the risks of toxicity for the neighboring tissues [85].

Due to their similar mode of action (i.e., ROS production), the rationale for CAP/PDT combination is relatively straightforward [86,87] (Figure 12.2). Karami-Gadallo et al. used LEDs mounted on their CAP applicator to simultaneously apply PDT and CAP to lung cancer cells previously incubated with a photosensitizer and found that the combination synergistically decreased cell viability [88]. The combination of PDT and CAP has been studied in cervical-cancer, colon-cancer, and glioblastoma cells and was found to be synergistic in all cases [89,90]. Songca et al. have also reported that using a combination of PDT with direct or indirect CAP treatment for antimicrobial applications can potentiate PDT in hypoxic regions where the lack of oxygen would limit its efficacy. CAP counters this problem by introducing RONS to the area, a reasoning that could be applied to the hypoxic cores of tumors [91].

Treating cancer cells with UV light could also be considered a phototherapy. UVC non-ionizing radiation (200 to 280 nm) has been used to treat cancer cells, leading to their death by apoptosis via the endoplasmic reticulum (ER) stress/unfolded protein response (UPR) pathway [92,93]. Like RT, irradiation by UVC can lead to immunogenic cell death [94]. Interestingly, plasma UV is one of the components of plasma, and depending on the parameters used, UVC can be emitted [95]. Wang et al. compared CAP-mediated phototherapy to UV-mediated phototherapy and found that using CAP increases the cytotoxicity of PDT more than using UV lights does [96].

12.4 SONODYNAMIC THERAPY

Sonodynamic therapy (SDT) is very similar to PDT: a sensitizing molecule called a sonosensitizer is injected into the patient before the tumor is exposed to low-intensity ultrasound. Separately, both treatments are harmless, but their combination is cytotoxic, mainly via the production of RONS [97,98] (Figure 12.2). The advantage of SDT compared to PDT is that US can penetrate deeper into the body, allowing treatment of less accessible tumors [97]. Like PDT, SDT has been successfully used with various other anti-cancer modalities [99].

To date, CAP has not been used in combination with SDT. One rationale for the combination is that CAP produces H_2O_2 in cancer cells, which transforms this H_2O_2 into OH- and most importantly •OH using the Fenton reaction. Ultrasonic waves have been shown to improve the efficacy of the Fenton reaction, leading to the production of more •OH and, therefore, more DNA damage [99].

12.5 PERMEABILIZATION

Broadly speaking, cell permeabilization is the transient or permanent creation of pores in the various membranes of a cell or of its organelles. Although cell permeabilization is often employed non-therapeutically in a research setting (e.g., immunohistochemistry, transfection), cell permeabilization as a physical therapeutic method for cancer is gaining momentum. In cancer therapy, permeabilization can be used in combination with other modalities to load cancer cells with active chemicals or antibodies or on its own to induce cell death, all of which can be achieved through various methods. This section will concentrate on transient permeabilization, designed to be combined with other modalities, rather than irreversible permeabilization that indiscriminately kills normal and cancer cells. All of the permeabilization methods described here can be used for both applications, depending on the applied parameters.

12.5.1 ELECTROMAGNETIC FIELDS

In cancer therapy, electric fields are primarily used to perform electroporation of cancer cells. Using short, high-voltage pulses of an external electric field, it is possible to create transient reversible openings in the cell membrane by overcoming its capacitance. Initially developed and optimized for gene transfer, electroporation allows fine-tuned delivery of drugs, ions, and antibodies into cells in an electrophoretically-driven or diffusion-driven fashion [100]. Optimization of protocols for transient cell permeabilization using electric fields has demonstrated a high permeabilization rate with low cell death. This low electroporation-induced cell death could help to limit injury to normal tissue in cancer therapy. In a clinical setting, electroporation has been successfully combined with chemotherapy in what was designated as electrochemotherapy (ECT) [101]. By using electroporation to improve intracellular delivery of drugs (especially hydrophilic ones), their cytotoxic potential can be improved from 2- to 5000-fold [100,101].

CAP-modified PBS has been combined with microsecond pulsed electric fields, resulting in significant increases in the level of cell membrane electroporation, even at very low electric field amplitude. Interestingly, the murine melanoma cell line was more sensitive to the treatment than the lung fibroblasts. The combined treatment with a lower electric field than for electroporation alone would allow for a reduction of side effects in normal cells [102]. The same approach was used to treat colorectal carcinoma spheroids treated with CAP-modified PBS and electric pulses. This combination has been shown to enhance CAP-modified PBS-associated toxicity, resulting in an earlier onset of DNA damage and caspase activation [103]. The combination of direct CAP treatment with microsecond pulsed electric fields has also been studied on lymphoma cells, and their cytotoxicity was found to be additive [104]. This combination is considered a good candidate for palliative care, where microsecond pulsed ECT has been used with encouraging results but still has room for improvement, given that 10–30% of patients are non-responders [105].

Additionally, the immunogenic effects of electroporation and electrochemotherapy have justified their combination with immunotherapy protocols, with potential benefits for metastatic disease [100,106,107]. Finally, in an analogous fashion to electroporation, magnetoporation is the use of magnetic fields to destabilize the cell membrane, with applications in drug delivery [108]. Both direct and indirect (CAP-modified medium) treatments with CAP have been combined with a static magnetic field (SMF), leading to a slight increase in toxicity compared to CAP alone. The low-frequency SMF was used to disrupt ion flux through the cellular membrane, increasing the influx of Ca^{2+} and leading to apoptosis [109].

12.5.2 Ultrasound-Mediated Microbubble Cavitation

Another physical modality for transient permeabilization of live cells consists of using sound waves, also known as sonoporation. As previously mentioned, HIFU is already being used for tissue ablation and hyperthermia generation in cancer treatment, but sonoporation-based treatment strategies using low-intensity ultrasound (US) waves have been gaining interest in recent years and have shown promising results [110]. It is now established that US waves can induce controlled cavitation of gas microbubbles (MB) *in vivo*, resulting in a variety of physiological effects depending on the type and size of the bubbles, as well as the parameters of the US pulse. Indeed, when placed in a driving pressure field, the surface of MBs exhibit non-spherical oscillations and volumetric oscillations in conjunction with a radiation force phenomenon, enhancing US-induced sonoporation and conferring various properties allowing drug delivery, gene delivery, and other clinical applications [111–113]. Mainly, when US-mediated MB cavitation increases shear stress beyond a threshold of a few kilopascals, a phenomenon of transient cell sonoporation synchronized with the US pulse is observed to drastically increase cellular membrane permeability [114,115]. The value of this threshold depends on US parameters, both frequency and oscillation cycles [116]. Furthermore, single sonoporation events have also been shown to generate intercellular gaps between endothelial cells, persisting over longer timescales than membrane pores [116,117]. Interestingly, the provascular effects of the combined use of US and MBs have also been demonstrated. Convective shear caused by therapeutic US results in increased tissue blood flow, amplified by cavitating MB [118,119].

CAP has not yet been used in combination with sonoporation. The rationale for using CAP with permeabilization methods is that large amounts of the RONS produced by CAP stay in the liquid

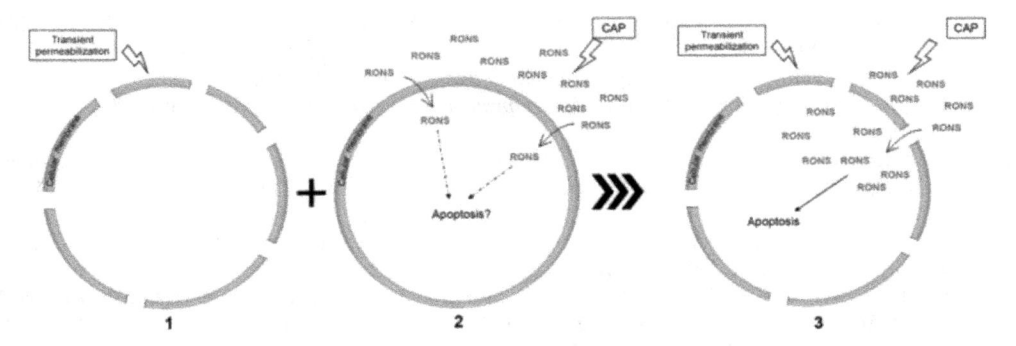

FIGURE 12.3 Simplified schematic representation of the combination of CAP and permeabilization methods at the cellular level. 1. Electromagnetic fields and ultrasonic waves are methods of transient permeabilization that lead to the formation of pores in the cellular membrane. 2. CAP-produced RONS do not penetrate the cell easily, but their accumulation can lead to apoptosis. 3. The combination of CAP with permeabilization treatments facilitates the entry of RONS into the cell and leads to apoptosis. This could allow for the use of a lower dose of CAP to obtain cytotoxic results. CAP: Cold atmospheric plasma; RONS: Reactive oxygen and nitrogen species.

phase surrounding the cells and do not penetrate the cells (Figure 12.3). As cell death correlates with the concentration of intracellular RONS and not with the concentration of produced RONS, the combination of CAP with permeabilization techniques would allow for higher quantities of RONS to enter the cells and, therefore, would potentiate CAP treatment [120,121].

12.5.3 CAP

Permeabilization of cells can also be achieved with the use of CAP. Plasma-assisted transfections have been accomplished, and molecules under 6.5 nm of radius have been proven to be able to enter the cells following permeabilization with CAP [122]. Gas-liquid interfacial atmospheric-pressure plasma jets have been used for cell permeabilization, primarily thanks to •OH radicals in the aqueous phase disrupting the membrane [123].

Cold atmospheric plasma is a versatile tool that can be combined with or replace various physical modalities for cancer treatment. The association of CAP with other physical therapies has been demonstrated to be primarily additive or synergistic *in vitro*, allowing for the use of smaller doses of both treatments and thereby leading to fewer side effects in normal cells. Additional research is warranted to evaluate the efficacy of these combinations *in vivo* and to better understand their interaction with the tumor microenvironment and vasculature, as well as better understand their immunogenicity. This would contribute to a better understanding of CAP and its place in the expanding arsenal of anti-cancer therapeutics.

REFERENCES

1. Jackson, S.E.; Chester, J.D. *Personalised cancer medicine*. Int J Cancer **2015**, 10.1002/ijc.28940.
2. Tsimberidou, A.M.; Fountzilas, E.; Nikanjam, M.; Kurzrock, R. *Review of precision cancer medicine: Evolution of the treatment paradigm*. Cancer Treat Rev **2020**, 10.1016/j.ctrv.2020.102019.
3. Son, S.; Kim, J.; Kim, J.; Kim, B.; Lee, J.; Kim, Y.; Li, M.; Kang, H.; Kim, J.S. *Cancer therapeutics based on diverse energy sources*. Chem Soc Rev **2022**, 10.1039/d2cs00102k.
4. *Basic clinical radiobiology*. 5th ed.; Taylor & Francis Group: 2018; p. 375.
5. Trikalinos, T.A.; Terasawa, T.; Ip, S.; et al. *Particle beam radiation therapies for cancer*. Agency for Healthcare Research and Quality (US): 2009; Vol. 1.
6. Bryant, A.K.; Banegas, M.P.; Martinez, M.E.; Mell, L.K.; Murphy, J.D. *Trends in radiation therapy among cancer survivors in the united states, 2000–2030*. Cancer Epidemiol Biomarkers Prev. A publication of the American Association for Cancer Research, cosponsored by the American Society of Preventive Oncology **2017**, 10.1158/1055-9965.epi-16-1023.
7. *Clinical radiation oncology*. 4th ed.; Elsevier Health Sciences: 2015; p. 1648.

8. Martin, A.; Gaya, A. *Stereotactic body radiotherapy: A review.* Clin Oncol (R Coll Radiol) **2010**, 10.1016/j.clon.2009.12.003.

9. *Radiotherapy in cancer care: Facing the global challenge.* IAEA: 2017; p. 578.

10. Lim, Y.K.; Kim, D. *Brachytherapy: A comprehensive review.* Prog Med Phys **2021**, 10.14316/pmp.2021.32.2.25.

11. Brewington, B.Y.; Shao, Y.F.; Davidorf, F.H.; Cebulla, C.M. *Brachytherapy for patients with uveal melanoma: Historical perspectives and future treatment directions.* Clin Ophthalmol **2018**, 10.2147/opth. S129645.

12. Stish, B.J.; Davis, B.J.; Mynderse, L.A.; McLaren, R.H.; Deufel, C.L.; Choo, R. *Low dose rate prostate brachytherapy.* Transl Androl Urol **2018**, 10.21037/tau.2017.12.15.

13. Arazi, L.; Cooks, T.; Schmidt, M.; Popovtzer, A.; Rosenfeld, E.; Mizrachi, A.; Ben-Hur, R.; Keisari, Y.; Kelson, I. *Alpha dart: Revolutionary alpha-emitters brachytherapy,* TAT11 **2019**; Sciences, J.o.M.I.a.R., Ed.

14. Bourhis, J.; Sozzi, W.J.; Jorge, P.G.; Gaide, O.; Bailat, C.; Duclos, F.; Patin, D.; Ozsahin, M.; Bochud, F.; Germond, J.F.; Moeckli, R.; Vozenin, M.C. *Treatment of a first patient with flash-radiotherapy.* Radiother Oncol **2019**, 10.1016/j.radonc.2019.06.019.

15. Symonds, P.; Jones, G.D.D. *Flash radiotherapy: The next technological advance in radiation therapy?* Clin Oncol (Royal College of Radiologists) **2019**, 10.1016/j.clon.2019.05.011.

16. Hughes, J.R.; Parsons, J.L. *Flash radiotherapy: Current knowledge and future insights using proton-beam therapy.* Int J Mol Sci **2020**, 10.3390/ijms21186492.

17. Volkert, W.A.; Hoffman, T.J. *Therapeutic radiopharmaceuticals.* Chem Rev **1999**, 10.1021/cr9804386.

18. Vermeulen, K.; Vandamme, M.; Bormans, G.; Cleeren, F. *Design and challenges of radiopharmaceuticals.* Semin Nucl Med **2019**, 10.1053/j.semnuclmed.2019.07.001.

19. *The basic science of oncology.* 5th ed.; McGraw Hill LLC: 2013; p. 576.

20. Brown, J.M. *Tumor hypoxia in cancer therapy.* Methods Enzymol **2007**, 10.1016/s0076-6879(07)35015-5.

21. Huang, J.L.; Chang, Y.T.; Hong, Z.Y.; Lin, C.S. *Targeting DNA damage response and immune checkpoint for anticancer therapy.* Int J Mol Sci **2022**, 10.3390/ijms23063238.

22. Vakifahmetoglu, H.; Olsson, M.; Zhivotovsky, B. *Death through a tragedy: Mitotic catastrophe.* Cell Death Differ **2008**, 10.1038/cdd.2008.47.

23. Hernandez-Segura, A.; Nehme, J.; Demaria, M. *Hallmarks of cellular senescence.* Trends Cell Biol **2018**, 10.1016/j.tcb.2018.02.001.

24. Herranz, N.; Gil, J. *Mechanisms and functions of cellular senescence.* J Clin Invest **2018**, 10.1172/JCI95148.

25. Castedo, M.; Perfettini, J.-L.; Roumier, T.; Andreau, K.; Medema, R.; Kroemer, G. *Cell death by mitotic catastrophe: A molecular definition.* Oncogene **2004**, 10.1038/sj.onc.1207528.

26. Vitale, I.; Galluzzi, L.; Castedo, M.; Kroemer, G. *Mitotic catastrophe: A mechanism for avoiding genomic instability.* Nat Rev Mol Cell Biol **2011**, 10.1038/nrm3115.

27. Heeran, A.B.; Berrigan, H.P.; O'Sullivan, J. *The radiation-induced bystander effect (ribe) and its connections with the hallmarks of cancer.* Radiat Res **2019**, 10.1667/rr15489.1.

28. Abdelrazzak, A.B.; Pottgiesser, S.J.; Hill, M.A.; O'Neill, P.; Bauer, G. *Enhancement of peroxidase release from non-malignant and malignant cells through low-dose irradiation with different radiation quality.* Radiat Res **2016**, 10.1667/RR14245.1.

29. Bogart, J.A.; Waqar, S.N.; Mix, M.D. *Radiation and systemic therapy for limited-stage small-cell lung cancer.* J Clin Oncol **2022**, 10.1200/jco.21.01639.

30. Liao, C.K.; Kuo, Y.T.; Lin, Y.C.; Chern, Y.J.; Hsu, Y.J.; Yu, Y.L.; Chiang, J.M.; Hsieh, P.S.; Yeh, C.Y.; You, J.F. *Neoadjuvant short-course radiotherapy followed by consolidation chemotherapy before surgery for treating locally advanced rectal cancer: A systematic review and meta-analysis.* Curr Oncol **2022**, 10.3390/curroncol29050297.

31. McAleavey, P.G.; Walls, G.M.; Chalmers, A.J. *Radiotherapy-drug combinations in the treatment of glioblastoma: A brief review.* CNS Oncol **2022**, 10.2217/cns-2021-0015.

32. Zhang, Z.; Liu, X.; Chen, D.; Yu, J. *Radiotherapy combined with immunotherapy: The dawn of cancer treatment.* Signal Transduct Target Ther **2022**, 10.1038/s41392-022-01102-y.

33. Yu, S.; Wang, Y.; He, P.; Shao, B.; Liu, F.; Xiang, Z.; Yang, T.; Zeng, Y.; He, T.; Ma, J.; Wang, X.; Liu, L. *Effective combinations of immunotherapy and radiotherapy for cancer treatment.* Front Oncol **2022**, 10.3389/fonc.2022.809304.

34. Roeder, F.; Krempien, R. *Intraoperative radiation therapy (iort) in soft-tissue sarcoma.* Radiat Oncol **2017**, 10.1186/s13014-016-0751-2.

35. Brown, A.; Buss, E.J.; Chin, C.; Liu, G.; Lee, S.; Rao, R.; Taback, B.; Wiechmann, L.; Horowitz, D.; Choi, J.C.; Katz, L.M.; Connolly, E.P. *Targeted intraoperative radiotherapy (targit-iort) for early-stage invasive breast cancer: A single institution experience.* Front Oncol **2022**, 10.3389/fonc.2022.788213.

36. Asha, W.; Geara, F.; Quinn, S.; Shah, C. *Intraoperative radiation therapy in the management of early stage breast cancer.* Brachytherapy **2022**, 10.1016/j.brachy.2022.09.003.
37. Giap, F.; O'Steen, L.; Liu, I.C.; Spiguel, L.E.; Shaw, C.M.; Morris, C.G.; Mailhot Vega, R.B.; Lightsey, J.L.; Bradley, J.A.; Mendenhall, N.P.; Okunieff, P.G.; Lockney, N.A. *Intraoperative radiation therapy for early-stage breast cancer: A single-institution experience.* Rep Pract Oncol Radiother **2022**, 10.5603/RPOR.a2022.0075.
38. Lin, L.; Wang, L.; Liu, Y.; Xu, C.; Tu, Y.; Zhou, J. *Nonthermal plasma inhibits tumor growth and proliferation and enhances the sensitivity to radiation in vitro and in vivo.* Oncol Rep **2018**, 10.3892/or.2018.6749.
39. Lafontaine, J.; Boisvert, J.S.; Glory, A.; Coulombe, S.; Wong, P. *Synergy between non-thermal plasma with radiation therapy and olaparib in a panel of breast cancer cell lines.* Cancers (Basel) **2020**, 10.3390/cancers12020348.
40. Moniruzzaman, R.; Rehman, M.U.; Zhao, Q.L.; Jawaid, P.; Takeda, K.; Ishikawa, K.; Hori, M.; Tomihara, K.; Noguchi, K.; Kondo, T.; Noguchi, M. *Cold atmospheric helium plasma causes synergistic enhancement in cell death with hyperthermia and an additive enhancement with radiation.* Sci Rep **2017**, 10.1038/s41598-017-11877-8.
41. Pasqual-Melo, G.; Sagwal, S.K.; Freund, E.; Gandhirajan, R.K.; Frey, B.; von Woedtke, T.; Gaipl, U.; Bekeschus, S. *Combination of gas plasma and radiotherapy has immunostimulatory potential and additive toxicity in murine melanoma cells in vitro.* Int J Mol Sci **2020**.
42. Kenari, A.J.; Siadati, S.N.; Abedian, Z.; Sohbatzadeh, F.; Amiri, M.; Gorji, K.E.; Babapour, H.; Zabihi, E.; Ghoreishi, S.M.; Mehraeen, R.; Monfared, A.S. *Therapeutic effect of cold atmospheric plasma and its combination with radiation as a novel approach on inhibiting cervical cancer cell growth (hela cells).* Bioorg Chem **2021**, 10.1016/j.bioorg.2021.104892.
43. Privat-Maldonado, A.; Schmidt, A.; Lin, A.; Weltmann, K.-D.; Wende, K.; Bogaerts, A.; Bekeschus, S. *Ros from physical plasmas: Redox chemistry for biomedical therapy.* Oxidative Med Cell Longev **2019**, 10.1155/2019/9062098.
44. Hirst, A.M.; Frame, F.M.; Arya, M.; Maitland, N.J.; O'Connell, D. *Low temperature plasmas as emerging cancer therapeutics: The state of play and thoughts for the future.* Tumour Biol **2016**, 10.1007/s13277-016-4911-7.
45. Bekeschus, S.; Schutz, C.S.; Niessner, F.; Wende, K.; Weltmann, K.D.; Gelbrich, N.; von Woedtke, T.; Schmidt, A.; Stope, M.B. *Elevated h2ax phosphorylation observed with kinpen plasma treatment is not caused by ros-mediated DNA damage but is the consequence of apoptosis.* Oxidative Med Cell Longev **2019**, 10.1155/2019/8535163.
46. Collet, G.; Robert, E.; Lenoir, A.; Vandamme, M.; Darny, T.; Dozias, S.; Kieda, C.; Pouvesle, J.M. *Plasma jet-induced tissue oxygenation: Potentialities for new therapeutic strategies.* Plasma Sources Sci Technol **2014**, 10.1088/0963-0252/23/1/012005.
47. van der Zee, J. *Heating the patient: A promising approach?* Ann Oncol **2002**.
48. Sugahara, T.; van der Zee, J.; Kampinga, H.H.; Vujaskovic, Z.; Kondo, M.; Ohnishi, T.; Li, G.; Park, H.J.; Leeper, D.B.; Ostapenko, V.; Repasky, E.A.; Watanabe, M.; Song, C.W. *Kadota fund international forum 2004. Application of thermal stress for the improvement of health, 15–18 june 2004, awaji yumebutai international conference center, awaji island, hyogo, japan. Final report.* Int J Hyperthermia **2008**, 10.1080/02656730701883675.
49. Fokas, E.; McKenna, W.G.; Muschel, R.J. *The impact of tumor microenvironment on cancer treatment and its modulation by direct and indirect antivascular strategies.* Cancer Metastasis Rev **2012**, 10.1007/s10555-012-9394-4.
50. Baronzio, G.; Gramaglia, A.; Fiorentini, G. *Review. Current role and future perspectives of hyperthermia for prostate cancer treatment. In Vivo* **2009**.
51. Bettaieb, A.; Wrzal, P.K.; Averill-Bates, D.A. Hyperthermia: Cancer treatment and beyond. In *Cancer treatment—conventional and innovative approaches*, Rangel, P.L., Ed. InTech: 2013; Vol. Book 2, pp. 257–283.
52. Richter, K.; Haslbeck, M.; Buchner, J. *The heat shock response: Life on the verge of death.* Molecular Cell **2010**, 10.1016/j.molcel.2010.10.006.
53. Sonna, L.A.; Fujita, J.; Gaffin, S.L.; Lilly, C.M. *Invited review: Effects of heat and cold stress on mammalian gene expression.* J Appl Physiol (1985) **2002**, 10.1152/japplphysiol.01143.2001.
54. Bettaieb, A.; Averill-Bates, D.A. *Thermotolerance induced at a fever temperature of 40 degrees c protects cells against hyperthermia-induced apoptosis mediated by death receptor signalling.* Biochem Cell Biol **2008**, 10.1139/O08-136.
55. Bettaieb, A.; Averill-Bates, D.A. *Thermotolerance induced at a mild temperature of 40 degrees c alleviates heat shock-induced er stress and apoptosis in hela cells.* Biochim Biophys Acta **2015**, 10.1016/j.bbamcr.2014.09.016.

56. Glory, A.; Bettaieb, A.; Averill-Bates, D.A. *Mild thermotolerance induced at 40 degrees c protects cells against hyperthermia-induced pro-apoptotic changes in bcl-2 family proteins.* Int J Hyperthermia **2014**, 10.3109/02656736.2014.968641.

57. Yuguchi, T.; Saito, M.; Yokoyama, Y.; Saito, T.; Nagata, T.; Sakamoto, T.; Tsukada, K. *Combined use of hyperthermia and irradiation cause antiproliferative activity and cell death to human esophageal cell carcinoma cells—mainly cell cycle examination.* Hum Cell **2002,**

58. Chae, S.Y.; Kim, Y.S.; Park, M.J.; Yang, J.; Park, H.; Namgung, M.S.; Rhim, H.; Lim, H.K. *High-intensity focused ultrasound-induced, localized mild hyperthermia to enhance anti-cancer efficacy of systemic doxorubicin: An experimental study.* Ultrasound Med Biol **2014**, 10.1016/j.ultrasmedbio.2014.01.005.

59. Malietzis, G.; Monzon, L.; Hand, J.; Wasan, H.; Leen, E.; Abel, M.; Muhammad, A.; Price, P.; Abel, P. *High-intensity focused ultrasound: Advances in technology and experimental trials support enhanced utility of focused ultrasound surgery in oncology.* Br J Radiol **2013**, 10.1259/bjr.20130044.

60. Maloney, E.; Hwang, J.H. *Emerging hifu applications in cancer therapy.* Int J Hyperthermia **2015**, 10.3109/02656736.2014.969789.

61. Wust, P.; Hildebrandt, B.; Sreenivasa, G.; Rau, B.; Gellermann, J.; Riess, H.; Felix, R.; Schlag, P.M. *Hyperthermia in combined treatment of cancer.* Lancet Oncol **2002**.

62. Cao, L.; Wu, Y.; Shan, Y.; Tan, B.; Liao, J. *A review: Potential application and outlook of photothermal therapy in oral cancer treatment.* Biomed Mater **2022**, 10.1088/1748-605X/ac5a23.

63. Blanco-Andujar, C.; Ortega, D.; Southern, P.; Nesbitt, S.A.; Thanh, N.T.; Pankhurst, Q.A. *Real-time tracking of delayed-onset cellular apoptosis induced by intracellular magnetic hyperthermia.* Nanomedicine (Lond) **2016**, 10.2217/nnm.15.185.

64. Kossatz, S.; Grandke, J.; Couleaud, P.; Latorre, A.; Aires, A.; Crosbie-Staunton, K.; Ludwig, R.; Dahring, H.; Ettelt, V.; Lazaro-Carrillo, A.; Calero, M.; Sader, M.; Courty, J.; Volkov, Y.; Prina-Mello, A.; Villanueva, A.; Somoza, A.; Cortajarena, A.L.; Miranda, R.; Hilger, I. *Efficient treatment of breast cancer xenografts with multifunctionalized iron oxide nanoparticles combining magnetic hyperthermia and anti-cancer drug delivery.* Breast Cancer Res **2015**, 10.1186/s13058-015-0576-1.

65. Bates, D.A.; Mackillop, W.J. *Hyperthermia, adriamycin transport, and cytotoxicity in drug-sensitive and -resistant chinese hamster ovary cells.* Cancer Res **1986**.

66. Song, C.W.; Park, H.J.; Lee, C.K.; Griffin, R. *Implications of increased tumor blood flow and oxygenation caused by mild temperature hyperthermia in tumor treatment.* Int J Hyperthermia **2005**, 10.1080/02656730500204487.

67. Wrzal, P.K.; Bettaieb, A.; Averill-Bates, D.A. *Molecular mechanisms of apoptosis activation by heat shock in multidrug-resistant chinese hamster cells.* Radiat Res **2008**.

68. Sakaguchi, Y.; Stephens, L.C.; Makino, M.; Kaneko, T.; Strebel, F.R.; Danhauser, L.L.; Jenkins, G.N.; Bull, J.M. *Apoptosis in tumors and normal tissues induced by whole body hyperthermia in rats.* Cancer Res **1995**.

69. Datta, N.R.; Puric, E.; Klingbiel, D.; Gomez, S.; Bodis, S. *Hyperthermia and radiation therapy in locoregional recurrent breast cancers: A systematic review and meta-analysis.* Int J Radiat Oncol Biol Phys **2016**, 10.1016/j.ijrobp.2015.12.361.

70. Raaphorst, G.; Mao, J.; Ng, C. *A comparison of hyperthermia inhibition of sublethal radiation damage recovery in four human cell lines with different radiosensitivity.* Int J Oncol **1996**, 10.3892/ijo.9.1.159.

71. Jolesch, A.; Elmer, K.; Bendz, H.; Issels, R.D.; Noessner, E. *Hsp70, a messenger from hyperthermia for the immune system.* Eur J Cell Biol **2012**, 10.1016/j.ejcb.2011.02.001.

72. Ishii, R.; Kamiya, T.; Hara, H.; Adachi, T. *Hyperthermia synergistically enhances cancer cell death by plasma-activated acetated ringer's solution.* Arch Biochem Biophys **2020**, 10.1016/j.abb.2020.108565.

73. Yazdani Z, Biparva P, Rafiei A, Kardan M, Hadavi S. *Combination effect of cold atmospheric plasma with green synthesized zero-valent iron nanoparticles in the treatment of melanoma cancer model.* PLoS One. **2022** Dec 19;17(12):e0279120. doi: 10.1371/journal.pone.0279120. PMID: 36534669; PMCID: PMC9762585.

74. Momeni S, Shanei A, Sazgarnia A, Attaran N, Aledavood SA. *The Synergistic Effect of Cold Atmospheric Plasma Mediated Gold Nanoparticles Conjugated with Indocyanine Green as An Innovative Approach to Cooperation with Radiotherapy.* Cell J. **2023** Jan 1;25(1):51–61. doi: 10.22074/cellj.2022.559078.1097. PMID: 36680484; PMCID: PMC9868434.

75. Qin, J.; Zhang, J.; Fan, G.; Wang, X.; Zhang, Y.; Wang, L.; Zhang, Y.; Guo, Q.; Zhou, J.; Zhang, W.; Ma, J. *Cold atmospheric plasma activates selective photothermal therapy of cancer.* Molecules **2022**, 10.3390/molecules27185941.

76. Dolmans, D.E.; Fukumura, D.; Jain, R.K. *Photodynamic therapy for cancer.* Nat Rev Cancer **2003**, 10.1038/nrc1071.

77. Agostinis, P.; Berg, K.; Cengel, K.A.; Foster, T.H.; Girotti, A.W.; Gollnick, S.O.; Hahn, S.M.; Hamblin, M.R.; Juzeniene, A.; Kessel, D.; Korbelik, M.; Moan, J.; Mroz, P.; Nowis, D.; Piette, J.; Wilson, B.C.; Golab, J. *Photodynamic therapy of cancer: An update.* CA Cancer J Clin **2011**, 10.3322/caac.20114.

78. Khalili, M.; Daniels, L.; Lin, A.; Krebs, F.C.; Snook, A.E.; Bekeschus, S.; Bowne, W.B.; Miller, V. *Non-thermal plasma-induced immunogenic cell death in cancer: A topical review.* J Phys D: Appl Phys **2019**, 10.1088/1361-6463/ab31c1.

79. Henderson, B.W.; Dougherty, T.J. *How does photodynamic therapy work?* Photochem Photobiol **1992**.

80. Lucena, S.R.; Salazar, N.; Gracia-Cazana, T.; Zamarron, A.; Gonzalez, S.; Juarranz, A.; Gilaberte, Y. *Combined treatments with photodynamic therapy for non-melanoma skin cancer.* Int J Mol Sci **2015**, 10.3390/ijms161025912.

81. Dougherty, T.J.; Gomer, C.J.; Henderson, B.W.; Jori, G.; Kessel, D.; Korbelik, M.; Moan, J.; Peng, Q. *Photodynamic therapy.* JNCI: J Natl Cancer Inst **1998**, 10.1093/jnci/90.12.889.

82. Wei, D.; Qi, J.; Hamblin, M.R.; Wen, X.; Jiang, X.; Yang, H. *Near-infrared photoimmunotherapy: Design and potential applications for cancer treatment and beyond.* Theranostics **2022**, 10.7150/thno.74820.

83. Adnane, F.; El-Zayat, E.; Fahmy, H.M. *The combinational application of photodynamic therapy and nanotechnology in skin cancer treatment: A review.* Tissue Cell **2022**, 10.1016/j.tice.2022.101856.

84. Tabosa, A.T.L.; Souza, M.G.; de Jesus, S.F.; Rocha, D.F.; Queiroz, L.; Santos, E.M.; Guimarães, V.H.D.; Andrade, L.A.A.; Santos, S.H.; de Paula, A.M.B.; de Souza, P.E.N.; Farias, L.C.; Guimarães, A.L.S. *Effect of low-level light therapy before radiotherapy in oral squamous cell carcinoma: An in vitro study.* Lasers Med Sci **2022**, 10.1007/s10103-022-03632-x.

85. Gustalik, J.; Aebisher, D.; Bartusik-Aebisher, D. *Photodynamic therapy in breast cancer treatment.* J Appl Biomed **2022**, 10.32725/jab.2022.013.

86. Tanaka, H.; Hori, M. *Medical applications of non-thermal atmospheric pressure plasma.* J Clin Biochem Nutr **2017**, 10.3164/jcbn.16-67.

87. Kajiyama, H.; Utsumi, F.; Nakamura, K.; Tanaka, H.; Toyokuni, S.; Hori, M.; Kikkawa, F. *Future perspective of strategic non-thermal plasma therapy for cancer treatment.* J Clin Biochem Nutr **2017**, 10.3164/jcbn.16-65.

88. Karami-Gadallo, L.; Ghoranneviss, M.; Ataie-Fashtami, L.; Pouladian, M.; Sardari, D. *Enhancement of cancerous cells treatment by applying cold atmospheric plasma and photo dynamic therapy simultaneously.* Clin Plasma Med **2017**.

89. Ha, J.H.; Kim, Y.J. *Photodynamic and cold atmospheric plasma combination therapy using polymeric nanoparticles for the synergistic treatment of cervical cancer.* Int J Mol Sci **2021**, 10.3390/ijms22031172.

90. Ara, E.S.; Noghreiyan, A.V.; Sazgarnia, A. *Evaluation of photodynamic effect of indocyanine green (icg) on the colon and glioblastoma cancer cell lines pretreated by cold atmospheric plasma.* Photodiagnosis Photodyn Ther **2021**, 10.1016/j.pdpdt.2021.102408.

91. Songca, S.P.; Adjei, Y. *Applications of antimicrobial photodynamic therapy against bacterial biofilms.* Int J Mol Sci **2022**, 10.3390/ijms23063209.

92. Murray, D.; Mirzayans, R. *Cellular responses to platinum-based anticancer drugs and uvc: Role of p53 and implications for cancer therapy.* Int J Mol Sci **2020**, 10.3390/ijms21165766.

93. Wang, S.C.; Chang, H.S.; Tang, J.Y.; Farooqi, A.A.; Kuo, Y.T.; Hsuuw, Y.D.; Lee, J.W.; Chang, H.W. *Combined treatment with cryptocaryone and ultraviolet c promotes antiproliferation and apoptosis of oral cancer cells.* Int J Mol Sci **2022**, 10.3390/ijms23062981.

94. Adkins, I.; Fucikova, J.; Garg, A.D.; Agostinis, P.; Spisek, R. *Physical modalities inducing immunogenic tumor cell death for cancer immunotherapy.* Oncoimmunology **2014**, 10.4161/21624011.2014.968434.

95. Kletschkus, K.; Gelbrich, N.; Burchardt, M.; Kramer, A.; Bekeschus, S.; Stope, M.B. *Emission of ultraviolet radiation from 220 to 280 nm by a cold physical plasma generating device.* Health Phys **2020**, 10.1097/HP.0000000000001276.

96. Wang, M.; Geilich, B.M.; Keidar, M.; Webster, T.J. *Killing malignant melanoma cells with protoporphyrin ix-loaded polymersome-mediated photodynamic therapy and cold atmospheric plasma.* Int J Nanomedicine **2017**, 10.2147/ijn.S129266.

97. McHale, A.P.; Callan, J.F.; Nomikou, N.; Fowley, C.; Callan, B. Sonodynamic therapy: Concept, mechanism and application to cancer treatment. In *Therapeutic ultrasound*, Escoffre, J.-M.; Bouakaz, A., Eds. Springer International Publishing: Cham, 2016; pp. 429–450.

98. Wu, N.; Fan, C.H.; Yeh, C.K. *Ultrasound-activated nanomaterials for sonodynamic cancer theranostics.* Drug Discov Today **2022**, 10.1016/j.drudis.2022.02.025.

99. Xu, M.; Zhou, L.; Zheng, L.; Zhou, Q.; Liu, K.; Mao, Y.; Song, S. *Sonodynamic therapy-derived multimodal synergistic cancer therapy.* Cancer Lett **2021**, 10.1016/j.canlet.2020.10.037.

100. Gehl, J. *Electroporation: Theory and methods, perspectives for drug delivery, gene therapy and research.* Acta Physiol Scand **2003**, 10.1046/j.1365-201X.2003.01093.x.

101. Larkin, J.O.; Collins, C.G.; Aarons, S.; Tangney, M.; Whelan, M.; O'Reily, S.; Breathnach, O.; Soden, D.M.; O'Sullivan, G.C. *Electrochemotherapy: Aspects of preclinical development and early clinical experience.* Ann Surg **2007**, 10.1097/01.sla.0000250419.36053.33.

102. Chung, T.H.; Stancampiano, A.; Sklias, K.; Gazeli, K.; André, F.M.; Dozias, S.; Douat, C.; Pouvesle, J.M.; Santos Sousa, J.; Robert, É.; Mir, L.M. *Cell electropermeabilisation enhancement by non-thermal-plasma-treated pbs.* Cancers (Basel) **2020**, 10.3390/cancers12010219.

103. Griseti, E.; Kolosnjaj-Tabi, J.; Gibot, L.; Fourquaux, I.; Rols, M.-P.; Yousfi, M.; Merbahi, N.; Golzio, M. *Pulsed electric field treatment enhances the cytotoxicity of plasma-activated liquids in a three-dimensional human colorectal cancer cell model.* Sci Rep **2019**, 10.1038/s41598-019-44087-5.

104. Wolff, C.M.; Kolb, J.F.; Weltmann, K.D.; von Woedtke, T.; Bekeschus, S. *Combination treatment with cold physical plasma and pulsed electric fields augments ros production and cytotoxicity in lymphoma.* Cancers (Basel) **2020**, 10.3390/cancers12040845.

105. Wolff, C.M.; Steuer, A.; Stoffels, I.; von Woedtke, T.; Weltmann, K.-D.; Bekeschus, S.; Kolb, J.F. *Combination of cold plasma and pulsed electric fields—a rationale for cancer patients in palliative care.* Clin Plasma Med **2019**, 10.1016/j.cpme.2020.100096.

106. Justesen, T.F.; Orhan, A.; Raskov, H.; Nolsoe, C.; Gögenur, I. *Electroporation and immunotherapy-unleashing the abscopal effect.* Cancers (Basel) **2022**, 10.3390/cancers14122876.

107. Aguilar, A.A.; Ho, M.C.; Chang, E.; Carlson, K.W.; Natarajan, A.; Marciano, T.; Bomzon, Z.; Patel, C.B. *Permeabilizing cell membranes with electric fields.* Cancers (Basel) **2021**, 10.3390/cancers13092283.

108. Yousefian, B.; Firoozabadi, S.M.; Mokhtari-Dizaji, M. *Magnetoporation: New method for permeabilization of cancerous cells to hydrophilic drugs.* J Biomed Phys Eng **2022**, 10.31661/jbpe.v0i0.1256.

109. Cheng, X.; Rajjoub, K.; Shashurin, A.; Yan, D.; Sherman, J.H.; Bian, K.; Murad, F.; Keidar, M. *Enhancing cold atmospheric plasma treatment of cancer cells by static magnetic field.* Bioelectromagnetics **2017**, 10.1002/bem.22014.

110. *Therapeutic ultrasound.* 1st ed.; Springer, Cham: 2016; pp. VIII, 465.

111. Kooiman, K.; Vos, H.J.; Versluis, M.; de Jong, N. *Acoustic behavior of microbubbles and implications for drug delivery.* Adv Drug Deliv Rev **2014**, 10.1016/j.addr.2014.03.003.

112. Shapiro, G.; Wong, A.W.; Bez, M.; Yang, F.; Tam, S.; Even, L.; Sheyn, D.; Ben-David, S.; Tawackoli, W.; Pelled, G.; Ferrara, K.W.; Gazit, D. *Multiparameter evaluation of in vivo gene delivery using ultrasound-guided, microbubble-enhanced sonoporation.* J Control Release **2016**, 10.1016/j.jconrel.2015.12.001.

113. Jain, A.; Tiwari, A.; Verma, A.; Jain, S.K. *Ultrasound-based triggered drug delivery to tumors.* Drug Deliv Transl Res **2018**, 10.1007/s13346-017-0448-6.

114. Lentacker, I.; De Cock, I.; Deckers, R.; De Smedt, S.C.; Moonen, C.T. *Understanding ultrasound induced sonoporation: Definitions and underlying mechanisms.* Adv Drug Deliv Rev **2014**, 10.1016/j.addr.2013.11.008.

115. Deng, C.X.; Sieling, F.; Pan, H.; Cui, J. *Ultrasound-induced cell membrane porosity.* Ultrasound Med Biol **2004**, 10.1016/j.ultrasmedbio.2004.01.005.

116. Hu, Y.; Wan, J.M.; Yu, A.C. *Membrane perforation and recovery dynamics in microbubble-mediated sonoporation.* Ultrasound Med Biol **2013**, 10.1016/j.ultrasmedbio.2013.08.003.

117. Helfield, B.; Chen, X.; Watkins, S.C.; Villanueva, F.S. *Biophysical insight into mechanisms of sonoporation.* Proc Natl Acad Sci U S A **2016**, 10.1073/pnas.1606915113.

118. Belcik, J.T.; Davidson, B.P.; Xie, A.; Wu, M.D.; Yadava, M.; Qi, Y.; Liang, S.; Chon, C.R.; Ammi, A.Y.; Field, J.; Harmann, L.; Chilian, W.M.; Linden, J.; Lindner, J.R. *Augmentation of muscle blood flow by ultrasound cavitation is mediated by atp and purinergic signaling.* Circulation **2017**, 10.1161/circulationaha.116.024826.

119. Belcik, J.T.; Mott, B.H.; Xie, A.; Zhao, Y.; Kim, S.; Lindner, N.J.; Ammi, A.; Linden, J.M.; Lindner, J.R. *Augmentation of limb perfusion and reversal of tissue ischemia produced by ultrasound-mediated microbubble cavitation.* Circ Cardiovasc Imaging **2015**, 10.1161/circimaging.114.002979.

120. Rehman, M.U.; Jawaid, P.; Uchiyama, H.; Kondo, T. *Comparison of free radicals formation induced by cold atmospheric plasma, ultrasound, and ionizing radiation.* Arch Biochem Biophys **2016**, 10.1016/j.abb.2016.04.005.

121. Ji, W.O.; Lee, M.H.; Kim, G.H.; Kim, E.H. *Quantitation of the ros production in plasma and radiation treatments of biotargets.* Sci Rep **2019**, 10.1038/s41598-019-56160-0.

122. Leduc, M.; Guay, D.; Leask, R.L.; Coulombe, S. *Cell permeabilization using a non-thermal plasma.* New J Phys **2009**, 10.1088/1367-2630/11/11/115021.

123. Kaneko, T.; Sasaki, S.; Takashima, K.; Kanzaki, M. *Gas-liquid interfacial plasmas producing reactive species for cell membrane permeabilization.* J Clin Biochem Nutr **2017**, 10.3164/jcbn.16-73.

13 Oncological Application of Medically Relevant Plasma-Treated Liquids

Hiromasa Tanaka, Shinya Toyokuni, Masaru Hori

13.1 INTRODUCTION

Plasma cancer treatment is among the most challenging topics in plasma medicine and biology [1–3]. Plasma cancer treatments can be direct or indirect [4,5]. The use of direct plasma cancer treatments has been reported in patients with head and neck cancer who were treated with low-temperature plasma (LTP) in Germany [6] and in patients with colorectal cancer who were treated with LTP in the United States [7]. Although these types of direct treatments are straightforward and powerful, the discovery of indirect treatments has broadened the ways in which LTP can be applied in medicine and biology.

Various liquids have been studied as potential medically relevant plasma-treated liquids for oncological application. Plasma-activated water is the simplest plasma-treated liquid, and its plasma-liquid interactions have been studied for various applications, such as the cleaning of polluted water and sterilization [8]. Since the anti-tumor effects of plasma-activated medium (PAM) were discovered [9], various liquids have been studied for oncological application. Phosphate-buffered saline (PBS) is a simple liquid when compared to culture medium, and plasma-treated PBS has been tested as a plasma cancer treatment [10,11]. Infusion fluids, such as Ringer's lactate solution, Ringer's acetate solution, and Ringer's bicarbonate solution, are commonly used clinically, and they are also simple liquids when compared to culture medium. Ringer's lactate solution has also been treated with LTP for oncological application [12,13].

Another discovery related to indirect plasma treatment is the plasma-induced immune response [14]. LTP induces immunogenic cell death (ICD) in tumor cells, and it is believed that reactive oxygen species (ROS) are needed to induce the ICD. *In vivo* experiments using mice also revealed that LTP activates immune cells. It was demonstrated that ICD could also be induced by plasma-treated liquids in melanoma and pancreatic cancer cells [15–19]. Plasma-treated liquids also potentially activate immune cells, such as T-cells and macrophages [17,18,20].

In this chapter, the plasma sources used to produce plasma-treated liquids for oncological application will be introduced. In plasma-treated solutions, various components are produced from plasma-gas interactions and plasma-liquid interactions. Reactive oxygen and nitrogen species (RONS) play key roles in physiological responses, as do organic compounds produced in plasma-treated liquids. LTP is expected to be useful as a novel tool for synthesizing anticancer drugs, and this potential has led to the development of a new field: plasma pharmacy. Over the last decade, the mechanisms of the PAM or plasma-activated Ringer's lactate solution (PAL)-induced cell death have been elucidated. Comprehensive gene expression analysis, proteomics approaches, and metabolomic analysis are useful for understanding the intracellular molecular mechanisms of the induced cell death in PAM/PAL-treated cells. In most cases, the intracellular molecular mechanisms were found to be related to redox biology concepts. Many animal experiments have been conducted to test the effectiveness and safety of plasma-treated liquids. In the future, clinical research on plasma-treated liquids for oncological application will most likely proceed through various regulations and procedures.

DOI: 10.1201/9781003328056-15

13.2   PLASMA SOURCES FOR PRODUCING PLASMA-TREATED LIQUIDS

Plasma jets have been developed to directly treat cancers. For example, the kINPen MED [21,22] was developed at the Leibniz Institute for Plasma Science and Technology (INP Greifswald) in Germany to treat head and neck cancers [6,23,24], and the European Cooperation in Science and Technology (COST) developed the COST Reference Microplasma jet (COST-jet) under COST Action MP1101 "Biomedical Applications of Atmospheric Pressure Plasmas" [25]. kINPen and COST-jet have been compared in terms of oxidation and nitrosylation for biological applications [26], and COST-jet was found to result in higher S-nitrosocystein production than kINPen.

Besides plasma jets, dielectric barrier discharges (DBD) are the most common plasma-source type used in plasma medicine. PAM and PAL were developed using LTP with a high electron density [27]. With LTP, the electron density is 3.2×10^{16} cm^{-3} at a 1% O_2/Ar flow rate of 5.0 standard liters per minute, and the oxygen radical concentration is 1.6×10^{15} cm^{-3} at a 1% O_2/Ar flow rate of 15 standard liters per minute; these values are higher than the ones in conventional DBD plasmas. These results suggested that LTP can produce a high concentration of reactive species and plasma-activated organic compounds. LTP-based machines have therefore been developed to produce plasma-treated liquids for clinical applications [28]. In these machines, the molecules surrounding the plasma in the air are replaced by argon gas in a chamber, and the proportions of oxygen and nitrogen are regulated by changing the feed gases of oxygen and nitrogen. Humidity is also removed by the argon gas purge process. Thus, these machines can be used to produce stable plasma-treated liquids.

13.3   PLASMA CHEMISTRY TO ELUCIDATE THE COMPONENTS OF PLASMA-TREATED LIQUIDS

LTP interacts with many chemical compounds in gases and liquids (see also chapters 2 and 3). The most commonly used gases are air (including nitrogen, oxygen, and water [humidity]) and noble gases such as helium, argon, or neon. As liquids, water (distilled water, tap water, or Milli-Q water), saline solutions (PBS or Ringer's solutions), or cell culture media (Dulbecco's Modified Eagle's Medium or Roswell Park Memorial Institute [RPMI] medium) have been investigated. A comparison of 6 clinically relevant liquids has shown the advantages and disadvantages of several types of liquids with regard to plasma treatment [29].

Interactions between LTP and oxygen, nitrogen, and water produce various short-lived RONS, such as hydroxyl (OH) and nitric oxide (NO) radicals. The spatial distributions of such RONS have been measured using vacuum ultraviolet absorption spectroscopy and laser-induced fluorescence spectroscopy [30]. Results showed that the NO density increased with increasing distance from the gas slit up to 10 mm, at which point it became saturated. The OH density decreased with increasing distance from the gas slit, and OH radicals were observed up to 5 mm from the plasma jet. These short-lived RONS are considered important since they react with other molecules, such as organic compounds in liquids, to produce secondary chemical compounds.

In plasma-treated liquids, it is well known that RONS, such as hydrogen peroxide, nitrite, and nitrate, are produced through plasma-liquid interactions. A high concentration of hydrogen peroxide generally damages cells. However, the results of many studies have suggested that not only hydrogen peroxide but also other components in PAM/PAL could contribute to cellular effects [5,31,32]. For example, a study showed that 63 μM hydrogen peroxide and 1890 μM nitrite were produced in PAM when the culture medium was treated with LTP for 3 min, and that the 63 μM hydrogen peroxide killed approximately 70% of the glioblastoma cells while the 1890 μM nitrite had no cytotoxic effect on the cells. However, when both 63 μM hydrogen peroxide and 1890 μM nitrite were added to the medium, approximately 90% of the glioblastoma cells were killed, and PAM killed almost 100% of the glioblastoma cells [33]. These results suggested that hydrogen peroxide and nitrite have synergistic effects in glioblastoma cells and that other components of PAM could also contribute to cellular effects induced by PAM. The mechanisms of the synergistic effects of hydrogen peroxide and nitrites have been discussed in more detail in a previous paper [34]. It was proposed that, first, peroxynitrite is produced from hydrogen peroxide and nitrite, then primary singlet oxygen is produced from the peroxynitrite and hydrogen peroxide, and the primary singlet oxygen subsequently attacks catalase around the plasma membrane. Hydrogen peroxide accumulates around the plasma membrane through this process, and peroxynitrite is also produced via plasma membrane-bound enzymes. Then, secondary singlet oxygen is produced and accumulates around the plasma membrane. This hypothesis suggests that the levels of catalase on tumor cell membranes would correspond to the sensitivity of tumor cells to plasma-derived RONS; however, this hypothesis was not confirmed in a screening of 36 plasma-treated human cell lines [35].

Recently, components other than RONS have been identified in PAL [36]. Ringer's lactate solution contains NaCl, KCl, CaCl$_2$, and L-sodium lactate. A previous study demonstrated that among these four components, L-sodium lactate is responsible for the anti-tumor effects of PAL against glioblastoma cells [12]. These results suggested that organic compounds produced through the interactions between the plasma and L-sodium lactate are key factors for the anti-tumor effects of PAL. Such organic compounds include pyruvate, acetate, formate, glyoxylate, and 2,3-dimethyl tartrate, as identified by nuclear magnetic resonance analysis and direct infusion-electrospray ionization with tandem mass spectroscopy analysis. No cell cytotoxicity was observed from pyruvate, acetate, or formate at 5, 10, or 20 mM. However, glyoxylate or 2,3-dimethyl tartrate at more than 10 mM had cytotoxic effects in glioblastoma cells. Interestingly, 2,3-dimethyl tartrate selectively killed cancer cells, while glyoxylate killed both cancer cells and normal cells (Figure 13.1). For PAL generated with the kINPen, the generation of acetic acid, pyruvic acid, and formic acid was found, which correlated to cytotoxicity and immunogenic effects in tumor cells [19].

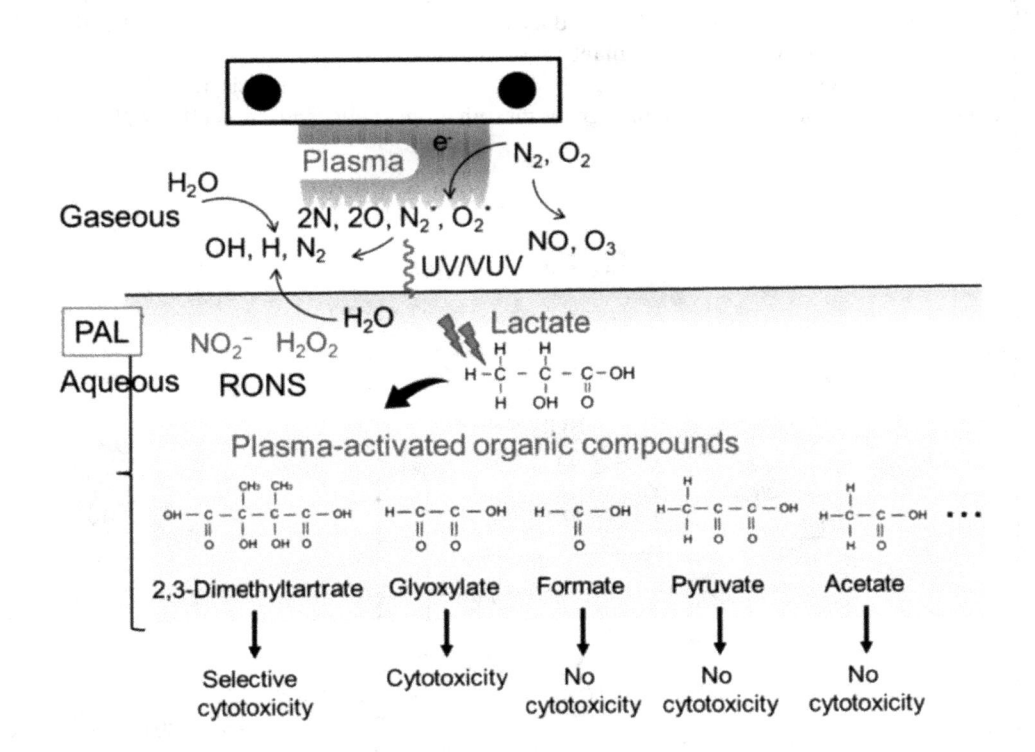

FIGURE 13.1 Components in PAL identified by nuclear magnetic resonance analysis and direct infusion-electrospray ionization with tandem mass spectroscopy analysis (Reproduced from [36]).

13.4   PLASMA BIOLOGY TO UNDERSTAND THE INTRACELLULAR MOLECULAR MECHANISMS OF PAM/PAL IN CELLS

It is well known that LTP and PAM/PAL can induce apoptosis, i.e., programmed cell death [37–39]. Over the last decade, the intracellular molecular mechanisms of LTP- and PAM/PAL-induced apoptosis in cancer cells have been well studied. LTP and PAM/PAL generally induce the production of intracellular RONS, which then interact with several biomolecules, such as proteins, nucleic acids (DNA and RNA), and lipids; these interactions affect intracellular signaling pathways and activate some signaling pathways that lead to apoptosis. For example, it was proposed that intracellular ROS induce DNA damage in cells, which activates the p53 signal transduction pathway to induce apoptosis [40,41].

Interestingly, PAM was found to induce more intracellular ROS than PAL in glioblastoma cells [42]. Comprehensive gene expression analysis using a microarray revealed that 61 genes were upregulated more than two-fold in PAM-treated glioblastoma cells when compared to the control (untreated Lactec-treated) cells. It is noteworthy that genes related to the GADD45 signaling pathway were upregulated in glioblastoma cells by PAM but not by PAL. The GADD45 signaling pathway mediates oxidative stress-dependent apoptosis. These results suggested that PAM induces oxidative stress-dependent cell death while PAL induces oxidative stress-independent cell death in glioblastoma cells. Previous studies have demonstrated that PAM downregulates survival and proliferation signaling pathways, such as the PI3K-AKT and RAS-MAPK signaling pathways [9,43]. It is known that the Gadd45α gene is downregulated by the PI3K-AKT signaling pathway. These results are consistent with the hypothesis that PAM induces oxidative stress-dependent apoptosis by downregulating survival and proliferation signaling networks (Figure 13.2). PAM and PAL have also been shown to affect the metabolites in glioblastoma cells [44]. Analysis using an extracellular flux analyzer revealed that PAL also affects the cellular respiratory system in HeLa cells [45].

Recently, it was demonstrated that PAL induces ferroptosis, a type of programmed cell death discovered in the early 2000s, in malignant mesothelioma cells [46]. Ferroptosis is defined as catalytic Fe (II)-dependent programmed cell death. A metabolome analysis revealed that PAL upregulates the citrulline-NO cycle in malignant mesothelioma cells, and intracellular NO induces

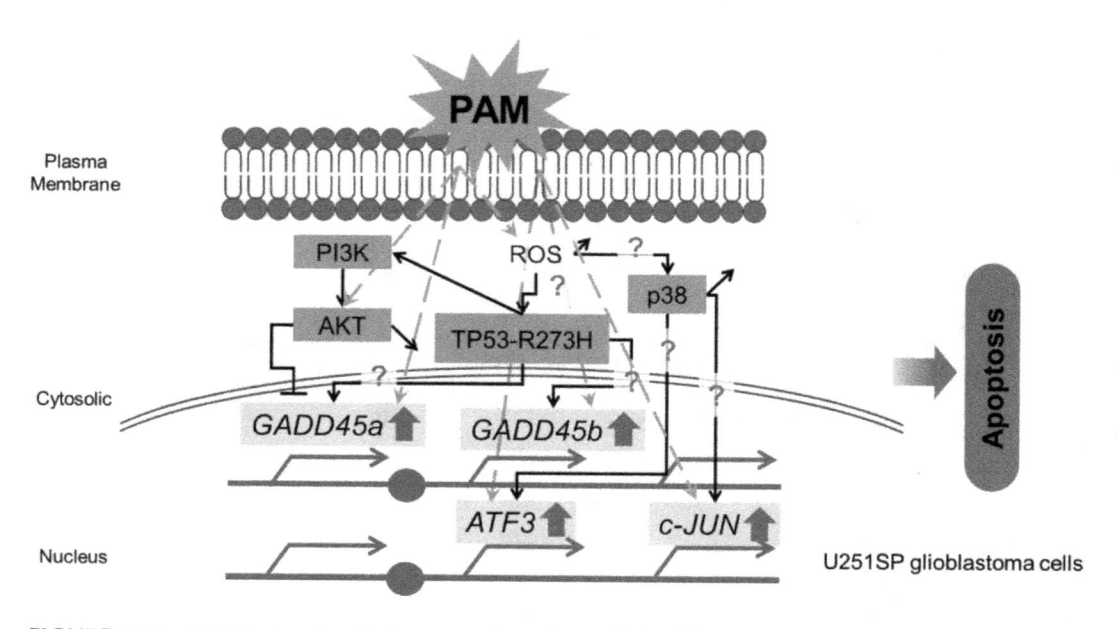

FIGURE 13.2   PAM-induced oxidative stress-dependent cell death in glioblastoma cells.

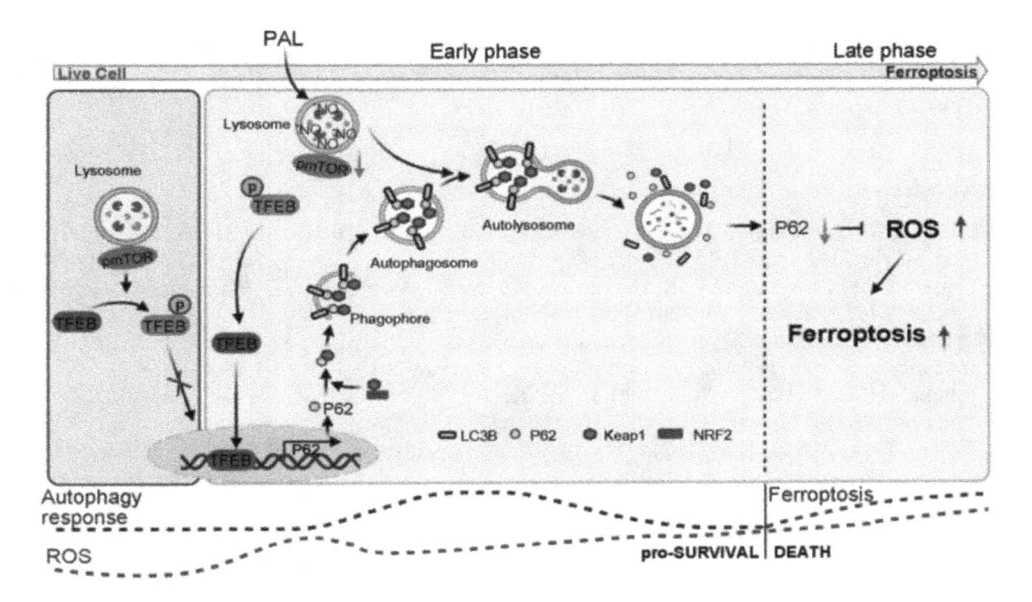

FIGURE 13.3 PAL-induced ferroptosis in malignant mesothelioma cells (Reproduced from [46]).

lysosome-dependent autophagy at the early phase in PAL-treated malignant mesothelioma cells. However, in the late phase, the autophagic process was terminated, p62-mediated antioxidant response was downregulated, and lysosomal dysfunction led to ferroptosis in PAL-treated malignant mesothelioma cells (Figure 13.3).

13.5 ANIMAL EXPERIMENTS LEADING TO THE CLINICAL APPLICATION OF PLASMA-TREATED LIQUIDS

Animal experiments are important for investigating the effectiveness and safety of plasma-treated liquids. Animal experiments using nude mice have demonstrated that LTP and plasma-treated liquids have anti-tumor effects in various cancer cells [47–51]. Plasma-treated liquids are expected to be useful in cancer patients to prevent peritoneal metastasis of cancer cells. A mouse model has been developed for visualizing the micrometastasis of gastric cancer cells, and the anti-tumor effects of PAM have been investigated in this mouse model [52]. The anti-tumor effects of the intraperitoneal administration of PAM against ovarian cancer cells have also been tested using a mouse model bearing ovarian cancer cells [53]. In addition, the anti-tumor effects of PAL have been evaluated using a peritoneal dissemination mouse model of pancreatic [54] and colorectal [55] cancer cells. Moreover, PAM produced with the kINPen was shown to be toxic in pancreatic cancer cells [56], while plasma-treated saline (PAS) reduced tumor burden in syngeneic models of colorectal cancer [17,18,57]. Recently, it was proposed that PAL washing therapy may be used to prevent intraperitoneal metastasis of cancers after surgical operations [58].

13.6 CONCLUSIONS

In this chapter, we introduced the current state-of-the-art in oncological applications of medically relevant plasma-treated liquids. Plasma sources have been developed for the clinical application of plasma-treated liquids. Studies on the physicochemical reactions of plasma-treated liquids have revealed that the anti-tumor effects are attributable to the RONS and plasma-activated organic compounds produced in plasma-treated liquids. In addition, studies on the biochemical reactions of plasma-treated liquids have revealed that plasma-treated liquids induce programmed cell death, such as apoptosis and ferroptosis. The anti-tumor effects of plasma-treated solutions have also been demonstrated in animal studies.

REFERENCES

[1] Laroussi, M. *From killing bacteria to destroying cancer cells: 20 years of plasma medicine.* Plasma Processes and Polymers **2014**, 10.1002/ppap.201400152.

[2] Von Woedtke, T.; Schmidt, A.; Bekeschus, S.; Wende, K.; Weltmann, K.D. *Plasma medicine: A field of applied redox biology.* In Vivo **2019**, 10.21873/invivo.11570.

[3] Keidar, M.; Shashurin, A.; Volotskova, O.; Stepp, M.A.; Srinivasan, P.; Sandler, A.; Trink, B. *Cold atmospheric plasma in cancer therapy.* Physics of Plasmas **2013**, 10.1063/1.4801516.

[4] Tanaka, H.; Bekeschus, S.; Yan, D.; Hori, M.; Keidar, M.; Laroussi, M. *Plasma-treated solutions (pts) in cancer therapy.* Cancers **2021**, 10.3390/cancers13071737.

[5] Tanaka, H.; Ishikawa, K.; Mizuno, M.; Toyokuni, S.; Kajiyama, H.; Kikkawa, F.; Metelmann, H.R.; Hori, M. *State of the art in medical applications using non-thermal atmospheric pressure plasma.* Reviews of Modern Plasma Physics **2017**, 10.1007/s41614-017-0004-3.

[6] Metelmann, H.R; Nedrelow, D.S.; Seebauer, C.; Schuster, M.; von Woedtke, T.; Weltmann, K.D.; Kindler, S.; Metelmann, P.H.; Finkelstein, S.E.; Von Hoff, D.D.; Podmelle, F. *Head and neck cancer treatment and physical plasma.* Clinical Plasma Medicine **2015**, 10.1016/j.cpme.2015.02.001.

[7] Canady, J.; Murthy, S.R.K.; Zhuang, T.; Gitelis, S.; Nissan, A.; Ly, L.; Jones, O.Z.; Cheng, X.; Adileh, M.; Blank, A.T.; Colman, M.W.; Millikan, K.; O'Donoghue, C.; Stenson, K.M.; Ohara, K.; Schtrechman, G.; Keidar, M.; Basadonna, G. *The first cold atmospheric plasma phase i clinical trial for the treatment of advanced solid tumors: A novel treatment arm for cancer.* Cancers **2023**, 10.3390/cancers15143688.

[8] Bruggeman, P.J.; Kushner, M.J.; Locke, B.R.; Gardeniers, J.G.E.; Graham, W.G.; Graves, D.B.; Hofman-Caris, R.C.H.M.; Maric, D.; Reid, J.P.; Ceriani, E.; Rivas, D.F.; Foster, J.E.; Garrick, S.C.; Gorbanev, Y.; Hamaguchi, S.; Iza, F.; Jablonowski, H.; Klimova, E.; Kolb, J.; Krcma, F.; Lukes, P.; Machala, Z.; Marinov, I.; Mariotti, D.; Thagard, S.M.; Minakata, D.; Neyts, E.C.; Pawlat, J.; Petrovic, Z.L.; Pflieger, R.; Reuter, S.; Schram, D.C.; Schroter, S.; Shiraiwa, M.; Tarabova, B.; Tsai, P.A.; Verlet, J.R.R.; von Woedtke, T.; Wilson, K.R.; Yasui, K.; Zvereva, G. *Plasma-liquid interactions: A review and roadmap.* Plasma Sources Science and Technology **2016**, 10.1088/0963-0252/25/5/053002.

[9] Tanaka, H.; Mizuno, M.; Ishikawa, K.; Nakamura, K.; Kajiyama, H.; Kano, H.; Kikkawa, F.; Hori, M. *Plasma-activated medium selectively kills glioblastoma brain tumor cells by down-regulating a survival signaling molecule, akt kinase.* Plasma Medicine **2011**, 10.1615/PlasmaMed.2012006275.

[10] Yan, D.; Nourmohammadi, N.; Bian, K.; Murad, F.; Sherman, J.H.; Keidar, M. *Stabilizing the cold plasma-stimulated medium by regulating medium's composition.* Scientific Reports **2016**, 10.1038/srep26016.

[11] Bekeschus, S.; Kading, A.; Schroder, T.; Wende, K.; Hackbarth, C.; Liedtke, K.R.; van der Linde, J.; von Woedtke, T.; Heidecke, C.D.; Partecke, L.I. *Cold physical plasma-treated buffered saline solution as effective agent against pancreatic cancer cells.* Anticancer Agents in Medicinal Chemistry **2018**, 10.21 74/1871520618666180507130243.

[12] Tanaka, H.; Nakamura, K.; Mizuno, M.; Ishikawa, K.; Takeda, K.; Kajiyama, H.; Utsumi, F.; Kikkawa, F.; Hori, M. *Non-thermal atmospheric pressure plasma activates lactate in ringer's solution for anti-tumor effects.* Scientific Reports **2016**, 10.1038/srep36282.

[13] Freund, E.; Bekeschus, S. *Gas plasma-oxidized liquids for cancer treatment: Preclinical relevance, immuno-oncology, and clinical obstacles.* IEEE Transactions on Radiation and Plasma Medical Sciences **2021**, 10.1109/trpms.2020.3029982.

[14] Khalili, M.; Daniels, L.; Lin, A.; Krebs, F.C.; Snook, A.E.; Bekeschus, S.; Bowne, W.B.; Miller, V. *Non-thermal plasma-induced immunogenic cell death in cancer: A topical review.* Journal of Physics D-Applied Physics **2019**, 10.1088/1361-6463/ab31c1.

[15] Azzariti, A.; Iacobazzi, R.M.; Di Fonte, R.; Porcelli, L.; Gristina, R.; Favia, P.; Fracassi, F.; Trizio, I.; Silvestris, N.; Guida, G.; Tommasi, S.; Sardella, E. *Plasma-activated medium triggers cell death and the presentation of immune activating danger signals in melanoma and pancreatic cancer cells.* Scientific Reports **2019**, 10.1038/s41598-019-40637-z.

[16] Liedtke, K.R.; Freund, E.; Hackbarth, C.; Heidecke, C.D.; Partecke, L.I.; Bekeschus, S. *A myeloid and lymphoid infiltrate in murine pancreatic tumors exposed to plasma-treated medium.* Clinical Plasma Medicine **2018**, 10.1016/j.cpme.2018.07.001.

[17] Freund, E.; Liedtke, K.R.; van der Linde, J.; Metelmann, H.R.; Heidecke, C.D.; Partecke, L.I.; Bekeschus, S. *Physical plasma-treated saline promotes an immunogenic phenotype in ct26 colon cancer cells in vitro and in vivo*. Scientific Reports **2019**, 10.1038/s41598-018-37169-3.

[18] Miebach, L.; Freund, E.; Clemen, R.; Kersting, S.; Partecke, L.I.; Bekeschus, S. *Gas plasma-oxidized sodium chloride acts via hydrogen peroxide in a model of peritoneal carcinomatosis*. Proceedings of the National Academy of Sciences of the United States of America (PNAS) **2022**, 10.1073/pnas.2200708119.

[19] Miebach, L.; Mohamed, H.; Wende, K.; Miller, V.; Bekeschus, S. *Pancreatic cancer cells undergo immunogenic cell death upon exposure to gas plasma-oxidized ringers lactate*. Cancers **2023**, 10.3390/cancers15010319.

[20] Bekeschus, S.; Scherwietes, L.; Freund, E.; Liedtke, K.R.; Hackbarth, C.; von Woedtke, T.; Partecke, L.I. *Plasma-treated medium tunes the inflammatory profile in murine bone marrow-derived macrophages*. Clinical Plasma Medicine **2018**, 10.1016/j.cpme.2018.06.001.

[21] Reuter, S.; von Woedtke, T.; Weltmann, K.D. *The kinpen—a review on physics and chemistry of the atmospheric pressure plasma jet and its applications*. Journal of Physics D: Applied Physics **2018**, 10.1088/1361-6463/aab3ad.

[22] Bekeschus, S.; Schmidt, A.; Weltmann, K.D.; von Woedtke, T. *The plasma jet kinpen—a powerful tool for wound healing*. Clinical Plasma Medicine **2016**, 10.1016/j.cpme.2016.01.001.

[23] Metelmann, H.R.; Seebauer, C.; Miller, V.; Fridman, A.; Bauer, G.; Graves, D.B.; Pouvesle, J.M.; Rutkowski, R.; Schuster, M.; Bekeschus, S.; Wende, K.; Masur, K.; Hasse, S.; Gerling, T.; Hori, M.; Tanaka, H.; Choi, E.; Weltemann, K.D.; von Woedtke, T. *Clinical experience with cold plasma in the treatment of locally advanced head and neck cancer*. Clinical Plasma Medicine **2018**, 10.1016/j.cpme.2017.09.001.

[24] Rutkowski, R.; Daeschlein, G.; von Woedtke, T.; Smeets, R.; Gosau, M.; Metelmann, H.R. *Long-term risk assessment for medical application of cold atmospheric pressure plasma*. Diagnostics **2020**, 10.3390/diagnostics10040210.

[25] Golda, J.; Held, J.; Redeker, B.; Konkowski, M.; Beijer, P.; Sobota, A.; Kroesen, G.; Braithwaite, N.S.J.; Reuter, S.; Turner, M.M.; Gans, T.; O'Connell, D.; Schulz-von der Gathen, V. *Concepts and characteristics of the 'cost reference microplasma jet'*. Journal of Physics D: Applied Physics **2016**, 10.1088/0022-3727/49/8/084003.

[26] Lackmann, J.W.; Bruno, G.; Jablonowski, H.; Kogelheide, F.; Offerhaus, B.; Held, J.; Schulz-von der Gathen, V.; Stapelmann, K.; von Woedtke, T.; Wende, K. *Nitrosylation vs. Oxidation—how to modulate cold physical plasmas for biological applications*. PloS One **2019**, 10.1371/journal.pone.0216606.

[27] Iwasaki, M.; Inui, H.; Matsudaira, Y.; Kano, H.; Yoshida, N.; Ito, M.; Hori, M. *Nonequilibrium atmospheric pressure plasma with ultrahigh electron density and high performance for glass surface cleaning*. Applied Physics Letters **2008**, 10.1063/1.2885084.

[28] Nakamura, K.; Yoshikawa, N.; Yoshihara, M.; Ikeda, Y.; Higashida, A.; Niwa, A.; Jindo, T.; Tanaka, H.; Ishikawa, K.; Mizuno, M.; Toyokuni, S.; Hori, M.; Kikkawa, F.; Kajiyama, H. *Adjusted multiple gases in the plasma flow induce differential antitumor potentials of plasma-activated solutions*. Plasma Processes and Polymers **2020**, 10.1002/ppap.201900259.

[29] Freund, E.; Liedtke, K.R.; Gebbe, R.; Heidecke, A.K.; Partecke, L.I.; Bekeschus, S. *In vitro anticancer efficacy of six different clinically approved types of liquids exposed to physical plasma*. IEEE Transactions on Radiation and Plasma Medical Sciences **2019**, 10.1109/trpms.2019.2902015.

[30] Takeda, K.; Ishikawa, K.; Tanaka, H.; Sekine, M.; Hori, M. *Spatial distributions of o, n, no, oh and vacuum ultraviolet light along gas flow direction in an ac-excited atmospheric pressure ar plasma jet generated in open air*. Journal of Physics D: Applied Physics **2017**, 10.1088/1361-6463/aa6555.

[31] Tanaka, H.; Mizuno, M.; Ishikawa, K.; Kondo, H.; Takeda, K.; Hashizume, H.; Nakamura, K.; Utsumi, F.; Kajiyama, H.; Kano, H.; Okazaki, Y.; Toyokuni, S.; Akiyama, S.; Maruyama, S.; Yamada, S.; Kodera, Y.; Kaneko, H.; Terasaki, H.; Hara, H.; Adachi, T.; Iida, M.; Yajima, I.; Kato, M.; Kikkawa, F.; Hori, M. *Plasma with high electron density and plasma-activated medium for cancer treatment*. Clinical Plasma Medicine **2015**, 10.1016/j.cpme.2015.09.001.

[32] Tanaka, H.; Mizuno, M.; Toyokuni, S.; Maruyama, S.; Kodera, Y.; Terasaki, H.; Adachi, T.; Kato, M.; Kikkawa, F.; Hori, M. *Cancer therapy using non-thermal atmospheric pressure plasma with ultra-high electron density*. Physics of Plasmas **2015**, 10.1063/1.4933366.

[33] Kurake, N.; Tanaka, H.; Ishikawa, K.; Kondo, T.; Sekine, M.; Nakamura, K.; Kajiyama, H.; Kikkawa, F.; Mizuno, M.; Hori, M. *Cell survival of glioblastoma grown in medium containing hydrogen peroxide and/or nitrite, or in plasma-activated medium*. Archives of Biochemistry and Biophysics **2016**, 10.1016/j.abb.2016.01.011.

[34] Bauer, G. *The synergistic effect between hydrogen peroxide and nitrite, two long-lived molecular species from cold atmospheric plasma, triggers tumor cells to induce their own cell death.* Redox Biology **2019**, 10.1016/j.redox.2019.101291.

[35] Bekeschus, S.; Liebelt, G.; Menz, J.; Berner, J.; Sagwal, S.K.; Wende, K.; Weltmann, K.D.; Boeckmann, L.; von Woedtke, T.; Metelmann, H.R.; Emmert, S.; Schmidt, A. *Tumor cell metabolism correlates with resistance to gas plasma treatment: The evaluation of three dogmas.* Free Radical Biology and Medicine **2021**, 10.1016/j.freeradbiomed.2021.02.035.

[36] Tanaka, H.; Hosoi, Y.; Ishikawa, K.; Yoshitake, J.; Shibata, T.; Uchida, K.; Hashizume, H.; Mizuno, M.; Okazaki, Y.; Toyokuni, S.; Nakamura, K.; Kajiyama, H.; Kikkawa, F.; Hori, M. *Low temperature plasma irradiation products of sodium lactate solution that induce cell death on u251sp glioblastoma cells were identified.* Scientific Reports **2021**, 10.1038/s41598-021-98020-w.

[37] Kieft, I.E.; Kurdi, M.; Stoffels, E. *Reattachment and apoptosis after plasma-needle treatment of cultured cells.* Ieee Transactions on Plasma Science **2006**, 10.1109/Tps.2006.876511.

[38] Fridman, G.; Shereshevsky, A.; Jost, M.M.; Brooks, A.D.; Fridman, A.; Gutsol, A.; Vasilets, V.; Friedman, G. *Floating electrode dielectric barrier discharge plasma in air promoting apoptotic behavior in melanoma skin cancer cell lines.* Plasma Chemistry and Plasma Processing **2007**, 10.1007/s11090-007-9048-4.

[39] Tanaka, H.; Mizuno, M.; Ishikawa, K.; Nakamura, K.; Kajiyama, H.; Kano, H.; Kikkawa, F.; Hori, M. *Plasma-activated medium selectively kills glioblastoma brain tumor cells by down-regulating a survival signaling molecule, akt kinase.* Plasma Medicine **2013**, 10.1615/PlasmaMed.2012006275

[40] Ma, Y.; Ha, C.S.; Hwang, S.W.; Lee, H.J.; Kim, G.C.; Lee, K.W.; Song, K. *Non-thermal atmospheric pressure plasma preferentially induces apoptosis in p53-mutated cancer cells by activating ros stress-response pathways.* PloS One **2014**, 10.1371/journal.pone.0091947.

[41] Chang, J.W.; Kang, S.U.; Shin, Y.S.; Kim, K.I.; Seo, S.J.; Yang, S.S.; Lee, J.S.; Moon, E.; Baek, S.J.; Lee, K.; Kim, C.H. *Non-thermal atmospheric pressure plasma induces apoptosis in oral cavity squamous cell carcinoma: Involvement of DNA-damage-triggering sub-g(1) arrest via the atm/p53 pathway.* Archives of Biochemistry and Biophysics **2014**, 10.1016/j.abb.2014.01.022.

[42] Tanaka, H.; Mizuno, M.; Katsumata, Y.; Ishikawa, K.; Kondo, H.; Hashizume, H.; Okazaki, Y.; Toyokuni, S.; Nakamura, K.; Yoshikawa, N.; Kajiyama, H.; Kikkawa, F.; Hori, M. *Oxidative stress-dependent and -independent death of glioblastoma cells induced by non-thermal plasma-exposed solutions.* Scientific Reports **2019**, 10.1038/s41598-019-50136-w.

[43] Tanaka, H.; Mizuno, M.; Ishikawa, K.; Nakamura, K.; Utsumi, F.; Kajiyama, H.; Kano, H.; Maruyama, S.; Kikkawa, F.; Hori, M. *Cell survival and proliferation signaling pathways are downregulated by plasma-activated medium in glioblastoma brain tumor cells.* Plasma Medicine **2012**, 10.1615/PlasmaMed.2013008267.

[44] Ishikawa, K.; Hosoi, Y.; Tanaka, H.; Jiang, L.; Toyokuni, S.; Nakamura, K.; Kajiyama, H.; Kikkawa, F.; Mizuno, M.; Hori, M. *Non-thermal plasma-activated lactate solution kills u251sp glioblastoma cells in an innate reductive manner with altered metabolism.* Archives of Biochemistry and Biophysics **2020**, 10.1016/j.abb.2020.108414.

[45] Tanaka, H.; Maeda, S.; Nakamura, K.; Hashizume, H.; Ishikawa, K.; Ito, M.; Ohno, K.; Mizuno, M.; Motooka, Y.; Okazaki, Y.; Toyokuni, S.; Kajiyama, H.; Kikkawa, F.; Hori, M. *Plasma-activated ringer's lactate solution inhibits the cellular respiratory system in hela cells.* Plasma Processes and Polymers **2021**, 10.1002/ppap.202100056.

[46] Jiang, L.; Zheng, H.; Lyu, Q.; Hayashi, S.; Sato, K.; Sekido, Y.; Nakamura, K.; Tanaka, H.; Ishikawa, K.; Kajiyama, H.; Mizuno, M.; Hori, M.; Toyokuni, S. *Lysosomal nitric oxide determines transition from autophagy to ferroptosis after exposure to plasma-activated ringer's lactate.* Redox Biology **2021**, 10.1016/j.redox.2021.101989.

[47] Vandamme, M.; Robert, E.; Pesnel, S.; Barbosa, E.; Dozias, S.; Sobilo, J.; Lerondel, S.; Le Pape, A.; Pouvesle, J.M. *Antitumor effect of plasma treatment on u87 glioma xenografts: Preliminary results.* Plasma Processes and Polymers **2010**, 10.1002/ppap.200900080.

[48] Keidar, M.; Walk, R.; Shashurin, A.; Srinivasan, P.; Sandler, A.; Dasgupta, S.; Ravi, R.; Guerrero-Preston, R.; Trink, B. *Cold plasma selectivity and the possibility of a paradigm shift in cancer therapy.* British Journal of Cancer **2011**, 10.1038/bjc.2011.386.

[49] Utsumi, F.; Kajiyama, H.; Nakamura, K.; Tanaka, H.; Mizuno, M.; Ishikawa, K.; Kondo, H.; Kano, H.; Hori, M.; Kikkawa, F. *Effect of indirect nonequilibrium atmospheric pressure plasma on anti-proliferative activity against chronic chemo-resistant ovarian cancer cells in vitro and in vivo.* PLoS ONE **2013**, 10.1371/journal.pone.0081576.

[50] Hattori, N.; Yamada, S.; Torii, K.; Takeda, S.; Nakamura, K.; Tanaka, H.; Kajiyama, H.; Kanda, M.; Fujii, T.; Nakayama, G.; Sugimoto, H.; Koike, M.; Nomoto, S.; Fujiwara, M.; Mizuno, M.; Hori, M.; Kodera, Y. *Effectiveness of plasma treatment on pancreatic cancer cells.* International Journal of Oncology **2015**, 10.3892/ijo.2015.3149.

[51] Miebach, L.; Melo-Zainzinger, G.; Freund, E.; Clemen, R.; Cecchini, A.L.; Bekeschus, S. *Medical gas plasma technology combines with antimelanoma therapies and promotes immune-checkpoint therapy responses.* Advanced Science **2023**, 10.1002/advs.202303183.

[52] Takeda, S.; Yamada, S.; Hattori, N.; Nakamura, K.; Tanaka, H.; Kajiyama, H.; Kanda, M.; Kobayashi, D.; Tanaka, C.; Fujii, T.; Fujiwara, M.; Mizuno, M.; Hori, M.; Kodera, Y. *Intraperitoneal administration of plasma-activated medium: Proposal of a novel treatment option for peritoneal metastasis from gastric cancer.* Annals of Surgical Oncology **2017**, 10.1245/s10434-016-5759-1.

[53] Nakamura, K.; Peng, Y.; Utsumi, F.; Tanaka, H.; Mizuno, M.; Toyokuni, S.; Hori, M.; Kikkawa, F.; Kajiyama, H. *Novel intraperitoneal treatment with non-thermal plasma-activated medium inhibits metastatic potential of ovarian cancer cells.* Scientific Reports **2017**, 10.1038/s41598-017-05620-6.

[54] Sato, Y.; Yamada, S.; Takeda, S.; Hattori, N.; Nakamura, K.; Tanaka, H.; Mizuno, M.; Hori, M.; Kodera, Y. *Effect of plasma-activated lactated ringer's solution on pancreatic cancer cells in vitro and in vivo.* Annals of Surgical Oncology **2018**, 10.1245/s10434-017-6239-y.

[55] Miebach, L.; Freund, E.; Cecchini, A.L.; Bekeschus, S. *Conductive gas plasma treatment augments tumor toxicity of ringer's lactate solutions in a model of peritoneal carcinomatosis.* Antioxidants **2022**, 10.3390/antiox11081439.

[56] Liedtke, K.R.; Bekeschus, S.; Kaeding, A.; Hackbarth, C.; Kuehn, J.P.; Heidecke, C.D.; von Bernstorff, W.; von Woedtke, T.; Partecke, L.I. *Non-thermal plasma-treated solution demonstrates antitumor activity against pancreatic cancer cells in vitro and in vivo.* Scientific Reports **2017**, 10.1038/s41598-017-08560-3.

[57] Freund, E.; Miebach, L.; Clemen, R.; Schmidt, M.; Heidecke, A.; von Woedtke, T.; Weltmann, K.D.; Kersting, S.; Bekeschus, S. *Large volume spark discharge and plasma jet-technology for generating plasma-oxidized saline targeting colon cancer in vitro and in vivo.* Journal of Applied Physics **2021**, 10.1063/5.0033406.

[58] Nakamura, K.; Yoshikawa, N.; Mizuno, Y.; Ito, M.; Tanaka, H.; Mizuno, M.; Toyokuni, S.; Hori, M.; Kikkawa, F.; Kajiyama, H. *Preclinical verification of the efficacy and safety of aqueous plasma for ovarian cancer therapy.* Cancers **2021**, 10.3390/cancers13051141.

14 Model Systems in Plasma Biomedicine

Angela Privat-Maldonado, Christophe Deben

14.1 INTRODUCTION

With only a few clinical trials ongoing or completed, the transition of biomedical plasmas from bench to bedside remains slow. Many of the preclinical studies still rely largely on the classic 2-dimensional (2D) monolayer culture model due to its low cost and ease of use. However, the 2D model does not accurately represent how cells are affected by disease and treatment. As the field moves forward, researchers need better tools and model systems. Taking the drug discovery industry as an example, one of the most promising areas for the development of the field is the use of preclinical models that better simulate *in vivo* biology and tissue microenvironment [1]. It is well accepted that 3-dimensional (3D) culture models represent the complex interactions between cells and their microenvironment better than standard 2D systems, as these interactions affect gene expression and the response to treatments [2]. The 3D system models present concentration gradients for oxygen, pH, nutrients, and metabolites found in real tissues, which can affect cell behavior, motility, and cell signaling. These factors play a key role in multiple processes, including wound healing and tumor progression [3]. In addition, 3D system models can incorporate different cell populations (e.g., stromal, endothelial, immune cells), making them a competitive alternative to expensive *in vivo* models for the development of plasma therapies.

14.2 RELEVANT 3D *IN VITRO* MODELS

14.2.1 SPHEROIDS

3D spheroids can be generated using immortalized, commercially available cell lines to reproduce key features of solid tumors *in vivo*. Compared to 2D cultures (Figure 14.1A), cells in spheroids interact in an environment with variable stiffness (extracellular matrix [ECM]), and this 3D organization recreates the gradients of nutrients, oxygen, and metabolites commonly found in solid tumors. Spheroids present heterogeneous populations, with proliferating cells in the outer layer, quiescent cells in the inner layers, and a hypoxic core. Spheroids are generated in a few days, starting from single-cell suspensions using ultra-low attachment plates (Figure 14.1B), hanging drops, magnetic levitation, microwells, or spinner cultures [4]. In addition, spheroids can be formed using cancer cells alone (homotypic) or together with other cells of the tumor microenvironment (TME), such as fibroblasts, immune cells, and endothelial cells (heterotypic). Although most of the work done with spheroids is in cancer research, the model is also valuable for the study of stem cells for tissue regeneration and transplantation therapy, as it provides a platform to study cell-to-cell interaction, maintenance of stemness, and differentiation potential upon treatment [5].

In all cases, optimal spheroids of 300–500 µm diameter can be generated with 2,000–10,000 cells per spheroid, best mimicking the *in vivo* hypoxic and proliferating gradients [6]. The initial cell aggregation occurs between the extracellular matrix fibers and integrins, after which spheroids are formed. In this second step, the level of compactness depends on the expression of E-cadherin. Highly compact spheroids are formed between cells with high levels of E-cadherin, whereas loose aggregates are observed when N-cadherin is favored, as the E-to-N cadherin switch is often a hallmark of aggressive

DOI: 10.1201/9781003328056-16

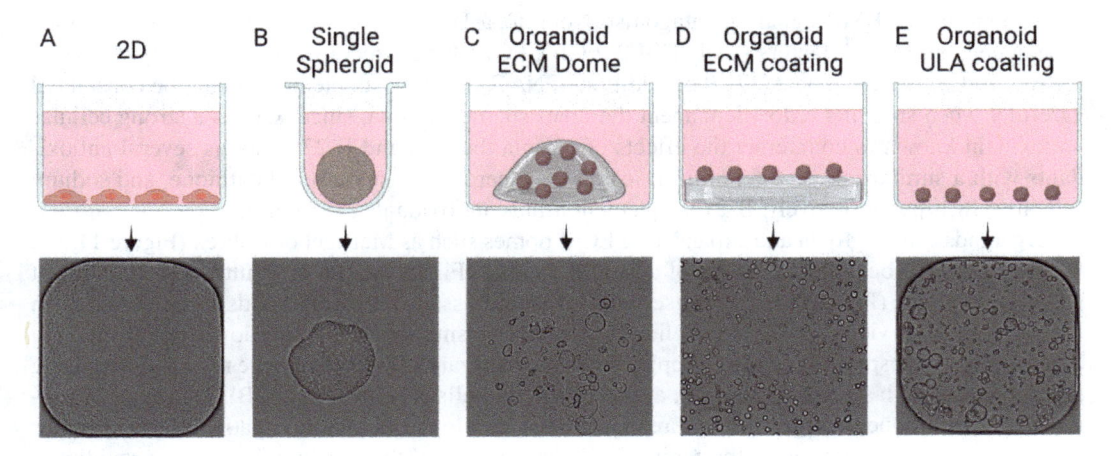

FIGURE 14.1 Overview of 2D and 3D *in vitro* culturing methods. (A) 2D cell culture in flat bottom microplates. The image shows A549 lung cancer cells grown in a 384-well microplate. (B) Single 3D spheroid culture in round-bottom ultra-low attachment (ULA) plates. The image shows A549 lung cancer cells grown in a 96-well microplate. (C) Patient-derived tumor organoids (PDTOs) grown in extracellular matrix (ECM) domes. The image shows pancreatic cancer organoids grown in a 6-well microplate. (D) PDTOs grown on top of an ECM coating in a flat-bottom microplate. The image shows head-and-neck carcinoma organoids grown in a 96-well microplate. (E) PDTOs grown in a flat bottom ULA microplate. The image shows pancreatic cancer organoids grown in a 384-well microplate. ECM: extracellular matrix. ULA: ultra-low attachment. Created with BioRender.com.

cancers [7,8]. Compact spheroids are usually more resistant to therapies than loosely aggregated cells, and smaller spheroids are more sensitive, as drugs or RONS can penetrate the tissue more easily. To improve spheroid compactness and uniformity in cells with low expression of E-cadherin, additives such as Matrigel (matrix extract from the Engelbreth-Holm-Swarm mouse sarcoma, rich in laminin, collagen IV, entactin, proteoglycans, and growth factors), collagen, and methylcellulose can be added to the medium during spheroid generation. Plasma has been shown to downregulate the expression of E-cadherin and disrupt cell-to-cell adhesion, resulting in the opening of intercellular channels to enhance the delivery of antineoplastic drugs [9,10] and exert an effect in other cells due to the bystander effect [11]. This makes spheroids a suitable model for studying the effect of plasma on cell-to-cell and cell-to-ECM interactions, combination therapies, and immune cell infiltration.

To support spheroid researchers, the platform MISpheroID shares the characteristics of more than 4,500 spheroid experiments, providing key technical and biological information for 3D cultures with the aim of increasing reproducibility and uniformity in reported parameters [12]. This information is highly important since the type of cell culture medium has already been shown to influence cellular response to plasma treatment [13].

14.2.2 Organoids

Organoids are multicellular 3D structures that can be derived from healthy and tumor tissue from patient's adult (cancer) stem cells, and that resemble the original tissue in terms of morphology, heterogeneity, and molecular characteristics. In addition, they can self-organize, differentiate, and self-renew [14–16]. These organoid models have already proven their value in preclinical, translational, and even clinical applications because they have been shown to predict patient therapy response, opening the way for next-generation precision medicine using these 'patients-in-the-lab'.

In general, the culture medium for patient-derived organoids is more standardized compared to conventional cell-line media. Typically, a basal medium adapted for serum-free conditions is used (advanced DMEM/F-12), supplemented with promotors of the Wnt/β-catenin pathway (Wnt3a and/

or R-Spondin-3), a BMP signaling antagonist (Noggin), a TGF-β pathway inhibitor (A83–01), tissue type dependent growth factors (EGF, FGF2, FGF7, FGF10, etc.), nicotinamide, N-Acetyl Cysteine (NAC), and B-27 supplement [17]. The addition of NAC and B-27 supplement should be considered carefully when studying redox-dependent therapies such as plasma, since NAC is a strong cellular antioxidant known to counteract the effects of plasma therapy, and B-27 contains several antioxidants with a similar outcome (vitamin E, catalase, superoxide dismutase, glutathione, and sodium selenite) [18,19]. Alternatively, B-27 supplement minus antioxidants can be used.

Organoids can be grown and expanded in ECM domes such as Matrigel or Cultrex (Figure 14.1C), but they can also be cultured on top of an ECM coating (Figure 14.1D) or in ultra-low attachment flat-bottom plates (Figure 14.1E). These multi-organoid assays allow organoids to be obtained in the same field of view, which is beneficial for direct plasma treatment and downstream imaging applications. Like spheroids, more complex heterotypic organoid cultures can be reconstituted using autologous fibroblasts, immune cells, and endothelial cells (Figures 14.2A-B). However, culturing methods unique to organoid cultures have been developed to generate patient-derived tumor organoids (PDTOs) that retain the native TME and tumor-infiltrating immune cells, providing a unique ex vivo model to study immunological responses and even predict immunotherapy response [20,21]. Multiple studies have already proven the immunogenicity of plasma therapy and its potential to target cancer-associated fibroblasts [22–24]. These complex ex vivo models can be an important alternative to studying the influence of plasma therapy on an animal-free TME.

The most widely used assay to characterize therapy response in organoids is the luminescent ATP-based CellTiter-Glo 3D cell viability assay (Promega). The limitation of this assay is that it is a bulk endpoint measurement and is unable to distinguish cytotoxic from cytostatic responses to obtain more in-depth insights into the therapy outcome. Live-cell imaging is a much more powerful tool since it allows for kinetic measurements and for studying therapy response on individual organoids or even at the (sub)cellular level. It is also possible to include fluorescent reporter genes to study cell death, apoptosis, oxidative stress, hypoxia, and senescence, among other factors, in real time. When studying redox-dependent therapies such as plasma, fluorescent viability dyes should be used with caution since photoexcited fluorophores can release ROS, which can damage nearby cellular structures [25]. A process known as phototoxicity can unwantedly enhance the effect of plasma therapy. Phototoxicity

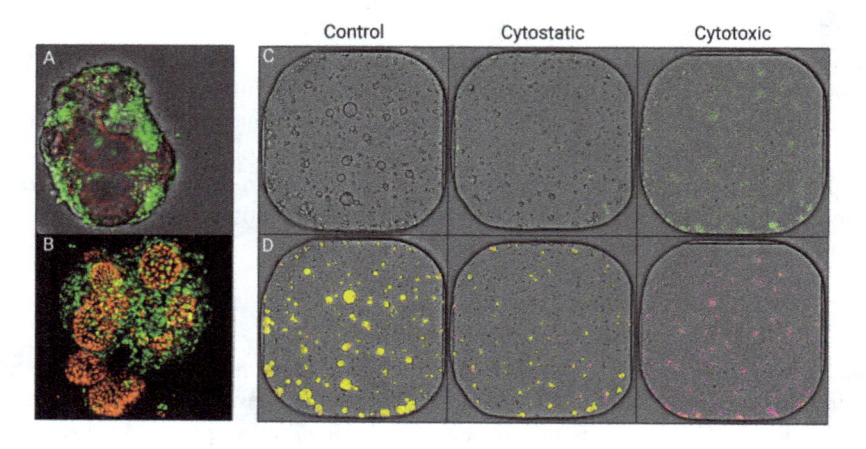

FIGURE 14.2 Live-cell imaging of patient-derived tumor organoids. (A) Widefield and (B) confocal image of a 'micro-tumor' composed of patient-derived pancreatic cancer organoids (red) and cancer-associated fibroblasts (green). (C) Example of whole 384-well brightfield and fluorescence imaging of patient-derived pancreatic cancer organoids treated with control (DMSO), a cytostatic drug (paclitaxel), and a cytotoxic drug (staurosporine). Cytotox green was used as a fluorescent cell-death marker. (D) Corresponding label-free organoid (yellow) and fluorescence segmentation (pink) using the Orbits image analysis software [26]. Created with BioRender.com.

can be reduced by lowering the intensity and exposure time of the light source, reducing the frequency of the measurements, or choosing fluorophores with excitation in the (far-)red spectrum and avoiding the blue range [25]. The implementation of artificial intelligence (AI) in live-cell image analysis is an exciting development, since the algorithm can be trained to detect organoid structures label-free, eliminating fluorophore-induced phototoxicity and improving the overall accuracy of the model. An example of label-free detection of organoids by the Orbits image analysis software is shown in Figures 14.2C-D, and the therapy response can be studied more accurately at a single-organoid level in combination with a fluorescent cell-death marker [26]. The choice of microscope mainly depends on the complexity of the model and the research question. Widefield live-cell imaging systems are mostly sufficient to study tumor-only organoids in a single field of view, mainly at the organoid level. More in-depth analysis at the cellular level can be obtained by high-content confocal imaging systems that can penetrate deeper into the 3D structure, allowing for Z-stacking and a higher resolution [27]. Confocal imaging is especially advised in co-culture models since it allows the study of cell-type specific responses (Figures 14.2A-B). A downside is that fluorescent dyes are still required for high-content imaging to distinguish different cell types and cellular responses.

An important benefit of PDTOs is that both normal and tumor organoids can be cultured in nearly identical conditions from the same patient, excluding the influence of culture medium on plasma treatment response and providing an ideal model to study the selectivity of plasma treatment towards cancer cells [13]. To date, however, there is only one publication on biomedical plasmas using organoids [28].

14.2.3 Tumor-on-a-chip/Microfluidics

The next level in mimicking the TME *in vitro* is the inclusion of cellular layers, air, and blood flow in so-called tumor-on-a-chip models in combination with microfluidics. Cell layers can be recreated by using a membrane on which, for example, endothelial cells are grown on one side to mimic blood vessels, and normal epithelial and/or tumor cells are grown on the other side, separating each layer in different compartments while retaining interaction between all cell types. These microfluidic devices can then be used to establish a well-defined gradient to simulate, for example, the recruitment of immune cells from the blood flow on the endothelial side to the epithelial/tumor side through inflammatory chemokine/cytokine gradients or by attraction towards the tumor cells [29,30]. Currently, this model has not been used in the field of plasma research, but it can be an interesting model for studying the immunogenicity of plasma-treated cancer cells or the effect of indirect plasma perfusion on both tumoral and healthy tissue.

14.2.4 Scaffolds

Scaffolds are useful for the study of plasma-derived ROS on growth and invasion of cancer cells, angiogenesis, cell differentiation, and wound healing. Scaffolds are low-cost and tuneable and can be generated using natural or synthetic polymers or a combination of both types. These polymers can be functionalized and mixed with other materials to improve their mechanical, biological, and chemical properties or loaded with nutrients, soluble factors, and ROS [31]. The cell behavior is influenced by the chemical, biological, and physical properties of the polymers [3], and the material to be used must be carefully chosen. The advantage of scaffolds is that they are highly reproducible. However, the architecture and properties are limited to the components present in them. To improve a scaffold, different materials can be combined. For example, hyaluronan can be combined with chitosan to improve the physical and mechanical properties of the scaffold and to promote cell binding [32], whereas the combination of collagen and hyaluronan favors water retention and gas exchange to promote wound healing [33]. The ability of some scaffolds, such as hydrogels, to retain large amounts of water makes them an ideal system for studying tissue regeneration [34]. Scaffolds can also be used for 3D bioprinting, which allows the fabrication of complex matrices with high resolution and controlled microarchitecture, with the option to combine polymers and cells (cellular bioprinting) [35].

Besides artificial scaffolds, natural decellularized scaffolds can also be obtained after the complete removal of cell content from a tissue or organ while preserving the native ECM 3D structure and organization [36]. Complete cell removal is achieved by using chemical agents (acids or bases), hypotonic or hypertonic solutions, detergents (ionic, non-ionic, and zwitterionic), physical methods, or enzymatic agents [37]. The selection of decellularization is critical because each agent can alter the different components of the d(decellularized)ECM. The composition and structure of the dECM must be assessed after decellularization to determine its integrity, as well as to confirm complete cell removal. Large organs such as porcine hearts or lungs can be perfused with decellularization agents to effectively remove cells [38]. In the case of mice and other small animals, organs can be subjected to washing steps in conical tubes [39], whereas delicate tissues such as brains should be sectioned prior to decellularization to avoid structural damage [40]. Alternatively, more than 80 commercially decellularized scaffolds based on intact dECMs from human and animal models [41] are available for use in more controlled studies on the effects of plasma treatments.

It has been shown that plasma can improve cell attachment to 3D scaffolds, and that the type of working gas has an impact on the number of functional groups found on the material [42]. Yet the level of attachment obtained is cell-dependent and must be optimized according to the system used. As plasma induces different responses in cells when in 2D or 3D systems [43] and can change the biophysical properties (polymerization, hydrophilicity, roughness, etc.) of the ECM components [44–46], these model systems are valuable for studying the response of the tissue microenvironment to plasma or combination therapies including plasma. Thus, we believe that it is paramount to consider the mechanophysical and biological properties of the tissue, as they can shape the response to therapies.

14.3   THE CHICKEN CHORIOALLANTOIC MEMBRANE MODEL (CAM)

This is a simple, robust model used to study angiogenesis, tissue transplantation, metastasis, and response to therapies. The highly vascularized membrane provides an ideal *in vivo* environment faster and at lower costs than other *in vivo* models. The CAM is easily accessible, and the vascular network develops quickly, in 3 to 10 days. Cell seeding or manipulation of the CAM is possible towards the end of this period, as the blood vessels are more developed. The response to treatment can be monitored in real-time throughout the experiment using non-invasive imaging methods such as bioluminescence and magnetic resonance imaging [47]. Since it uses chicken embryos, the CAM model does not require animal protocol approval in most countries, which facilitates research.

In ex ovo experiments, the egg contents are carefully transferred into a petri dish or weight boat, offering an increased CAM surface to work with. In this scenario, the survival of the embryos is often reduced (30–50% on day 14 due to humidity stress, damaged yolks or membranes, and contamination), and more careful manipulation is necessary [48]. However, this method allows the performance of multiple tests on individual CAMs, which reduces the inter-embryo variability and number of eggs required. In contrast, the survival is significantly improved in in ovo experiments to up to 93% on day 14, as the shell protects from dehydration and mechanical damage [49].

The CAM model is ideal for assessing the effect of plasma treatments and combination therapies for cancer [50–53], angiogenesis [54], and analysis of the biocompatibility of biomaterials used for wound healing and regenerative therapies [55]. In addition, the CAM model can be used to test tissue tolerability to physical plasmas to determine its effect on coagulation, vascular stability, and irritation [56].

Although the system is convenient and easy to use, it offers only a short post-treatment observation time (up to 3 days after treatment, ending on day 14), which is a limitation for studying longer biological processes after plasma treatment [57] or immune responses [58]. The need for 2–10 million cells to generate a single in ovo tissue and the limited selection of reagents compatible with the model (antibodies, cytokines) should also be considered when choosing this model for plasma research. The CAM model cannot fully replace the current preclinical models, but it could bridge the gap between 2D cultures and more complex mammalian models.

14.4 ANIMAL MODELS

The animal models are valuable for imitating disease progression, studying the response to treatments, identifying optimal doses, and determining the side effects of drugs before their use in humans. The mammals most commonly used in research are mice, rats, rabbits, pigs, and monkeys, due to their phylogenetic resemblance to humans, but approx. 95% of the research is done in mice and rats [59]. The model must be carefully chosen based on the characteristics of the study, the genetic properties of the model, sex, similarity of signaling pathways between the model and humans, ability to emulate the disease/pathology, lifespan, and feasibility.

In oncology, tumors can be transplanted subcutaneously or orthotopically (in the normal place in the body). In some cases, immunodeficient animal models that lack some or all cells of the immune system are required, and the level of immunodeficiency must be chosen according to the needs of the study [60]. Immunocompromised animals can be engrafted with functional human cells (patient-derived cells or cell lines) or transduced with transgenes encoding for human factors. When engrafted with human peripheral blood mononuclear cells (humanized models), these animals can be used to study the human immune response in disease and tumor therapy [61]. Alternatively, allogeneic murine cancer cell lines can be used in immunocompetent mice to study the involvement of the (murine) immune system. In addition, transgenic or knockout mouse models can be used to study the role of specific enzymes.

The translation of treatment outcomes from animal models to humans is unpredictable and varied due to the difference in metabolic pathways, injury or illness induced, and treatment schedules, among other factors. For example, the skin anatomy of rodents differs from that of humans, and their immune system is stronger. Yet rodents are still widely used for wound healing studies because of their ease of use, cost, and reproducibility [62]. The conclusions obtained from animal studies should be carefully drawn, as they cannot be directly transferred to humans.

14.5 LIMITATIONS OF 3D MODELS FOR PLASMA BIOMEDICINE

One of the main challenges is the choice of relevant system models that are compatible with the method of administration of plasma. While most of the models presented in this chapter can be used for direct or indirect plasma treatments, it is important to consider the interference of the excessive amount of liquid required to avoid dehydration of samples during the experiments [63]. Plasma sources with an active gas flow suffer from evaporation and disturbance of 3D tissues due to the constant gas flow towards the samples, depending on the flow rate, distance to sample, and volume in the well. Moreover, the composition of culture media and buffer solutions used during plasma treatments *in vitro* are not representative of the organic and inorganic components of fluids found in living tissues. Lastly, the presence of RONS scavengers in the culture medium and the phototoxicity from fluorescent probes commonly used for live imaging *in vitro* can interfere with the assessment of pro-oxidant therapies like plasma and must be considered in each case. Despite these limitations, these system models provide more representative conditions than conventional 2D cultures and better mimic the response in patients.

14.6 OPPORTUNITIES FOR THE ADVANCEMENT OF PLASMA BIOMEDICINE

Recent studies have rightly pointed out the difference between human diseases and animal models of disease, and these differences could explain why almost 90% of the results from *in vivo* animal work fail to match the results obtained in human clinical trials [2]. To date, multiple 3D cellular models that resemble human disease have been developed and are currently being used in drug discovery and therapy development, with positive results. Scientists can benefit from the existing databases of non-animal technologies that facilitate their access to protocols and validated technologies [64,65], improving technology transfer into the clinic and contributing to the reduction of animal use in research.

All the 3D model systems described in this chapter can better represent the complex interactions taking place in real tissues, whether between cells of the same or different type or with molecules making up the tissue microenvironment. The gradients in nutrients, metabolites, oxygen and proliferation, compartmentalization, mechanophysical properties, and differential gene/protein expression in 3D cultures make them suitable candidates for the study of plasma therapies for oncology, wound healing, and regenerative medicine. Furthermore, these models present a unique opportunity to study the response to treatment when multiple cell types are present in the tissue of interest (immune, stromal, endothelial cells, etc.). Yet not all the system models described here are widely used by the plasma community (see Table 14.1). While 3D spheroids are becoming more common, other systems have yet to be adopted. 3D model systems, combined with proper imaging tools, can provide a significant

TABLE 14.1
Non-Exhaustive List of Model Systems Used for the Study of Biomedical Plasmas.

Model system	Device used	Malignancy	Treatment	References
Spheroids	kINPen devices	Cancer	Direct	[52,53,66–73]
		Cancer	Indirect	[69,74,75]
	DBD plasma jets	Cancer	Direct	[76]
		Cancer	Indirect	[77–82]
	Other plasma jets	Cancer	Indirect	[83,84]
	DBD plasmas	Cancer	Indirect	[85,86]
	Other reactors	Cancer	Direct	[87,88]
	Surface DBD	Cancer	Indirect	[89]
Organoids	Endoscopic plasma jet	Cancer	Direct	[28]
Scaffolds	Plasma jets	Tissue regeneration	Direct	[46,90]
	kINPen devices	Cancer	Indirect	[43,91–95]
		Cancer/Tissue regeneration		[95,96]
	GlidArc (GAD) plasma	Tissue regeneration/Wound	Direct	[97]
	DBD plasma spray	healing	Indirect	[98]
CAM model	kINPen devices	Cancer	Direct	[52,53,57,73,99,100]
		Mutagenicity		[101]
		Angiogenesis		[102–104]
		Cancer	Indirect	[50,51,75]
	miniFlatPlaSter	Cancer	Direct	[54]
Mouse model	kINPen devices	Safety/Risk assessment	Direct	[105,106]
		Cancer	Direct	[107–109]
		Coagulation	Direct	[110]
		Wound healing	Direct	[105,111–115]
		Cancer	Indirect	[116–118]
	DBD CINOGY	Cancer	Direct	[107]
	Plasma jets	Skin diseases	Direct	[119]
		Wound healing/Angiogenesis	Direct	[120,121]
		Cancer	Direct	[122,123]
	DBD plasmas	Cancer	Direct	[124,125]
		Cancer	Indirect	[86,126,127]
	Hollow-structured microneedle	Cancer	Direct	[128]
	Plasma needle	Wound healing	Direct	[129]
	Plasma streamer	Cancer	Direct	[130]
Other animal models	kINPen device	Cancer/Tissue regeneration	Indirect	[95]
	DBD plasma spray	Wound healing	Indirect	[98]
	FE-DBD plasma	Wound sterilization	Direct	[131]
	CAP Med II	Wound sterilization	Direct	[132]

amount of informative data about the value of the therapies being developed. Although there is no perfect system model that can perfectly represent the response in real patients, these physiologically relevant model systems can bring us closer to developing safe and effective plasma therapies.

REFERENCES

1. Wang, Y.; Jeon, H. *3D cell cultures toward quantitative high-throughput drug screening.* Trends in Pharmacological Sciences **2022**, 10.1016/j.tips.2022.03.014.
2. Brancato, V.; Oliveira, J.M.; Correlo, V.M.; Reis, R.L.; Kundu, S.C. *Could 3d models of cancer enhance drug screening?* Biomaterials **2020**, 10.1016/j.biomaterials.2019.119744.
3. Langhans, S.A. *Three-dimensional in vitro cell culture models in drug discovery and drug repositioning.* Frontiers in Pharmacology **2018**, 10.3389/fphar.2018.00006.
4. Liu, D.; Chen, S.; Win Naing, M. *A review of manufacturing capabilities of cell spheroid generation technologies and future development.* Biotechnology & Bioengineering **2021**, 10.1002/bit.27620.
5. Ryu, N.E.; Lee, S.H.; Park, H. *Spheroid culture system methods and applications for mesenchymal stem cells.* Cells **2019**, 10.3390/cells8121620.
6. Pinto, B.; Henriques, A.C.; Silva, P.M.A.; Bousbaa, H. *Three-dimensional spheroids as in vitro preclinical models for cancer research.* Pharmaceutics **2020**, 10.3390/pharmaceutics12121186.
7. Wheelock, M.J.; Shintani, Y.; Maeda, M.; Fukumoto, Y.; Johnson, K.R. *Cadherin switching.* Journal of Cell Science **2008**, 10.1242/jcs.000455.
8. Han, S.J.; Kwon, S.; Kim, K.S. *Challenges of applying multicellular tumor spheroids in preclinical phase.* Cancer Cell International **2021**, 10.1186/s12935-021-01853-8.
9. Lee, H.Y.; Choi, J.H.; Hong, J.W.; Kim, G.C.; Lee, H.J. *Comparative study of the Ar and He atmospheric pressure plasmas on e-cadherin protein regulation for plasma-mediated transdermal drug delivery.* Journal of Physics D-Applied Physics **2018**, 10.1088/1361-6463/aabd8c.
10. Choi, J.H.; Nam, S.H.; Song, Y.S.; Lee, H.W.; Lee, H.J.; Song, K.; Hong, J.W.; Kim, G.C. *Treatment with low-temperature atmospheric pressure plasma enhances cutaneous delivery of epidermal growth factor by regulating e-cadherin-mediated cell junctions.* Archives of Dermatological Research **2014**, 10.1007/s00403-014-1463-9.
11. Bauer, G. *Intercellular singlet oxygen-mediated bystander signaling triggered by long-lived species of cold atmospheric plasma and plasma-activated medium.* Redox Biology **2019**, 10.1016/j.redox.2019.101301.
12. Peirsman, A.; Blondeel, E.; Ahmed, T.; Anckaert, J.; Audenaert, D.; Boterberg, T.; Buzas, K.; Carragher, N.; Castellani, G.; Castro, F.; Dangles-Marie, V.; Dawson, J.; De Tullio, P.; De Vlieghere, E.; Dedeyne, S.; Depypere, H.; Diosdi, A.; Dmitriev, R.I.; Dolznig, H.; Fischer, S.; Gespach, C.; Goossens, V.; Heino, J.; Hendrix, A.; Horvath, P.; Kunz-Schughart, L.A.; Maes, S.; Mangodt, C.; Mestdagh, P.; Michlikova, S.; Oliveira, M.J.; Pampaloni, F.; Piccinini, F.; Pinheiro, C.; Rahn, J.; Robbins, S.M.; Siljamaki, E.; Steigemann, P.; Sys, G.; Takayama, S.; Tesei, A.; Tulkens, J.; Van Waeyenberge, M.; Vandesompele, J.; Wagemans, G.; Weindorfer, C.; Yigit, N.; Zablowsky, N.; Zanoni, M.; Blondeel, P.; De Wever, O. *Misspheroid: A knowledgebase and transparency tool for minimum information in spheroid identity.* Nature Methods **2021**, 10.1038/s41592-021-01291-4.
13. Biscop, E.; Lin, A.; Boxem, W.V.; Loenhout, J.V.; Backer, J.; Deben, C.; Dewilde, S.; Smits, E.; Bogaerts, A.A. *Influence of cell type and culture medium on determining cancer selectivity of cold atmospheric plasma treatment.* Cancers **2019**, 10.3390/cancers11091287.
14. Clevers, H. *Modeling development and disease with organoids.* Cell **2016**, 10.1016/j.cell.2016.05.082.
15. Veninga, V.; Voest, E.E. *Tumor organoids: Opportunities and challenges to guide precision medicine.* Cancer Cell **2021**, 10.1016/j.ccell.2021.07.020.
16. Sato, T.; Stange, D.E.; Ferrante, M.; Vries, R.G.; Van Es, J.H.; Van den Brink, S.; Van Houdt, W.J.; Pronk, A.; Van Gorp, J.; Siersema, P.D.; Clevers, H. *Long-term expansion of epithelial organoids from human colon, adenoma, adenocarcinoma, and Barrett's epithelium.* Gastroenterology **2011**, 10.1053/j.gastro.2011.07.050.
17. Driehuis, E.; Gracanin, A.; Vries, R.G.J.; Clevers, H.; Boj, S.F. *Establishment of pancreatic organoids from normal tissue and tumors.* STAR Protocols **2020**, 10.1016/j.xpro.2020.100192.
18. Van Loenhout, J.; Freire Boullosa, L.; Quatannens, D.; De Waele, J.; Merlin, C.; Lambrechts, H.; Lau, H.W.; Hermans, C.; Lin, A.; Lardon, F.; Peeters, M.; Bogaerts, A.; Smits, E.; Deben, C. *Auranofin and cold atmospheric plasma synergize to trigger distinct cell death mechanisms and immunogenic responses in glioblastoma.* Cells **2021**, 10.3390/cells10112936.

19. Shaw, P.; Kumar, N.; Privat-Maldonado, A.; Smits, E.; Bogaerts, A. *Cold atmospheric plasma increases temozolomide sensitivity of three-dimensional glioblastoma spheroids via oxidative stress-mediated DNA damage.* Cancers **2021**, 10.3390/cancers13081780.

20. Dao, V.; Yuki, K.; Lo, Y.H.; Nakano, M.; Kuo, C.J. *Immune organoids: From tumor modeling to precision oncology.* Trends in Cancer **2022**, 10.1016/j.trecan.2022.06.001.

21. Yuki, K.; Cheng, N.; Nakano, M.; Kuo, C.J. *Organoid models of tumor immunology.* Trends in Immunology **2020**, 10.1016/j.it.2020.06.010.

22. Van Loenhout, J.; Flieswasser, T.; Freire Boullosa, L.; De Waele, J.; Van Audenaerde, J.; Marcq, E.; Jacobs, J.; Lin, A.; Lion, E.; Dewitte, H.; Peeters, M.; Dewilde, S.; Lardon, F.; Bogaerts, A.; Deben, C.; Smits, E. *Cold atmospheric plasma-treated PBS eliminates immunosuppressive pancreatic stellate cells and induces immunogenic cell death of pancreatic cancer cells.* Cancers **2019**, 10.3390/cancers11101597.

23. Khalili, M.; Daniels, L.; Lin, A.; Krebs, F.C.; Snook, A.E.; Bekeschus, S.; Bowne, W.B.; Miller, V. *Non-thermal plasma-induced immunogenic cell death in cancer: A topical review.* Journal of Physics D-Applied Physics **2019**, 10.1088/1361-6463/ab31c1.

24. Lin, A.; De Backer, J.; Quatannens, D.; Cuypers, B.; Verswyvel, H.; De La Hoz, E.C.; Ribbens, B.; Siozopoulou, V.; Van Audenaerde, J.; Marcq, E.; Lardon, F.; Laukens, K.; Vanlanduit, S.; Smits, E.; Bogaerts, A. *The effect of local non-thermal plasma therapy on the cancer-immunity cycle in a melanoma mouse model.* Bioengineering & Translational Medicine **2022**, 10.1002/btm2.10314.

25. Icha, J.; Weber, M.; Waters, J.C.; Norden, C. *Phototoxity in live fluorescence microscopy, and how to avoid it.* Bioessays **2017**, 10.1002/bies.201700003.

26. Deben, C.; De La Hoz, E.C.; Le Compte, M.; Van Schil, P.; Hendriks, J.M.; Lauwers, P.; Yogeswaran, S.K.; Lardon, F.; Pauwels, P.; Bogaerts, A.; Smits, E.; Vanlanduit, S.; Lin, A. *Orbits: A high-throughput, time-lapse, and label-free drug screening platform for patient-derived 3d organoids.* bioRxiv **2021**, 10.1101/2021.09.09.459656.

27. Rios, A.C.; Clevers, H. *Imaging organoids: A bright future ahead.* Nature Methods **2018**, 10.1038/nmeth.4537.

28. Hadefi, A.; Leprovots, M.; Thulliez, M.; Bastin, O.; Lefort, A.; Libert, F.; Nonclercq, A.; Delchambre, A.; Reniers, F.; Deviere, J.; Garcia, M.I. *Cold atmospheric plasma differentially affects cell renewal and differentiation of stem cells and APC-deficient-derived tumor cells in intestinal organoids.* Cell Death and Discovery **2022**, 10.1038/s41420-022-00835-7.

29. Liu, X.; Fang, J.; Huang, S.; Wu, X.; Xie, X.; Wang, J.; Liu, F.; Zhang, M.; Peng, Z.; Hu, N. *Tumor-on-a-chip: From bioinspired design to biomedical application.* Microsystems & Nanoengineering **2021**, 10.1038/s41378-021-00277-8.

30. Zhang, J.; Tavakoli, H.; Ma, L.; Li, X.; Han, L.; Li, X. *Immunotherapy discovery on tumor organoid-on-a-chip platforms that recapitulate the tumor microenvironment.* Advanced Drug Delivery Reviews **2022**, 10.1016/j.addr.2022.114365.

31. Hamouda, I.; Labay, C.; Ginebra, M.P.; Nicol, E.; Canal, C. Poster: Evaluation of a photo-crosslinked hydrogel as carrier of reactive species generated by plasma jets. In *6th International Workshop on Plasma for Cancer Treatment—IWPCT 2019*, Antwerp, Belgium, 2019.

32. Lee, E.J.; Kang, E.; Kang, S.-W.; Huh, K.M. *Thermo-irreversible glycol chitosan/hyaluronic acid blend hydrogel for injectable tissue engineering.* Carbohydrate Polymers **2020**.

33. Ying, H.; Zhou, J.; Wang, M.; Su, D.; Ma, Q.; Lv, G.; Chen, J. *In situ formed collagen-hyaluronic acid hydrogel as biomimetic dressing for promoting spontaneous wound healing.* Materials Science and Engineering: C **2019**.

34. Spicer, C.D. *Hydrogel scaffolds for tissue engineering: The importance of polymer choice.* Polymer Chemistry **2020**, 10.1039/c9py01021a.

35. Antezana, P.E.; Municoy, S.; Alvarez-Echazu, M.I.; Santo-Orihuela, P.L.; Catalano, P.N.; Al-Tel, T.H.; Kadumudi, F.B.; Dolatshahi-Pirouz, A.; Orive, G.; Desimone, M.F. *The 3D bioprinted scaffolds for wound healing.* Pharmaceutics **2022**, 10.3390/pharmaceutics14020464.

36. Costa, A.; Naranjo, J.D.; Londono, R.; Badylak, S.F. *Biologic scaffolds.* Cold Spring Harbor Perspectives in Medicine **2017**, 10.1101/cshperspect.a025676.

37. Mendibil, U.; Ruiz-Hernandez, R.; Retegi-Carrion, S.; Garcia-Urquia, N.; Olalde-Graells, B.; Abarrategi, A. *Tissue-specific decellularization methods: Rationale and strategies to achieve regenerative compounds.* International Journal of Molecular Sciences **2020**, 10.3390/ijms21155447.

38. Hillebrandt, K.H.; Everwien, H.; Haep, N.; Keshi, E.; Pratschke, J.; Sauer, I.M. *Strategies based on organ decellularization and recellularization.* Transplant International **2019**, 10.1111/tri.13462.

39. Guller, A.; Kuschnerus, I.; Rozova, V.; Nadort, A.; Yao, Y.; Khabir, Z.; Garcia-Bennett, A.; Liang, L.O.; Polikarpova, A.; Qian, Y.; Goldys, E.M.; Zvyagin, A.V. *Chick embryo experimental platform for micrometastases research in a 3D tissue engineering model: Cancer biology, drug development, and nanotechnology applications.* Biomedicines **2021**, 10.3390/biomedicines9111578.

40. De Waele, J.; Reekmans, K.; Daans, J.; Goossens, H.; Berneman, Z.; Ponsaerts, P. *3D culture of murine neural stem cells on decellularized mouse brain sections.* Biomaterials **2015**, 10.1016/j.biomaterials.2014.11.025.

41. Neishabouri, A.; Soltani Khaboushan, A.; Daghigh, F.; Kajbafzadeh, A.M.; Majidi Zolbin, M. *Decellularization in tissue engineering and regenerative medicine: Evaluation, modification, and application methods.* Frontiers in Bioengineering and Biotechnology **2022**, 10.3389/fbioe.2022.805299.

42. Prasertsung, I.; Kanokpanont, S.; Mongkolnavin, R.; Wong, C.S.; Panpranot, J.; Damrongsakkul, S. *Plasma enhancement of in vitro attachment of rat bone-marrow-derived stem cells on cross-linked gelatin films.* Journal of Biomaterials Science, Polymer Edition **2012**, 10.1163/092050611X584900.

43. Mateu-Sanz, M.; Ginebra, M.P.; Tornin, J.; Canal, C. *Cold atmospheric plasma enhances doxorubicin selectivity in metastasic bone cancer.* Free Radical Biology and Medicine **2022**, 10.1016/j.freeradbiomed.2022.07.007.

44. Przekora, A. *Current trends in fabrication of biomaterials for bone and cartilage regeneration: Materials modifications and biophysical stimulations.* International Journal of Molecular Sciences **2019**, 10.3390/ijms20020435.

45. Bhushan, B.; Kumar, R. Plasma treated and untreated thermoplastic biopolymers/biocomposites in tissue engineering and biodegradable implants. In *Materials for biomedical engineering: Hydrogels and polymer-based scaffolds*, Holban, A.-M.; Grumezescu, A., Eds. Elsevier: Amsterdam, Netherlands, 2019.

46. Lee, S.J.; Yan, D.; Zhou, X.; Cui, H.; Esworthy, T.; Hann, S.Y.; Keidar, M.; Zhang, L.G. *Integrating cold atmospheric plasma with 3D printed bioactive nanocomposite scaffold for cartilage regeneration.* Materials Science & Engineering C-Materials for Biological Applications **2020**, 10.1016/j.msec.2020.110844.

47. Winter, G.; Koch, A.B.F.; Loffler, J.; Jelezko, F.; Linden, M.; Li, H.; Abaei, A.; Zuo, Z.; Beer, A.J.; Rasche, V. *In vivo PET/MRI imaging of the chorioallantoic membrane.* Frontiers in Physics **2020**, 10.3389/fphy.2020.00151.

48. Dohle, D.S.; Pasa, S.D.; Gustmann, S.; Laub, M.; Wissler, J.H.; Jennissen, H.P.; Dünker, N. *Chick ex ovo culture and ex ovo CAM assay: How it really works.* Journal of Visualized Experiments **2009**, 10.3791/1620.

49. Li, M.; Pathak, R.R.; Lopez-Rivera, E.; Friedman, S.L.; Aguirre-Ghiso, J.A.; Sikora, A.G. *The in ovo chick chorioallantoic membrane (CAM) assay as an efficient xenograft model of hepatocellular carcinoma.* Journal of Visualized Experiments **2015**, 10.3791/52411.

50. Kumar, N.; Perez-Novo, C.; Shaw, P.; Logie, E.; Privat-Maldonado, A.; Dewilde, S.; Smits, E.; Berghe, W.V.; Bogaerts, A. *Physical plasma-derived oxidants sensitize pancreatic cancer cells to ferroptotic cell death.* Free Radical Biology and Medicine **2021**.

51. Shaw, P.; Kumar, N.; Hammerschmid, D.; Privat-Maldonado, A.; Dewilde, S.; Bogaerts, A. *Synergistic effects of melittin and plasma treatment: A promising approach for cancer therapy.* Cancers **2019**.

52. Liedtke, K.R.; Diedrich, S.; Pati, O.; Freund, E.; Flieger, R.; Heidecke, C.D.; Partecke, L.I.; Bekeschus, S. *Cold physical plasma selectively elicits apoptosis in murine pancreatic cancer cells in vitro and in ovo.* Anticancer Research **2018**, 10.21873/anticanres.12901.

53. Freund, E.; Spadola, C.; Schmidt, A.; Privat-Maldonado, A.; Bogaerts, A.; von Woedtke, T.; Weltmann, K.-D.; Heidecke, C.-D.; Partecke, L.-I.; Käding, A.; Bekeschus, S. *Risk evaluation of EMT and inflammation in metastatic pancreatic cancer cells following plasma treatment.* Frontiers in Physics **2020**, 10.3389/fphy.2020.569618.

54. Kugler, P.; Becker, S.; Welz, C.; Wiesmann, N.; Sax, J.; Buhr, C.R.; Thoma, M.H.; Brieger, J.; Eckrich, J. *Cold atmospheric plasma reduces vessel density and increases vascular permeability and apoptotic cell death in solid tumors.* Cancers **2022**, 10.3390/cancers14102432.

55. Intasa-ard, S.; Birault, A. Chapter nine—nanoparticles characterization using the CAM assay. In *The enzymes*, Tamanoi, F., Ed. Academic Press, 2019; Vol. 46, pp. 129–160.

56. Harnoss, J.C.; Abu Elrub, Q.M.; Jung, J.O.; Koburger, T.; Assadian, O.; Dissemond, J.; Baguhl, R.; Papke, R.; Kramer, A.; Canc, I.S.C.I. *Irritative potency of selected wound antiseptics in the hen's egg test on chorioallantoic membrane to predict their compatibility to wounds.* Wound Repair and Regeneration **2019**, 10.1111/wrr.12689.

57. Privat-Maldonado, A.; Verloy, R.; Cardenas Delahoz, E.; Lin, A.; Vanlanduit, S.; Smits, E.; Bogaerts, A. *Cold atmospheric plasma does not affect stellate cells phenotype in pancreatic cancer tissue in ovo.* International Journal of Molecular Sciences **2022**.

58. Miebach, L.; Berner, J.; Bekeschus, S. *In ovo model in cancer research and tumor immunology.* Frontiers in Immunology **2022**, 10.3389/fimmu.2022.1006064.

59. *Mice & rats, the essential need for animals in medical research.* Foundation for Biomedical Research, 2016.

60. Lampreht Tratar, U.; Horvat, S.; Cemazar, M. *Transgenic mouse models in cancer research.* Frontiers in Oncology **2018**, 10.3389/fonc.2018.00268.

61. Kersten, K.; de Visser, K.E.; van Miltenburg, M.H.; Jonkers, J. *Genetically engineered mouse models in oncology research and cancer medicine.* EMBO Molecular Medicine **2017**, 10.15252/emmm.201606857.

62. Sami, D.G.; Heiba, H.H.; Abdellatif, A. *Wound healing models: A systematic review of animal and non-animal models.* Wound Medicine **2019**, 10.1016/j.wndm.2018.12.001.

63. Privat-Maldonado, A.; Bengtson, C.; Razzokov, J.; Smits, E.; Bogaerts, A. *Modifying the tumour micro-environment: Challenges and future perspectives for anticancer plasma treatments.* Cancers **2019**, 10.3390/cancers11121920.

64. Re-place: Database. https://re-place.be/.

65. Nat database. www.nat-database.org.

66. Privat-Maldonado, A.; Gorbanev, Y.; Dewilde, S.; Smits, E.; Bogaerts, A. *Reduction of human glioblastoma spheroids using cold atmospheric plasma: The combined effect of short- and long-lived reactive species.* Cancers **2018**.

67. Bekeschus, S.; Freund, E.; Spadola, C.; Privat-Maldonado, A.; Hackbarth, C.; Bogaerts, A.; Schmidt, A.; Wende, K.; Weltmann, K.D.; von Woedtke, T.; Heidecke, C.D.; Partecke, L.I.; Kading, A. *Risk assessment of kINPen plasma treatment of four human pancreatic cancer cell lines with respect to metastasis.* Cancers **2019**, 10.3390/cancers11091237.

68. Bekeschus, S.; Lippert, M.; Diepold, K.; Chiosis, G.; Seufferlein, T.; Azoitei, N. *Physical plasma-triggered ROS induces tumor cell death upon cleavage of HSP90 chaperone.* Scientific Reports **2019**, 10.1038/s41598-019-38580-0.

69. Hasse, S.; Meder, T.; Freund, E.; von Woedtke, T.; Bekeschus, S. *Plasma treatment limits human melanoma spheroid growth and metastasis independent of the ambient gas composition.* Cancers **2020**, 10.3390/cancers12092570.

70. Shaw, P.; Kumar, N.; Privat-Maldonado, A.; Smits, E.; Bogaerts, A. *Cold atmospheric plasma increases temozolomide sensitivity of three-dimensional glioblastoma spheroids via oxidative stress-mediated DNA damage.* Cancers **2021**.

71. Miebach, L.; Freund, E.; Horn, S.; Niessner, F.; Sagwal, S.K.; von Woedtke, T.; Emmert, S.; Weltmann, K.D.; Clemen, R.; Schmidt, A.; Gerling, T.; Bekeschus, S. *Tumor cytotoxicity and immunogenicity of a novel v-jet neon plasma source compared to the kINPen.* Scientific Reports **2021**, 10.1038/s41598-020-80512-w.

72. Mahdikia, H.; Saadati, F.; Freund, E.; Gaipl, U.S.; Majidzadeh-a, K.; Shokri, B.; Bekeschus, S. *Gas plasma irradiation of breast cancers promotes immunogenicity, tumor reduction, and an abscopal effect in vivo.* OncoImmunology **2021**, 10.1080/2162402X.2020.1859731.

73. Gelbrich, N.; Miebach, L.; Berner, J.; Freund, E.; Saadati, F.; Schmidt, A.; Stope, M.; Zimmermann, U.; Burchardt, M.; Bekeschus, S. *Medical gas plasma augments bladder cancer cell toxicity in pre-clinical models and patient-derived tumor tissues.* Journal of Advanced Research **2023**, 10.1016/j.jare.2022.07.012.

74. Freund, E.; Liedtke, K.R.; van der Linde, J.; Metelmann, H.R.; Heidecke, C.D.; Partecke, L.I.; Bekeschus, S. *Physical plasma-treated saline promotes an immunogenic phenotype in CT26 colon cancer cells in vitro and in vivo.* Scientific Reports **2019**, 10.1038/s41598-018-37169-3.

75. Miebach, L.; Freund, E.; Clemen, R.; Kersting, S.; Partecke, L.-I.; Bekeschus, S. *Gas plasma-oxidized sodium chloride acts via hydrogen peroxide in a model of peritoneal carcinomatosis.* Proceedings of the National Academy of Sciences **2022**, 10.1073/pnas.2200708119.

76. Plewa, J.M.; Yousfi, M.; Frongia, C.; Eichwald, O.; Ducommun, B.; Merbahi, N.; Lobjois, V. *Low-temperature plasma-induced antiproliferative effects on multi-cellular tumor spheroids.* New Journal of Physics **2014**, 10.1088/1367-2630/16/4/043027.

77. Judee, F.; Fongia, C.; Ducommun, B.; Yousfi, M.; Lobjois, V.; Merbahi, N. *Short and long time effects of low temperature plasma activated media on 3D multicellular tumor spheroids.* Scientific Reports **2016**, 10.1038/srep21421.

78. Florian, J.; Merbahi, N.; Yousfi, M. *Genotoxic and cytotoxic effects of plasma-activated media on multicellular tumor spheroids.* **2016**, 10.1615/PlasmaMed.2016015823.

79. Chauvin, J.; Gibot, L.; Griseti, E.; Golzio, M.; Rols, M.P.; Merbahi, N.; Vicendo, P. *Elucidation of in vitro cellular steps induced by antitumor treatment with plasma-activated medium.* Scientific Reports **2019**, 10.1038/s41598-019-41408-6.

80. Griseti, E.; Kolosnjaj-Tabi, J.; Gibot, L.; Fourquaux, I.; Rols, M.P.; Yousfi, M.; Merbahi, N.; Golzio, M. *Pulsed electric field treatment enhances the cytotoxicity of plasma-activated liquids in a three-dimensional human colorectal cancer cell model.* Scientific Reports **2019**, 10.1038/s41598-019-44087-5.

81. Griseti, E.; Merbahi, N.; Golzio, M. *Anti-cancer potential of two plasma-activated liquids: Implication of long-lived reactive oxygen and nitrogen species.* Cancers **2020**, 10.3390/cancers12030721.

82. Lee, Y.J.; Kim, S.W.; Jung, M.H.; Kim, Y.S.; Kim, K.S.; Suh, D.S.; Kim, K.H.; Choi, E.H.; Kim, J.; Kwon, B.S. *Plasma-activated medium inhibits cancer stem cell-like properties and exhibits a synergistic effect in combination with cisplatin in ovarian cancer.* Free Radical Biology and Medicine **2022**, 10.1016/j.freeradbiomed.2022.03.001.

83. Li, Y.; Lv, Y.; Tang, M.X.; Choi, E.H.; Wang, J.J.; Lv, G.Q.; Zhu, Y.; Wang, S.B.; Liu, Y.J. *Low-temperature plasma-jet-activated medium inhibited tumorigenesis of lung adenocarcinoma in a 3D in vitro culture model.* Plasma Processes and Polymers **2021**, 10.1002/ppap.202100049.

84. Li, Y.; Lv, Y.; Zhu, Y.; Yang, X.; Lin, B.; Li, M.; Zhou, Y.; Tan, Z.; Choi, E.H.; Wang, J.; Wang, S.; Liu, Y. *Low-temperature plasma-activated medium inhibited proliferation and progression of lung cancer by targeting the PI3K/AKT and MAPK pathways.* Oxidative Medicine and Cellular Longevity **2022**, 10.1155/2022/9014501.

85. Mihai, C.T.; Mihaila, I.; Pasare, M.A.; Pintilie, R.M.; Ciorpac, M.; Topala, I. *Cold atmospheric plasma-activated media improve paclitaxel efficacy on breast cancer cells in a combined treatment model.* Current Issues in Molecular Biology **2022**, 10.3390/cimb44050135.

86. Zhang, H.; Xu, S.; Zhang, J.; Wang, Z.; Liu, D.; Guo, L.; Cheng, C.; Cheng, Y.; Xu, D.; Kong, M.G.; Rong, M.; Chu, P.K. *Plasma-activated thermosensitive biogel as an exogenous ROS carrier for post-surgical treatment of cancer.* Biomaterials **2021**, 10.1016/j.biomaterials.2021.121057.

87. Zhen, X.; Sun, H.N.; Liu, R.; Choi, H.S.; Lee, D.S. *Non-thermal plasma-activated medium induces apoptosis of AsPC1 cells through the ros-dependent autophagy pathway.* In Vivo **2020**, 10.21873/invivo.11755.

88. Wanigasekara, J.; Barcia, C.; Cullen, P.J.; Tiwari, B.; Curtin, J.F. *Plasma induced reactive oxygen species-dependent cytotoxicity in glioblastoma 3D tumourspheres.* Plasma Processes and Polymers **2022**, 10.1002/ppap.202100157.

89. Zhang, H.; Zhang, J.; Xu, S.; Wang, Z.; Xu, D.; Guo, L.; Liu, D.; Kong, M.G.; Rong, M. *Antitumor effects of hyperthermia with plasma-treated solutions on 3D bladder tumor spheroids.* Plasma Processes and Polymers **2021**, 10.1002/ppap.202100070.

90. Wang, M.; Favi, P.; Cheng, X.; Golshan, N.H.; Ziemer, K.S.; Keidar, M.; Webster, T.J. *Cold atmospheric plasma (CAP) surface nanomodified 3D printed polylactic acid (PLA) scaffolds for bone regeneration.* Acta Biomaterialia **2016**, 10.1016/j.actbio.2016.09.030.

91. Labay, C.; Hamouda, I.; Tampieri, F.; Ginebra, M.-P.; Canal, C. *Production of reactive species in alginate hydrogels for cold atmospheric plasma-based therapies.* Scientific Reports **2019**, 10.1038/s41598-019-52673-w.

92. Labay, C.; Roldán, M.; Tampieri, F.; Stancampiano, A.; Bocanegra, P.E.; Ginebra, M.-P.; Canal, C. *Enhanced generation of reactive species by cold plasma in gelatin solutions for selective cancer cell death.* ACS Applied Materials & Interfaces **2020**, 10.1021/acsami.0c12930.

93. Tornin, J.; Villasante, A.; Sole-Marti, X.; Ginebra, M.P.; Canal, C. *Osteosarcoma tissue-engineered model challenges oxidative stress therapy revealing promoted cancer stem cell properties.* Free Radical Biology and Medicine **2021**, 10.1016/j.freeradbiomed.2020.12.437.

94. Sole-Marti, X.; Vilella, T.; Labay, C.; Tampieri, F.; Ginebra, M.P.; Canal, C. *Thermosensitive hydrogels to deliver reactive species generated by cold atmospheric plasma: A case study with methylcellulose.* Biomaterial Science **2022**, 10.1039/d2bm00308b.

95. Solé-Martí, X.; Labay, C.; Raymond, Y.; Franch, J.; Benitez, R.; Ginebra, M.P.; Canal, C. *Ceramic-hydrogel composite as carrier for cold-plasma reactive-species: Safety and osteogenic capacity in vivo.* Plasma Processes and Polymers **2022**, 10.1002/ppap.202200155.

96. Fischer, M.; Bortel, E.; Schoon, J.; Behnke, E.; Hesse, B.; Weitkamp, T.; Bekeschus, S.; Pichler, M.; Wassilew, G.I.; Schulze, F. *Cold physical plasma treatment optimization for improved bone allograft processing.* Frontiers in Bioengineering and Biotechnology **2023**, 10.3389/fbioe.2023.1264409.

97. Przekora, A.; Audemar, M.; Pawlat, J.; Canal, C.; Thomann, J.-S.; Labay, C.; Wojcik, M.; Kwiatkowski, M.; Terebun, P.; Ginalska, G.; Hermans, S.; Duday, D. *Positive effect of cold atmospheric nitrogen plasma on the behavior of mesenchymal stem cells cultured on a bone scaffold containing iron oxide-loaded silica nanoparticles catalyst.* International Journal of Molecular Sciences **2020**.

98. Lee, H.R.; Lee, H.-Y.; Heo, J.; Jang, J.Y.; Shin, Y.S.; Kim, C.-H. *Liquid-type nonthermal atmospheric plasma enhanced regenerative potential of silk—fibrin composite gel in radiation-induced wound failure.* Materials Science and Engineering: C **2021**, 10.1016/j.msec.2021.112304.

99. Partecke, L.I.; Evert, K.; Haugk, J.; Doering, F.; Normann, L.; Diedrich, S.; Weiss, F.-U.; Evert, M.; Huebner, N.O.; Guenther, C.; Heidecke, C.D.; Kramer, A.; Bussiahn, R.; Weltmann, K.-D.; Pati, O.; Bender, C.; von Bernstorff, W. *Tissue tolerable plasma (TTP) induces apoptosis in pancreatic cancer cells in vitro and in vivo.* BMC Cancer **2012**, 10.1186/1471-2407-12-473.

100. Miebach, L.; Freund, E.; Clemen, R.; Weltmann, K.D.; Metelmann, H.R.; von Woedtke, T.; Gerling, T.; Wende, K.; Bekeschus, S. *Conductivity augments ROS and RNS delivery and tumor toxicity of an argon plasma jet.* Free Radical Biology and Medicine **2022**, 10.1016/j.freeradbiomed.2022.01.014.

101. Kluge, S.; Bekeschus, S.; Bender, C.; Benkhai, H.; Sckell, A.; Below, H.; Stope, M.B.; Kramer, A. *Investigating the mutagenicity of a cold argon-plasma jet in an HET-MN model.* PLoS One **2016**, 10.1371/journal.pone.0160667.

102. Haertel, B.; Eiden, K.; Deuter, A.; Wende, K.; von Woedtke, T.; Lindequist, U. *Differential effect of non thermal atmospheric pressure plasma on angiogenesis.* Letters in Applied NanoBioscience **2014**.

103. Bender, C.; Partecke, L.I.; Kindel, E.; Doring, F.; Lademann, J.; Heidecke, C.D.; Kramer, A.; Hubner, N.O. *The modified HET-CAM as a model for the assessment of the inflammatory response to tissue tolerable plasma.* Toxicology *In Vitro* **2011**, 10.1016/j.tiv.2010.11.012.

104. Bender, C.; Matthes, R.; Kindel, E.; Kramer, A.; Lademann, J.; Weltmann, K.D.; Eisenbeiss, W.; Hubner, N.O. *The irritation potential of nonthermal atmospheric pressure plasma in the HET-CAM.* Plasma Processes and Polymers **2010**, 10.1002/ppap.200900119.

105. Schmidt, A.; Woedtke, T.V.; Stenzel, J.; Lindner, T.; Polei, S.; Vollmar, B.; Bekeschus, S. *One year follow-up risk assessment in SKH-1 mice and wounds treated with an argon plasma jet.* International Journal of Molecular Sciences **2017**.

106. Evert, K.; Kocher, T.; Schindler, A.; Müller, M.; Müller, K.; Pink, C.; Holtfreter, B.; Schmidt, A.; Dombrowski, F.; Schubert, A.; von Woedtke, T.; Rupf, S.; Calvisi, D.F.; Bekeschus, S.; Jablonowski, L. *Repeated exposure of the oral mucosa over 12 months with cold plasma is not carcinogenic in mice.* Scientific Reports **2021**, 10.1038/s41598-021-99924-3.

107. Daeschlein, G.; Scholz, S.; Lutze, S.; Arnold, A.; von Podewils, S.; Kiefer, T.; Tueting, T.; Hardt, O.; Haase, H.; Grisk, O.; Langner, S.; Ritter, C.; von Woedtke, T.; Junger, M. *Comparison between cold plasma, electrochemotherapy and combined therapy in a melanoma mouse model.* Experimental Dermatology **2013**, 10.1111/exd.12201.

108. Bekeschus, S.; Clemen, R.; Niessner, F.; Sagwal, S.K.; Freund, E.; Schmidt, A. *Medical gas plasma jet technology targets murine melanoma in an immunogenic fashion.* Advanced Science **2020**, 10.1002/advs.201903438.

109. Miebach, L.; Melo-Zainzinger, G.; Freund, E.; Clemen, R.; Cecchini, A.L.; Bekeschus, S. *Medical gas plasma technology combines with antimelanoma therapies and promotes immune-checkpoint therapy responses.* Advanced Science **2023**, 10.1002/advs.202303183.

110. Bekeschus, S.; Bruggemeier, J.; Hackbarth, C.; von Woedtke, T.; Partecke, L.I.; van der Linde, J. *Platelets are key in cold physical plasma-facilitated blood coagulation in mice.* Clinical Plasma Medicine **2017**, 10.1016/j.cpme.2017.10.001.

111. Schmidt, A.; Bekeschus, S.; Wende, K.; Vollmar, B.; von Woedtke, T. *A cold plasma jet accelerates wound healing in a murine model of full-thickness skin wounds.* Experimental Dermatology **2017**, 10.1111/exd.13156.

112. Schmidt, A.; Liebelt, G.; Niessner, F.; von Woedtke, T.; Bekeschus, S. *Gas plasma-spurred wound healing is accompanied by regulation of focal adhesion, matrix remodeling, and tissue oxygenation.* Redox Biology **2021**, 10.1016/j.redox.2020.101809.

113. Schmidt, A.; Niesner, F.; von Woedtke, T.; Bekeschus, S. *Hyperspectral imaging of wounds reveals augmented tissue oxygenation following cold physical plasma treatment in vivo.* IEEE Transactions on Radiation and Plasma Medical Sciences **2021**, 10.1109/trpms.2020.3009913.

114. Schmidt, A.; von Woedtke, T.; Vollmar, B.; Hasse, S.; Bekeschus, S. *Nrf2 signaling and inflammation are key events in physical plasma-spurred wound healing.* Theranostics **2019**, 10.7150/thno.29754.

115. Schmidt, A.; Woedtke, T.V.; Stenzel, J.; Lindner, T.; Polei, S.; Vollmar, B.; Bekeschus, S. *One year follow-up risk assessment in skh-1 mice and wounds treated with an argon plasma jet.* International Journal of Molecular Sciences **2017**, 10.3390/ijms18040868.

116. Liedtke, K.R.; Bekeschus, S.; Kaeding, A.; Hackbarth, C.; Kuehn, J.P.; Heidecke, C.D.; von Bernstorff, W.; von Woedtke, T.; Partecke, L.I. *Non-thermal plasma-treated solution demonstrates antitumor activity against pancreatic cancer cells in vitro and in vivo.* Scientific Reports **2017**, 10.1038/s41598-017-08560-3.

117. Miebach, L.; Freund, E.; Cecchini, A.L.; Bekeschus, S. *Conductive gas plasma treatment augments tumor toxicity of Ringer's lactate solutions in a model of peritoneal carcinomatosis.* Antioxidants **2022**, 10.3390/antiox11081439.

118. Miebach, L.; Freund, E.; Clemen, R.; Kersting, S.; Partecke, L.I.; Bekeschus, S. *Gas plasma-oxidized sodium chloride acts via hydrogen peroxide in a model of peritoneal carcinomatosis.* Proceedings of the National Academy of Sciences of the United States of America (PNAS) **2022**, 10.1073/pnas.2200708119.

119. Sun, T.; Zhang, X.; Hou, C.; Yu, S.; Zhang, Y.; Yu, Z.; Kong, L.; Liu, C.; Feng, L.; Wang, D.; Ni, G. *Cold plasma irradiation attenuates atopic dermatitis via enhancing HIF-1alpha-induced MANF transcription expression.* Frontiers in Immunology **2022**, 10.3389/fimmu.2022.941219.

120. Nasruddin; Nakajima, Y.; Mukai, K.; Rahayu, H.S.E.; Nur, M.; Ishijima, T.; Enomoto, H.; Uesugi, Y.; Sugama, J.; Nakatani, T. *Cold plasma on full-thickness cutaneous wound accelerates healing through promoting inflammation, re-epithelialization and wound contraction.* Clinical Plasma Medicine **2014**, 10.1016/j.cpme.2014.01.001.

121. Dang, C.P.; Weawseetong, S.; Charoensappakit, A.; Sae-Khow, K.; Thong-Aram, D.; Leelahavanichkul, A. *Non-thermal atmospheric pressure argon-sourced plasma flux promotes wound healing of burn wounds and burn wounds with infection in mice through the anti-inflammatory macrophages.* Applied Sciences **2021**.

122. Yajima, I.; Iida, M.; Kumasaka, M.Y.; Omata, Y.; Ohgami, N.; Chang, J.; Ichihara, S.; Hori, M.; Kato, M. *Non-equilibrium atmospheric pressure plasmas modulate cell cycle-related gene expressions in melanocytic tumors of RET-transgenic mice.* Experimental Dermatology **2014**, 10.1111/exd.12415.

123. Mirpour, S.; Piroozmand, S.; Soleimani, N.; Jalali Faharani, N.; Ghomi, H.; Fotovat Eskandari, H.; Sharifi, A.M.; Mirpour, S.; Eftekhari, M.; Nikkhah, M. *Utilizing the micron sized non-thermal atmospheric pressure plasma inside the animal body for the tumor treatment application.* Scientific Reports **2016**, 10.1038/srep29048.

124. Vandamme, M.; Robert, E.; Pesnel, S.; Barbosa, E.; Dozias, S.; Sobilo, J.; Lerondel, S.; Le Pape, A.; Pouvesle, J.-M. *Antitumor effect of plasma treatment on U87 glioma xenografts: Preliminary results.* Plasma Processes and Polymers **2010**, 10.1002/ppap.200900080.

125. Brullé, L.; Vandamme, M.; Riès, D.; Martel, E.; Robert, E.; Lerondel, S.; Trichet, V.; Richard, S.; Pouvesle, J.-M.; Le Pape, A. *Effects of a non thermal plasma treatment alone or in combination with gemcitabine in a Mia PaCa2-luc orthotopic pancreatic carcinoma model.* PLoS One **2012**, 10.1371/journal.pone.0052653.

126. Ye, F.; Kaneko, H.; Nagasaka, Y.; Ijima, R.; Nakamura, K.; Nagaya, M.; Takayama, K.; Kajiyama, H.; Senga, T.; Tanaka, H.; Mizuno, M.; Kikkawa, F.; Hori, M.; Terasaki, H. *Plasma-activated medium suppresses choroidal neovascularization in mice: A new therapeutic concept for age-related macular degeneration.* Scientific Reports **2015**, 10.1038/srep07705.

127. Wahyuningtyas, E.S.; Iswara, A.; Sari, Y.; Kamal, S.; Santosa, B.; Ishijima, T.; Nakatani, T.; Putri, I.K.; Nasruddin, N. *Comparative study on Manuka and Indonesian honeys to support the application of plasma jet during proliferative phase on wound healing.* Clinical Plasma Medicine **2018**, 10.1016/j.cpme.2018.08.001.

128. Chen, G.; Chen, Z.; Wen, D.; Wang, Z.; Li, H.; Zeng, Y.; Dotti, G.; Wirz, R.E.; Gu, Z. *Transdermal cold atmospheric plasma-mediated immune checkpoint blockade therapy.* Proceedings of the National Academy of Sciences **2020**, 10.1073/pnas.1917891117.

129. Garcia-Alcantara, E.; Lopez-Callejas, R.; Morales-Ramirez, P.R.; Pena-Eguiluz, R.; Fajardo-Munoz, R.; Mercado-Cabrera, A.; Barocio, S.R.; Valencia-Alvarado, R.; Rodriguez-Mendez, B.G.; Munoz-Castro, A.E.; de la Piedad-Beneitez, A.; Rojas-Olmedo, I.A. *Accelerated mice skin acute wound healing in vivo by combined treatment of argon and helium plasma needle.* Archives of Medical Research **2013**, 10.1016/j.arcmed.2013.02.001.

130. Jinno, R.; Komuro, A.; Yanai, H.; Ono, R. *Antitumor abscopal effects in mice induced by normal tissue irradiation using pulsed streamer discharge plasma.* Journal of Physics D-Applied Physics **2022**, 10.1088/1361-6463/ac6a24.

131. Dobrynin, D.; Wasko, K.; Friedman, G.; Fridman, A.A.; Fridman, G. *Cold plasma sterilization of open wounds: Live rat model.* Plasma Medicine **2011**, 10.1615/PlasmaMed.2011002698.

132. Li, G.L.; Li, D.H.; Li, J.; Jia, Y.N.; Zhu, C.Q.; Zhang, Y.; Li, H.P. *Promotion of the wound healing of in vivo rabbit wound infected with methicillin-resistant staphylococcus aureus treated by a cold atmospheric plasma jet.* IEEE Transactions on Plasma Science **2021**, 10.1109/Tps.2021.3092946.

15 Genotoxicity Risk Assessment of Plasma Exposure

Sander Bekeschus, Anke Schmidt

15.1 INTRODUCTION

Physical plasma is a partially ionized gas that produces various components. In the form of medical devices, gas plasmas are approved for several clinical applications in the field of dermatology with a focus on wound healing [1]. Here, several treatment cycles are usually required to achieve desired treatment outcomes. Other non-approved applications of plasma devices in medicine are found in dentistry and oncology. In dentistry, there are several putative medical applications, such as the treatment of mycosis, periodontitis, peri-implantitis, and tooth biofilm removal in general [2,3]. In oncology, clinical evidence is scarce with regard to plasma treatment. The only relevant cohort described so far is head and neck cancer patients [4]. In experimental pre-clinical tumor models, usually investigated in small rodents such as mice, a range of plasma anticancer effects across more than a dozen tumor types have been studied so far [5,6]. Here, most models also required repetitive plasma treatment regimens.

The main trait of plasma treatment is its versatile production of a range of reactive oxygen and nitrogen species simultaneously. These species are abbreviated as ROS in this chapter because RNS also contain oxygen. There is a consensus that most biomedical effects observed after plasma exposure are due to these species having specific or non-specific effects on the treated cells and tissues, depending on their concentration [7]. This is also referred to as hormesis, placing plasma biology and medicine at the heart of applied redox biology [8]. Instead of using various terms (cold atmospheric plasma, medical gas plasma, non-thermal plasma, etc.) or abbreviations (CAP, CAPP, NTP, etc.), in this chapter we use the term "plasma" and discuss cold physical plasma, not blood plasma. Plasma-derived ROS are mostly of short-lived nature when expelled within or in the immediate vicinity of the genuine plasma or afterglow discharge [9]. By contrast, long-lived oxidants and secondary products dominate with greater distances from the active gas plasma zone or when plasma is used to treat liquids (i.e., the short-lived reactive species have time and/or plenty of partners to react with and/or deteriorate) [10]. Therefore, plasma-treated liquids a priori are dominated by long-lived plasma-derived products, such as hydrogen peroxide (H_2O_2), nitrite (NO_2^-), and nitrate (NO_3^-) [11]. These agents have been tested for genotoxicity in recent decades [12,13]. Hence, this chapter focuses on direct plasma applications that include short-lived plasma-derived ROS chemistry, as these are unique and novel in contrast to the long-lived oxidants already tested in different contexts. Nevertheless, studies using plasma-treated liquids will be included as it is proposed that the combination of, e.g., H_2O_2 and NO_2^- and/or NO_3^- provide individual effects by themselves [14].

15.2 METHODS FOR ASSESSING GENOTOXICITY

Several methods are available to test genotoxicity under standardized *in vitro* culture conditions. Some of these tests have been approved by the Organization for Economic Co-Operation and Development (OECD) and thus serve as standard test assays for, e.g., the pharmaceutical industry during drug development as well as the government and independent laboratories including

DOI: 10.1201/9781003328056-17

academia for genotoxicity risk assessment of substances and processes. At the same time, other test systems that claim to be sensitive for indicating genotoxic effects have emerged from academia. However, these assays are not OECD-approved, because they either fail to be standardizable or lack a clear rationale and unambiguous identification of genotoxic effects; for example, the assay may be sensitive to genotoxicity but also to other cellular responses unrelated to genotoxicity. To prevent later harm in humans, the OECD appoints expert groups that perform research on the reproducibility and practicability of these methods before they are included in the catalog of recommended testing of substances or processes [15]. Moreover, it needs to be stressed that standardizable assays providing unambiguous information on genotoxicity are primarily employed *in vitro*, although a few *in vivo* assays exist. The latter are, however, increasingly discouraged within the 3R principles to reduce, refine, and replace animal experiments according to the ARRIVE (Animal Research. Reporting of *In Vivo* Experiments; https://arriveguidelines.org) guidelines, mainly if alternative assays exist that provide a similar degree of information without using animals.

15.2.1 OECD-Approved Methods

This section provides a brief summary of genotoxicity assays used in plasma medicine. More assays are available, but these have not been used in plasma medicine. For more details, the reader is referred to the Genetic Toxicology Guidance Document in its most recent version provided by the OECD. Full assay details can be easily obtained online as free resources at the OECD website. Generally, genotoxicity assays fall into three categories of analysis: i) gene mutations, ii) chromosome aberration or breakage, and iii) chromosome loss or gain.

In the hypoxanthine-guanine phosphoribosyl transferase (HPRT) gene test (OECD test 476), cell lines are used to measure forwarded mutations in the HPRT reporter gene. It is, therefore, an *in vitro* assay. The guideline describes the principle: "Mutant cells deficient in Hprt enzyme activity (. . .) are resistant to the cytostatic effects of the purine analogue 6-thioguanine (TG). The Hprt (. . .) proficient cells are sensitive to TG, which causes the inhibition of cellular metabolism and halts further cell division. Thus, mutant cells are able to proliferate in the presence of TG, whereas normal cells, which contain the Hprt (. . .) enzyme, are not." In other words, if a substance or process is genotoxic, it may by chance hit a gene encoding for an enzyme that produces a cytotoxic product from an externally added chemical. If this gene mutates, the enzyme no longer functions properly, and the externally added chemical is not converted into the cytotoxic product. Therefore, mutated cells survive and generate colonies *in vitro*, which can be counted using chromatic staining at the end of the experiment. Typical cell lines used for this assay are Chinese ovary (CHO), lung (CHL and V79) hamster cells, and human TK6 lymphoblastoid cells. The final determined number of mutants is then corrected with a factor for the toxicity of the tested agent. Usually, mutants are observed even in untreated conditions, and the question is whether this number increases significantly after the treatment. As positive controls, UV-B radiation and Ethyl methanesulfonate (EMS) are frequently used.

The *in vitro* mammalian cell micronucleus (MN) test (OECD test 487) is based on the principle of identifying small nuclei in the cytoplasm of cells in the interphase of the cell cycle. Among the reasons for MN formation is the lack of centromere, which leads to acentric chromosome fragments or to entire chromosomes being unable to migrate to the poles during the anaphase cell cycle stage. Hence, the assay allows for the identification of agents with chromosome-damaging potential in cells that have undergone at least one cell division. The test's specificity is increased when using the actin polymerization inhibitor cytochalasin B, effectively creating cells incapable of completing the cell cycle by an equal division of the cytosol. Instead, so-called binucleated cells (i.e., cells containing two nuclei) are formed. These are proof of proliferation and, therefore, indicate the validity of MN being a product of chromosomal damage that appears only during proliferation. This specificity, the relative speed of the procedure (only a few days), and the possibility of automating the cellular analysis by flow cytometry, image cytometry, and high-content imaging to score thousands of cells within only minutes make the assay one of the most popular genotoxicity assessments in the OECD test battery. The assay was adapted to nucleated erythrocytes in a different form but based on the same principle of identifying micronuclei in cells. In the 1990s, these assays were invented to study genotoxicity in readily available chicken embryos whose erythrocytes contain nuclei and are, therefore, capable of producing micronuclei [16]. The OECD has launched an addition to this assay employing bone-marrow-derived erythrocytes or nucleated peripheral blood erythrocytes of mammalians (OECD test 474), usually rodents. Blood smears on, e.g., microscopy slides can then be analyzed manually or automatically.

The *in vivo* mammalian alkaline (pH>13) Comet assay (OECD test 489) is used to detect DNA strand breaks in nuclei isolated from cells from tissues, usually of rodent origin, that have been exposed to potentially genotoxic noxae. The assay is not specific regarding the consequences of damage, i.e., whether the DNA damage would lead to cell death, mutations, or other dysfunctions. This is because the analysis is supposed to be performed relatively soon after exposure to prevent results on DNA damage from being disguised or artificially amplified by cellular responses (e.g., death or repair). The guideline's main scope is to use tissue-derived cells and not cells from *in vitro* experiments, but the latter are undoubtedly feasible, as shown with, e.g., oxidative stress [17,18]. The assay principle is to lyse the cells and decondense nuclear DNA before it is added to agarose gels, followed by electrophoresis. In the gel, intact nucleic DNA travels slowly because of a lack of fragmentation, and the DNA is made visible by, e.g., fluorescent dyes, indicating round nuclei. By contrast, fragmented DNA is much smaller and, therefore, travels faster in the gel, creating "Comets," i.e., tails that emerge from nucleus-shaped DNA towards the cathode. Recent criticism of the assay suggests that—despite innovations in automatic sample analysis— it is at best only semi-quantitative and complex to interpret [19].

15.2.2 NON-APPROVED METHODS

A range of assays on genotoxicity analysis are used in research but not approved by the OECD [20,21]. Perhaps the most famous member of this family is the γH2A.X assay. Here, the serin 139 phosphorylation (γ) of the X variant of the histone A2 (H2A, i.e., H2A.X) via several phosphoinositol three kinases (e.g., ATM and ATR) serves signaling purposes as a molecular marker of DNA double-strand breaks. However, because γH2A.X also appears in conditions other than DNA damage, it has not been included in the genotoxicity testing battery of, e.g., the OECD. Nevertheless, γH2A.X serves as a well-investigated marker in ionizing radiation research [22], where low-dose neutron or gamma radiation directly damages DNA. This role of γH2A.X, in turn, does not imply that each γH2A.X-positive result is automatically related to DNA damage in cells exposed to agents other than ionizing radiation. For instance, we have shown that H2A.X can be abrogated by pharmacological inhibition of caspases following plasma-induced oxidative stress or exposure to oxidants such as H_2O_2 and hypochlorous acid but not following UV-radiation (attacking DNA directly), suggesting that γH2A.X foci appearance is linked to apoptotic cell death signaling pathways [23]. The benefits and caveats of using the γH2A.X assay, especially in the field of plasma medicine (Table 15.1), have been extensively described in a recent review the interested reader is referred to [24].

In the following table, we non-comprehensively summarize results in plasma medicine using OECD-approved assays and list the main findings, with a focus on *in vitro* studies. There are several OECD-approved *in vivo* genotoxicity assays, such as the Earthworm Reproduction Test (OECD test 222), the Avian Acute Oral Toxicity Test (OECD test 223), and the Mammalian Erythrocyte Micronucleus test in rodents (OECD test 474). However, to our knowledge, no findings on using any OECD-approved *in vivo* genotoxicity test in plasma medicine have been published so far, except for a modified version of the OECD test 474 in chicken embryos [25], as described further below.

TABLE 15.1

Selection of γH2A.X Studies in Plasma Medicine.

mildly malignant cell lines	reference
HEK-293 (human embryonic kidney cells)	Kaushik et al. [26]
MRC-5 (human lung fibroblasts)	Kaushik et al. [26]
human primary fibroblasts	Lazović et al. [27]
TK6 cells (B lymphoblastoid)	Bekeschus et al. [23]
HaCaT cells (human keratinocytes)	Gaur et al. [28]
	Schmidt et al. [29]
MCF-10A (human breast epithelial cells)	Kalghatgi et al. [30]
highly malignant cell lines	
A549 (human lung adenocarcinoma)	Kaushik et al. [26]
T98G (human glioblastoma)	Kaushik et al. [26]
U87MG (human glioblastoma)	Gjika et al. [31]
MSK QLL1, SCC1483, SCC15, SCC25 (human oral squamous cell carcinoma)	Chang et al. [32]
SCC25 (human squamous cell carcinoma)	Han et al. [33]
Mel (human melanoma)	Arndt et al. [34]
SK-MEL 28 (human melanoma)	Sagwal et al. [35]
A549 (human lung adenocarcinoma)	Karki et al. [36]
A2058 (human melanoma), B16F10 (murine melanoma)	Sensing et al. [37]
	Kim et al. [38]
HCT116 (human colon carcinoma)	Judée et al. [39]
	Plewa et al. [40]
	Choi et al. [41]

All studies have provided evidence of γH2A.X induction following plasma treatment. The table is adapted from [24].

15.3 GENOTOXICITY AND LONG-TERM STUDIES IN PLASMA MEDICINE

Medical therapies must show good efficacy and safety, irrespective of being within local or systemic modalities. A sufficient genotoxic safety profile is part of safety assessments. To this end, several studies were conducted during and after development and approval of the first medical plasma devices in Europe to assess the absence or presence of genotoxic events following plasma exposure.

15.3.1 *In Vitro* Genotoxicity Plasma Studies: Micronucleus Assay

Many studies have used the cytokinesis-block micronucleus assay (CBMN) assay to investigate MN formation in *in vitro* plasma-treated cells. However, the quality of quantification schemes and the degree of adherence to the OECD guidelines vary. Several studies are available for the best-investigated plasma source worldwide, the kINPen atmospheric pressure argon plasma jet [42]. In HaCaT keratinocytes, MN formation was assessed, closely following the OECD guidelines, after several plasma treatment times and using several UV exposures as well as mitomycin C (MMC) as positive control groups [42]. However, plasma-generated ROS were used by applying plasma-treated cell culture medium, mainly including hydrogen peroxide, nitrite, and nitrate. MN identification was made manually. The results showed significantly increased MN formation in UV- and MMC-treated cells, while keratinocytes exposed to plasma-treated medium did not show elevated MN numbers. Two studies were performed using TK6 lymphocytes [43,44], a cell line frequently recommended in different OECD-based genotoxicity assays, and image cytometry to investigate MN formation after plasma treatment. The cytometry-generated data can be processed with appropriate image-analysis software and mask generation to generate unbiased results on MN formation strictly in binucleated cells. In one study, direct kINPen plasma treatment did produce elevated MN levels (Figure 15.1). in contrast to UV and the positive control methyl methanesulfonate (MMS) treatment [23]. In another study, the feed gas mixture added to the kINPen for plasma generation was varied, generating a varying cocktail of ROS with different biological consequences [45]. Against ionizing radiation as a positive control that induced significantly increased MN levels, none of the ten feed gas conditions investigated produced significantly higher MN counts in directly treated TK6 cells studied using image cytometry and algorithm-driven image analysis [43]. Using the same setup, another plasma jet, the v-jet, was tested against a positive MMS control, and no increased MN formation was found following the plasma treatment [44]. Another study, investigating a non-certified experimental argon plasma device and plasma-treated medium, reported results on MN formation in WIL2-NS cells, a human lymphoblast cell line frequently used in genotoxicity research. Performed according to OECD guidelines, the study showed significantly increased MN formation at acceptable toxicity profiles and only low (25 μM) H_2O_2 levels produced by the plasma source [46]. This is in agreement with an earlier study reporting markedly elevated MN levels in WIL2-NS cells following exposure to 20 μM H_2O_2, although the exact exposure regimens (cell culture medium, cell density, cell vessels) were not identical [47].

Several studies also report MN results using protocols or methods not compliant with the OECD protocol [35,48–51]. For example, one report mentioned increased MN frequency in Jurkat cells exposed to plasma-treated medium. However, no data were shown besides the results mentioned in the text [51], possible because MN were assessed using regular flow cytometry. This data acquisition and analysis procedure is complex and error-prone and quickly produces false positive results [52], especially at high toxicity rates, as seen in the plasma study [51]. For instance, MN formation was increased eight-fold in dexamethasone-treated cells using flow cytometry but not in microscopy investigations [53], demonstrating the need to be careful when interpreting flow cytometry-based results on MN frequencies. As another example, a report used human liver L02 cells and exposed them directly to a homemade helium plasma jet. Up to a seven-fold increase in MN formation was observed for the longest plasma treatment time of 30 min [54]. However, the procedure deviated

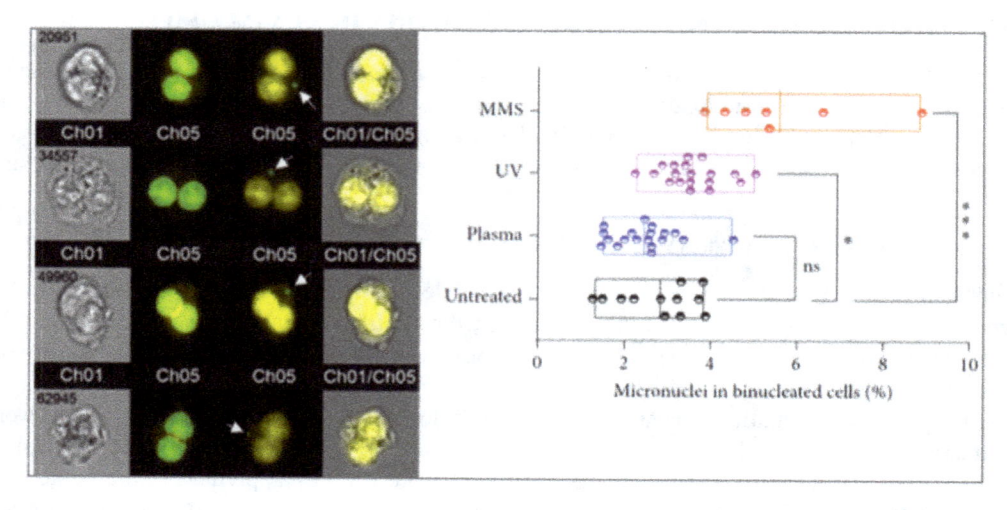

FIGURE 15.1 Representative images of four out of several thousand TK6 cells acquired using image cytometry (Amnis Image Stream ISX Mark II) indicated micronuclei in binucleated cells filtered by algorithm-driven image masking (left), and quantification of the percentage of binucleated cells carrying micronuclei in plasma-treated, UV-treated, and MMS-treated cells. Adapted from [23].

from the recommendations of the OECD protocol, and positive controls were missing. In addition, the liberal flow cytometry gating analysis suggests an underestimation of cytotoxicity, which is known to lead to false-positive results. This doubt on enhanced MN positivity is supported by the findings of an interesting experimental approach: after plasma treatment, living cells were sub-cultured for another seven passages before MN formation was investigated again. In these cells, elevated MN formation vanished, indicating that the cells with chromosomal aberrations had died or become senescent. In addition, the authors performed the HPRT assay (hypoxanthine-guanine-phosphoribosyl-transferase, see below), which showed no elevated mutation frequencies either in the directly treated cell or in the long-term subculture cells, suggesting a systematic error in the MN assessment had given rise to false-positive results.

The only quantitative *in vivo* report on genotoxicity in the field of plasma medicine used the hen's egg test (HET) on the chorioallantoic membrane (CAM) model (HET-CAM). In chicken embryos up to day 15 post-fertilization, the vasculature but not the nerve system is well developed, which means this model can be used for *in vivo* vertebrate testing without requiring ethical approval [55]. Neither the prototype kINPen 09 nor its clinical version kINPen MED exerted genotoxic effects at short or long (10 min) treatment times directly onto the CAM in either dry or wet (addition of sodium chloride to mimic wound fluid) conditions, compared to highly genotoxic positive controls cyclophosphamide and methotrexate when investigating nuclear aberrations in chicken red blood cells [25]. This provides quantitative evidence of the absence of genotoxic action of plasma treatment in vertebrates, at least for the kINPen jets.

Collectively, most studies strictly adhering to the CBMN assay OECD guidelines have not identified increased MN formation following plasma treatment. In general, care must be taken when using this assay because it is susceptible to false-positive results, especially as consequence of cytotoxic events such as apoptosis and beginning nuclear blebbing when analyzed manually by microscopy. Current research trends lean towards imaging or image cytometry-based systems capable of acquiring thousands of cells and using algorithm-driven imaging masks to identify binucleated cells and their micronuclei [56]. Recently, such methods were combined with neural networks [57] and deep learning [58] to increase automation for unbiased data analysis [59]. Recent inter-laboratory comparison of such methods will help generate a new industry (and, in the future, perhaps medical) standard [60] based on rapid training and classification of imaging cytometry data to remove human bias and error-prone manual analysis [61]. This structured methodology should eventually be applied to plasma medicine studies to define a clearer picture of MN formation with different cell types, plasma devices, and plasma settings.

15.3.2 *In Vitro* Genotoxicity Plasma Studies: HPRT Assay

While the CBMN and γH2A.X assays have been used in numerous publications in plasma medicine, reports on the more labor-intensive HPRT assay are infrequent. For the kINPen, neither plasma-treated medium nor direct plasma treatment led to increased colony formation indicative of mutants in HaCaT keratinocytes, while positive controls (UV radiation and EMS) did [42]. Interestingly, HaCaT exposure to H_2O_2 at the same concentration as produced by the plasma jet (in this instance, 80 μM) also did not produce more mutants. Similar findings were made for a portable plasma source using V79 cells, showing an absence of genotoxicity for plasma treatment at bactericidal dosages and genotoxic effects for EMS [62]. For the MicroPlaster β, a clinically certified device, plasma treatment of V79 cells did not show elevated numbers of mutants, even in the case of repetitive plasma treatment (2 min treatment over five days, i.e., five treatments), while UV radiation was strongly genotoxic [63]. A more recent study investigated plasma-treated Ringer's lactate and found that HeLa cells exposed to this liquid increased mutant numbers, although appropriate positive controls as indicated by OECD guidelines were missing. However, adding the H_2O_2-decomposing enzyme catalase entirely abrogated plasma-treated-liquid-enhanced colony formation, which led the authors to conclude that their observations were based on H_2O_2 produced by the plasma rather than the plasma treatment itself [64]. Unfortunately, two reports claiming increased HPRT frequencies following incubation of cells with plasma-treated fetal bovine serum or plasma-treated biomolecules had not included such catalase control or any other positive control [65,66] as recommended per OECD guidelines. In the first study, however, H_2O_2 exposure of CHO-K1 cells at lower micromolar range produced much greater mutant formation, which was indicative of the observations related to H_2O_2 [65]. In addition, in both studies, many of the plates investigated several weeks after initial seeding did not produce any colonies, which is not in line with general observations of other authors using these cells and the HPRT assay [67–70]. Finally, as indicated in the previous section, Ma and colleagues performed the HPRT assay in plasma-treated L02 liver cells. Against strongly genotoxic EMS as a positive control, no increased mutation frequency was found in either the directly treated cells or their subclones after seven additional passages [54].

In summary, none of the studies strictly adhering to OECD guidelines provided evidence of plasma-induced enhancement of mutation frequencies, and the other studies' potential genotoxic effects were proven to be or likely due to the effects of H_2O_2.

15.4 IMPLICATIONS OF GENOTOXIC SAFETY ASSESSMENTS FOR THE MODE OF ACTION OF PLASMAS

The findings in genotoxicity described above have implications for general considerations about the mode of action in plasma medicine and its redox biology-related effects. It is known that radiation, such as UV light and ionizing radiation, can penetrate cells and tissues, depending on wavelength and intensity, to produce gradual effects. Among these are, for instance, hydrolysis and hydroxyl radical (.OH) production, which can appear immediately within the nucleus, causing oxidative DNA damage either directly or indirectly by transformation to the much less reactive H_2O_2. Therefore, it is clear that reactive species can cause DNA damage. However, the conclusion that reactive species always cause DNA damage would be false parallelism. Due to the short half-life of most reactive species, their point of origin is essential. With ROS production externally from a cell, as in the case of plasma treatment, highly reactive short-lived species would have to diffuse long distances before reaching the nucleus, which their short half-lives and abundance of reaction partners disallow. Apart from the extracellular biomolecules surrounding the cells, cell membranes are the first to be attacked by plasma-derived reactive species, as indicated by lipid peroxidation studies in plasma medicine [71–74]. Accordingly, most primary plasma-derived short-lived species will react at this compartment (Figure 15.2). If non-lethal plasma membrane pore formation or secondary reactions from lipid peroxidation allow short-lived reactive species to enter the cytosol, they will meet abundant thiol-containing reaction partners that will decrease their further diffusion. Accordingly, it is impossible for plasma-derived short-lived species to directly affect nuclear DNA, especially considering the dozens of lipid bilayers of the endoplasmic reticulum surrounding the nuclear membrane, through which such species would have to travel without losing their reactivity. However, more long-lived oxidants, such as primary or secondary plasma-generated H_2O_2, may penetrate cells via aquaporins [75]. Yet it was recently shown that surface membrane aquaporin expression levels in 36 cell lines do not correlate with the cytotoxicity of plasma treatment [76], discouraging a more extensive role of aquaporins and, hence, H_2O_2 transported through this channel. From a redox biology perspective, it is more helpful to ask to what extent reactive-species-induced cell death occurs in cells. The lower such a threshold is, the higher the cytotoxicity will be with short treatment times. Hence, it appears logical that the probability of oxidative DNA damage is higher the more resistant a cell type is to plasma-induced cytotoxicity and the longer testable plasma treatment times or dosages are. Nevertheless, it would remain unlikely for primary plasma-derived species to travel from the plasma gas phase through the membranes of the cell, ER, nucleus, protein, and thiol-rich cytosol directly to the DNA to exert oxidative damage.

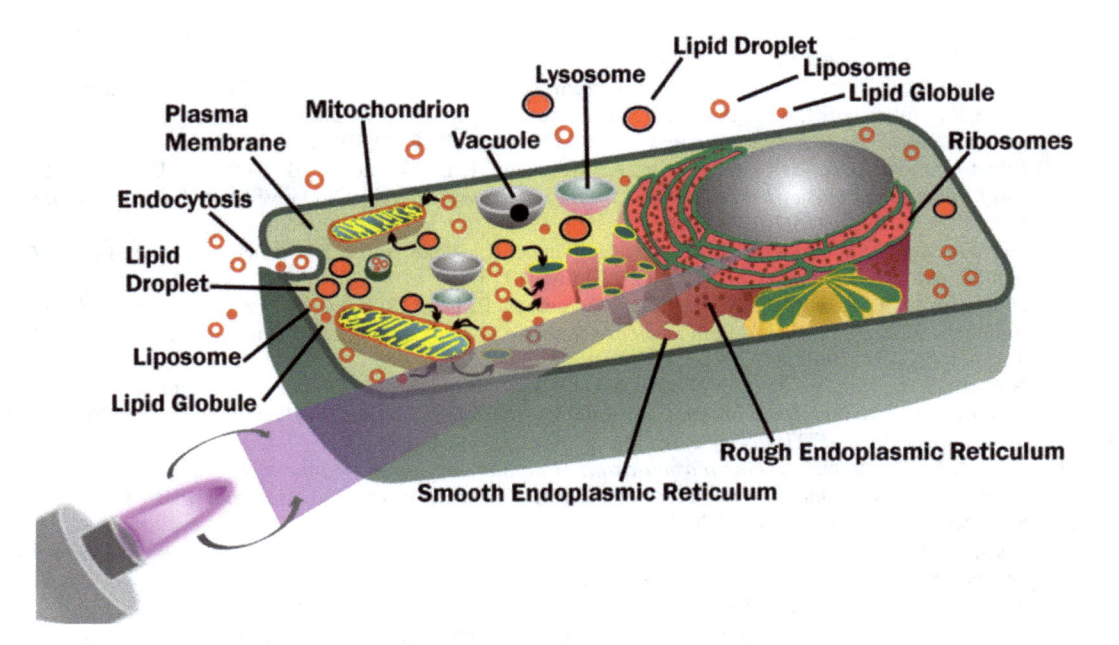

FIGURE 15.2 Cartoon of plasma jet (left) derived reactive species (triangle gradient) reaching cells. Most short-lived reactive species likely deteriorate at the cell membrane (faded violet color of the triangle), and with every compartment (cytosol rich in thiol-proteins, multiple ER membranes, nuclear membrane), it becomes less likely that primary plasma-derived reactive species directly reach into the nucleus to exert oxidative DNA damage and genotoxic effects. Adapted from [77].

15.5 CONCLUSION AND OUTLOOK

Based on published evidence from OECD guideline-based and controlled genotoxicity assessments, plasma treatment has been reported to have no genotoxic activity in several cell lines and for several plasma devices tested. These *in vitro* results were supported by findings of one *in vivo* work using chicken embryos. While clinical observations on the lack of long-term side effects [1] are in line with those data, it would be interesting from a redox biology point of view to capitalize on the battery of existing *in vivo* OECD genotoxic assays to further infer the relation between multiplex species production, diffusion, and biological consequences.

REFERENCES

[1] Bekeschus, S.; von Woedtke, T.; Emmert, S.; Schmidt, A. *Medical gas plasma-stimulated wound healing: Evidence and mechanisms.* Redox Biology **2021**, 10.1016/j.redox.2021.102116.

[2] Lata, S.; Chakravorty, S.; Mitra, T.; Pradhan, P.K.; Mohanty, S.; Patel, P.; Jha, E.; Panda, P.K.; Verma, S.K.; Suar, M. *Aurora borealis in dentistry: The applications of cold plasma in biomedicine.* Materials Today Bio **2022**, 10.1016/j.mtbio.2021.100200.

[3] Milhan, N.V.M.; Chiappim, W.; Sampaio, A.D.G.; Vegian, M.; Pessoa, R.S.; Koga-Ito, C.Y. *Applications of plasma-activated water in dentistry: A review.* International Journal of Molecular Sciences **2022**, 10.3390/ijms23084131.

[4] Metelmann, H.-R.; Seebauer, C.; Miller, V.; Fridman, A.; Bauer, G.; Graves, D.B.; Pouvesle, J.-M.; Rutkowski, R.; Schuster, M.; Bekeschus, S.; Wende, K.; Masur, K.; Hasse, S.; Gerling, T.; Hori, M.; Tanaka, H.; Ha Choi, E.; Weltmann, K.-D.; Metelmann, P.H.; Von Hoff, D.D.; Woedtke, T.V. *Clinical experience with cold plasma in the treatment of locally advanced head and neck cancer.* Clinical Plasma Medicine **2018**, 10.1016/j.cpme.2017.09.001.

[5] Bekeschus, S.; Clemen, R. *Plasma, cancer, immunity.* Journal of Physics D: Applied Physics **2022**, 10.1088/1361-6463/ac9398.

[6] Boeckmann, L.; Schafer, M.; Bernhardt, T.; Semmler, M.L.; Jung, O.; Ojak, G.; Fischer, T.; Peters, K.; Nebe, B.; Muller-Hilke, B.; Seebauer, C.; Bekeschus, S.; Emmert, S. *Cold atmospheric pressure plasma in wound healing and cancer treatment.* Applied Sciences **2020**, 10.3390/app10196898.

[7] Privat-Maldonado, A.; Schmidt, A.; Lin, A.; Weltmann, K.D.; Wende, K.; Bogaerts, A.; Bekeschus, S. *Ros from physical plasmas: Redox chemistry for biomedical therapy.* Oxidative Medicine and Cellular Longevity **2019**, 10.1155/2019/9062098.

[8] von Woedtke, T.; Schmidt, A.; Bekeschus, S.; Wende, K.; Weltmann, K.D. *Plasma medicine: A field of applied redox biology. In Vivo* **2019**, 10.21873/invivo.11570.

[9] Nikiforov, A.; Li, L.; Britun, N.; Snyders, R.; Vanraes, P.; Leys, C. *Influence of air diffusion on the oh radicals and atomic o distribution in an atmospheric ar (bio)plasma jet.* Plasma Sources Science and Technology **2014**, 10.1088/0963-0252/23/1/015015.

[10] Norberg, S.A.; Tian, W.; Johnsen, E.; Kushner, M.J. *Atmospheric pressure plasma jets interacting with liquid covered tissue: Touching and not-touching the liquid.* Journal of Physics D-Applied Physics **2014**, 10.1088/0022-3727/47/47/475203.

[11] Freund, E.; Bekeschus, S. *Gas plasma-oxidized liquids for cancer treatment: Preclinical relevance, immuno-oncology, and clinical obstacles.* IEEE Transactions on Radiation and Plasma Medical Sciences **2021**, 10.1109/trpms.2020.3029982.

[12] Romero, M.P.; Osuna, C.; Garcia-Perganeda, A.; Carrillo-Vico, A.; Guerrero, J.M. *The pineal secretory product melatonin reduces hydrogen peroxide-induced DNA damage in u-937 cells.* Journal of Pineal Research **1999**, 10.1111/j.1600-079x.1999.tb00588.x.

[13] Stivaktakis, P.; Vlastos, D.; Giannakopoulos, E.; Matthopoulos, D.P. *Differential micronuclei induction in human lymphocyte cultures by imidacloprid in the presence of potassium nitrate.* Scientific World Journal **2010**, 10.1100/tsw.2010.9.

[14] Tanaka, H.; Bekeschus, S.; Yan, D.; Hori, M.; Keidar, M.; Laroussi, M. *Plasma-treated solutions (pts) in cancer therapy.* Cancers **2021**, 10.3390/cancers13071737.

[15] Jacobs, M.N.; Colacci, A.; Corvi, R.; Vaccari, M.; Aguila, M.C.; Corvaro, M.; Delrue, N.; Desaulniers, D.; Ertych, N.; Jacobs, A.; Luijten, M.; Madia, F.; Nishikawa, A.; Ogawa, K.; Ohmori, K.; Paparella, M.; Sharma, A.K.; Vasseur, P. *Chemical carcinogen safety testing: Oecd expert group international consensus on the development of an integrated approach for the testing and assessment of chemical non-genotoxic carcinogens.* Archives of Toxicology **2020**, 10.1007/s00204-020-02784-5.

[16] Wolf, T.; Luepke, N.P. *Formation of micronuclei in incubated hen's eggs as a measure of genotoxicity.* Mutation Research—Genetic Toxicology and Environmental Mutagenesis **1997**, 10.1016/S1383-5718(97)00136-8.

[17] Fiskus, W.; Coothankandaswamy, V.; Chen, J.; Ma, H.; Ha, K.; Saenz, D.T.; Krieger, S.S.; Mill, C.P.; Sun, B.; Huang, P.; Mumm, J.S.; Melnick, A.M.; Bhalla, K.N. *Sirt2 deacetylates and inhibits the peroxidase activity of peroxiredoxin-1 to sensitize breast cancer cells to oxidant stress-inducing agents.* Cancer Research **2016**, 10.1158/0008-5472.CAN-16-0126.

[18] La Maestra, S.; Micale, R.T.; Ferretti, M.; Izzotti, A.; Gaggero, L. *Attenuation of oxidative stress and chromosomal aberrations in cultured macrophages and pulmonary cells following self-sustained high temperature synthesis of asbestos.* Scientific Reports **2020**, 10.1038/s41598-020-65620-x.

[19] Cordelli, E.; Bignami, M.; Pacchierotti, F. *Comet assay: A versatile but complex tool in genotoxicity testing.* Toxicology Research **2021**, 10.1093/toxres/tfaa093.

[20] Sponchiado, G.; Adam, M.L.; Silva, C.D.; Soley, B.S.; de Mello-Sampayo, C.; Cabrini, D.A.; Correr, C.J.; Otuki, M.F. *Quantitative genotoxicity assays for analysis of medicinal plants: A systematic review.* Journal of Ethnopharmacol **2016**, 10.1016/j.jep.2015.10.026.

[21] Knight, A.W.; Little, S.; Houck, K.; Dix, D.; Judson, R.; Richard, A.; McCarroll, N.; Akerman, G.; Yang, C.; Birrell, L.; Walmsley, R.M. *Evaluation of high-throughput genotoxicity assays used in profiling the us epa toxcast chemicals.* Regulatory Toxicology and Pharmacology **2009**, 10.1016/j.yrtph.2009.07.004.

[22] Mah, L.J.; El-Osta, A.; Karagiannis, T.C. *Gammah2ax: A sensitive molecular marker of DNA damage and repair.* Leukemia **2010**, 10.1038/leu.2010.6.

[23] Bekeschus, S.; Schutz, C.S.; Niessner, F.; Wende, K.; Weltmann, K.D.; Gelbrich, N.; von Woedtke, T.; Schmidt, A.; Stope, M.B. *Elevated h2ax phosphorylation observed with kinpen plasma treatment is not caused by ros-mediated DNA damage but is the consequence of apoptosis.* Oxidative Medicine and Cellular Longevity **2019**, 10.1155/2019/8535163.

[24] Schutz, C.S.; Stope, M.B.; Bekeschus, S. *H2a.X phosphorylation in oxidative stress and risk assessment in plasma medicine.* Oxidative Medicine and Cellular Longevity **2021**, 10.1155/2021/2060986.

[25] Kluge, S.; Bekeschus, S.; Bender, C.; Benkhai, H.; Sckell, A.; Below, H.; Stope, M.B.; Kramer, A. *Investigating the mutagenicity of a cold argon-plasma jet in an het-mn model.* PLoS One **2016**, 10.1371/journal.pone.0160667.

[26] Kaushik, N.; Uddin, N.; Sim, G.B.; Hong, Y.J.; Baik, K.Y.; Kim, C.H.; Lee, S.J.; Kaushik, N.K.; Choi, E.H. *Responses of solid tumor cells in dmem to reactive oxygen species generated by non-thermal plasma and chemically induced ros systems.* Scientific Reports **2015**, 10.1038/srep08587.

[27] Lazovic, S.; Maletic, D.; Leskovac, A.; Filipovic, J.; Puac, N.; Malovic, G.; Joksic, G.; Petrovic, Z.L. *Plasma induced DNA damage: Comparison with the effects of ionizing radiation.* Applied Physics Letters **2014**, 10.1063/1.4896626.

[28] Gaur, N.; Kurita, H.; Oh, J.S.; Miyachika, S.; Ito, M.; Mizuno, A.; Cowin, A.J.; Allinson, S.; Short, R.D.; Szili, E.J. *On cold atmospheric-pressure plasma jet induced DNA damage in cells.* Journal of Physics D-Applied Physics **2021**, 10.1088/1361-6463/abb8ab.

[29] Schmidt, A.; Bekeschus, S.; Jarick, K.; Hasse, S.; von Woedtke, T.; Wende, K. *Cold physical plasma modulates p53 and mitogen-activated protein kinase signaling in keratinocytes.* Oxidative Medicine and Cellular Longevity **2019**, 10.1155/2019/7017363.

[30] Kalghatgi, S.; Kelly, C.M.; Cerchar, E.; Torabi, B.; Alekseev, O.; Fridman, A.; Friedman, G.; Azizkhan-Clifford, J. *Effects of non-thermal plasma on mammalian cells.* PLoS One **2011**, 10.1371/journal.pone.0016270.

[31] Gjika, E.; Pal-Ghosh, S.; Kirschner, M.E.; Lin, L.; Sherman, J.H.; Stepp, M.A.; Keidar, M. *Combination therapy of cold atmospheric plasma (cap) with temozolomide in the treatment of u87mg glioblastoma cells.* Scientific Reports **2020**, 10.1038/s41598-020-73457-7.

[32] Chang, J.W.; Kang, S.U.; Shin, Y.S.; Kim, K.I.; Seo, S.J.; Yang, S.S.; Lee, J.S.; Moon, E.; Baek, S.J.; Lee, K.; Kim, C.H. *Non-thermal atmospheric pressure plasma induces apoptosis in oral cavity squamous cell carcinoma: Involvement of DNA-damage-triggering sub-g(1) arrest via the atm/p53 pathway.* Archives of Biochemistry and Biophysics **2014**, 10.1016/j.abb.2014.01.022.

[33] Han, X.; Klas, M.; Liu, Y.Y.; Stack, M.S.; Ptasinska, S. *DNA damage in oral cancer cells induced by nitrogen atmospheric pressure plasma jets.* Applied Physics Letters **2013**, 10.1063/1.4809830.

[34] Arndt, S.; Wacker, E.; Li, Y.F.; Shimizu, T.; Thomas, H.M.; Morfill, G.E.; Karrer, S.; Zimmermann, J.L.; Bosserhoff, A.K. *Cold atmospheric plasma, a new strategy to induce senescence in melanoma cells.* Experimental Dermatology **2013**, 10.1111/exd.12127.

[35] Sagwal, S.K.; Pasqual-Melo, G.; Bodnar, Y.; Gandhirajan, R.K.; Bekeschus, S. *Combination of chemotherapy and physical plasma elicits melanoma cell death via upregulation of slc22a16.* Cell Death and Diseases **2018**, 10.1038/s41419-018-1221-6.

[36] Karki, S.B.; Gupta, T.T.; Yildirim-Ayan, E.; Eisenmann, K.M.; Ayan, H. *Miniature non-thermal plasma induced cell cycle arrest and apoptosis in lung carcinoma cells.* Plasma Chemistry and Plasma Processing **2019**, 10.1007/s11090-019-10037-2.

[37] Sensenig, R.; Kalghatgi, S.; Cerchar, E.; Fridman, G.; Shereshevsky, A.; Torabi, B.; Arjunan, K.P.; Podolsky, E.; Fridman, A.; Friedman, G.; Azizkhan-Clifford, J.; Brooks, A.D. *Non-thermal plasma induces apoptosis in melanoma cells via production of intracellular reactive oxygen species.* Annals of Biomedical Engineering **2011**, 10.1007/s10439-010-0197-x.

[38] Kim, G.J.; Kim, W.; Kim, K.T.; Lee, J.K. *DNA damage and mitochondria dysfunction in cell apoptosis induced by nonthermal air plasma.* Applied Physics Letters **2010**, 10.1063/1.3292206.

[39] Judee, F.; Fongia, C.; Ducommun, B.; Yousfi, M.; Lobjois, V.; Merbahi, N. *Short and long time effects of low temperature plasma activated media on 3d multicellular tumor spheroids.* Scientific Reports **2016**, 10.1038/srep21421.

[40] Plewa, J.M.; Yousfi, M.; Frongia, C.; Eichwald, O.; Ducommun, B.; Merbahi, N.; Lobjois, V. *Low-temperature plasma-induced antiproliferative effects on multi-cellular tumor spheroids.* New Journal of Physics **2014**, 10.1088/1367-2630/16/4/043027.

[41] Choi, J.S.; Kim, J.; Hong, Y.J.; Bae, W.Y.; Choi, E.H.; Jeong, J.W.; Park, H.K. *Evaluation of non-thermal plasma-induced anticancer effects on human colon cancer cells.* Biomedical Optical Express **2017**, 10.1364/BOE.8.002649.

[42] Wende, K.; Bekeschus, S.; Schmidt, A.; Jatsch, L.; Hasse, S.; Weltmann, K.D.; Masur, K.; von Woedtke, T. *Risk assessment of a cold argon plasma jet in respect to its mutagenicity.* Mutation Research—Genetic Toxicology and Environmental Mutagenesis **2016**, 10.1016/j.mrgentox.2016.02.003.

[43] Bekeschus, S.; Schmidt, A.; Kramer, A.; Metelmann, H.R.; Adler, F.; von Woedtke, T.; Niessner, F.; Weltmann, K.D.; Wende, K. *High throughput image cytometry micronucleus assay to investigate the presence or absence of mutagenic effects of cold physical plasma.* Environmental and Molecular Mutagenesis **2018**, 10.1002/em.22172.

[44] Miebach, L.; Freund, E.; Horn, S.; Niessner, F.; Sagwal, S.K.; von Woedtke, T.; Emmert, S.; Weltmann, K.D.; Clemen, R.; Schmidt, A.; Gerling, T.; Bekeschus, S. *Tumor cytotoxicity and immunogenicity of a novel v-jet neon plasma source compared to the kinpen.* Scientific Reports **2021**, 10.1038/s41598-020-80512-w.

[45] Bekeschus, S.; Schmidt, A.; Niessner, F.; Gerling, T.; Weltmann, K.D.; Wende, K. *Basic research in plasma medicine—a throughput approach from liquids to cells.* Journal of Visualized Experiments **2017**, 10.3791/56331.

[46] Hong, S.H.; Szili, E.J.; Fenech, M.; Gaur, N.; Short, R.D. *Genotoxicity and cytotoxicity of the plasma jet-treated medium on lymphoblastoid wil2-ns cell line using the cytokinesis block micronucleus cytome assay.* Scientific Reports **2017**, 10.1038/s41598-017-03754-1.

[47] Saito, A.; Sugisawa, A.; Umegaki, K.; Sunagawa, H. *Protective effects of quercetin and its metabolites on h2o2-induced chromosomal damage to wil2-ns cells.* Bioscience, Biotechnology, and Biochemistry **2004**, 10.1271/bbb.68.271.

[48] Kaushik, N.K.; Uhm, H.; Choi, E.H. *Micronucleus formation induced by dielectric barrier discharge plasma exposure in brain cancer cells.* Applied Physics Letters **2012**, 10.1063/1.3687172.

[49] Pasqual-Melo, G.; Sagwal, S.K.; Freund, E.; Gandhirajan, R.K.; Frey, B.; von Woedtke, T.; Gaipl, U.; Bekeschus, S. *Combination of gas plasma and radiotherapy has immunostimulatory potential and additive toxicity in murine melanoma cells in vitro.* International Journal of Molecular Sciences **2020**, 10.3390/ijms21041379.

[50] Padureanu, S.; Burlica, R.; Stoleru, V.; Beniuga, O.; Dirlau, D.; Cretu, D.E.; Astanei, D.; Patras, A. *Non-thermal plasma-activated water: A cytogenotoxic potential on triticum aestivum.* Agronomy **2023**, 10.3390/agronomy13020459.

[51] Turrini, E.; Laurita, R.; Simoncelli, E.; Stancampiano, A.; Catanzaro, E.; Calcabrini, C.; Carulli, G.; Rousseau, M.; Gherardi, M.; Maffei, F.; Cocchi, V.; Lenzi, M.; Pellicioni, V.; Hrelia, P.; Colombo, V.; Fimognari, C. *Plasma-activated medium as an innovative anticancer strategy: Insight into its cellular and molecular impact on in vitro leukemia cells.* Plasma Processes and Polymers **2020**, 10.1002/ppap.202000007.

[52] Avlasevich, S.; Bryce, S.; De Boeck, M.; Elhajouji, A.; Van Goethem, F.; Lynch, A.; Nicolette, J.; Shi, J.; Dertinger, S. *Flow cytometric analysis of micronuclei in mammalian cell cultures: Past, present and future.* Mutagenesis **2011**, 10.1093/mutage/geq058.

[53] Collins, J.E.; Ellis, P.C.; White, A.T.; Booth, A.E.; Moore, C.E.; Burman, M.; Rees, R.W.; Lynch, A.M. *Evaluation of the litron in vitro microflow kit for the flow cytometric enumeration of micronuclei (mn) in mammalian cells.* Mutation Research **2008**, 10.1016/j.mrgentox.2008.05.003.

[54] Ma, M.; Duan, J.; Lu, X.; He, G. *Genotoxic and mutagenic properties of atmospheric pressure plasma jet on human liver cell line l02.* Physics of Plasmas **2019**, 10.1063/1.5087148.

[55] Miebach, L.; Berner, J.; Bekeschus, S. *In ovo model in cancer research and tumor immunology.* Frontiers in Immunology **2022**, 10.3389/fimmu.2022.1006064.

[56] Saadati, F.; da Silva Brito, W.A.; Emmert, S.; Bekeschus, S. *Optimized high-content imaging screening quantifying micronuclei formation in polymer-treated hacat keratinocytes.* Nanomaterials **2022**, 10.3390/nano12244463.

[57] Wei, W.; Tao, H.; Chen, W.; Wu, X. *Automatic recognition of micronucleus by combining attention mechanism and alexnet.* BMC Medical Informatics and Decision Making **2022**, 10.1186/s12911-022-01875-w.

[58] Rodrigues, M.A.; Probst, C.E.; Zayats, A.; Davidson, B.; Riedel, M.; Li, Y.; Venkatachalam, V. *The in vitro micronucleus assay using imaging flow cytometry and deep learning.* Systems Biology and Applications **2021**, 10.1038/s41540-021-00179-5.

[59] Wang, Q.; Rodrigues, M.A.; Repin, M.; Pampou, S.; Beaton-Green, L.A.; Perrier, J.; Garty, G.; Brenner, D.J.; Turner, H.C.; Wilkins, R.C. *Automated triage radiation biodosimetry: Integrating imaging flow cytometry with high-throughput robotics to perform the cytokinesis-block micronucleus assay.* Radiation Research **2019**, 10.1667/RR15243.1.

[60] Wills, J.W.; Verma, J.R.; Rees, B.J.; Harte, D.S.G.; Haxhiraj, Q.; Barnes, C.M.; Barnes, R.; Rodrigues, M.A.; Doan, M.; Filby, A.; Hewitt, R.E.; Thornton, C.A.; Cronin, J.G.; Kenny, J.D.; Buckley, R.; Lynch, A.M.; Carpenter, A.E.; Summers, H.D.; Johnson, G.E.; Rees, P. *Inter-laboratory automation of the in vitro micronucleus assay using imaging flow cytometry and deep learning.* Archives of Toxicology **2021**, 10.1007/s00204-021-03113-0.

[61] Eulenberg, P.; Kohler, N.; Blasi, T.; Filby, A.; Carpenter, A.E.; Rees, P.; Theis, F.J.; Wolf, F.A. *Reconstructing cell cycle and disease progression using deep learning.* Nature Communication **2017**, 10.1038/s41467-017-00623-3.

[62] Boxhammer, V.; Li, Y.F.; Koritzer, J.; Shimizu, T.; Maisch, T.; Thomas, H.M.; Schlegel, J.; Morfill, G.E.; Zimmermann, J.L. *Investigation of the mutagenic potential of cold atmospheric plasma at bactericidal dosages.* Mutation Research **2013**, 10.1016/j.mrgentox.2012.12.015.

[63] Maisch, T.; Bosserhoff, A.K.; Unger, P.; Heider, J.; Shimizu, T.; Zimmermann, J.L.; Morfill, G.E.; Landthaler, M.; Karrer, S. *Investigation of toxicity and mutagenicity of cold atmospheric argon plasma.* Environmental and Molecular Mutagenesis **2017**, 10.1002/em.22086.

[64] Liu, Y.; Nakatsu, Y.; Tanaka, H.; Koga, K.; Ishikawa, K.; Shiratani, M.; Hori, M. *Effects of plasma-activated ringer's lactate solution on cancer cells: Evaluation of genotoxicity.* Genes and Environment **2023**, 10.1186/s41021-023-00260-x.

[65] Boehm, D.; Heslin, C.; Cullen, P.J.; Bourke, P. *Cytotoxic and mutagenic potential of solutions exposed to cold atmospheric plasma.* Scientific Reports **2016**, 10.1038/srep21464.

[66] Heslin, C.; Boehm, D.; Gilmore, B.F.; Megaw, J.; Freeman, T.A.; Hickok, N.J.; Cullen, P.J.; Bourke, P. *Biomolecules as model indicators of in vitro and in vivo cold plasma safety.* Frontiers in Physics **2021**, 10.3389/fphy.2020.613046.

[67] Ding, G.R.; Wake, K.; Taki, M.; Miyakoshi, J. *Increase in hypoxanthine-guanine phosphoribosyl transferase gene mutations by exposure to electric field.* Life Science **2001**, 10.1016/s0024-3205(00)01007-9.

[68] Stearns, D.M.; Silveira, S.M.; Wolf, K.K.; Luke, A.M. *Chromium(iii) tris(picolinate) is mutagenic at the hypoxanthine (guanine) phosphoribosyltransferase locus in chinese hamster ovary cells.* Mutation Research **2002**, 10.1016/s1383-5718(01)00301-1.

[69] Yang, J.L.; Yeh, S.C.; Chang, C.Y. *Lead acetate mutagenicity and mutational spectrum in the hypoxanthine guanine phosphoribosyltransferase gene of chinese hamster ovary k1 cells.* Molecular Carcinogenesis **1996**, 10.1002/1098-2744(199612)17:4<181::aid-mc2940170402>3.0.co;2-f.

[70] Schwartz, J.L.; Rotmensch, J.; Sun, J.; An, J.; Xu, Z.; Yu, Y.; Hsie, A.W. *Multiplex polymerase chain reaction-based deletion analysis of spontaneous, gamma ray- and alpha-induced hprt mutants of cho-k1 cells.* Mutagenesis **1994**, 10.1093/mutage/9.6.537.

[71] Tani, A.; Fukui, S.; Ikawa, S.; Kitano, K. *Evaluation of fatty acid oxidation by reactive oxygen species induced in liquids using atmospheric-pressure nonthermal plasma jets.* Journal of Physics D-Applied Physics **2015**, 10.1088/0022-3727/48/42/424010.

[72] Wolff, C.M.; Kolb, J.F.; Weltmann, K.D.; von Woedtke, T.; Bekeschus, S. *Combination treatment with cold physical plasma and pulsed electric fields augments ros production and cytotoxicity in lymphoma.* Cancers **2020**, 10.3390/cancers12040845.

[73] Van der Paal, J.; Neyts, E.C.; Verlackt, C.C.W.; Bogaerts, A. *Effect of lipid peroxidation on membrane permeability of cancer and normal cells subjected to oxidative stress.* Chemical Sciences **2016**, 10.1039/c5sc02311d.

[74] Striesow, J.; Lackmann, J.W.; Ni, Z.; Wenske, S.; Weltmann, K.D.; Fedorova, M.; von Woedtke, T.; Wende, K. *Oxidative modification of skin lipids by cold atmospheric plasma (cap): A standardizable approach using rp-lc/ms(2) and di-esi/ms(2).* Chemistry and Physics of Lipids **2020**, 10.1016/j.chemphyslip.2019.104786.

[75] Bienert, G.P.; Chaumont, F. *Aquaporin-facilitated transmembrane diffusion of hydrogen peroxide.* Biochimica et Biophysica Acta **2014**, 10.1016/j.bbagen.2013.09.017.

[76] Bekeschus, S.; Liebelt, G.; Menz, J.; Berner, J.; Sagwal, S.K.; Wende, K.; Weltmann, K.D.; Boeckmann, L.; von Woedtke, T.; Metelmann, H.R.; Emmert, S.; Schmidt, A. *Tumor cell metabolism correlates with resistance to gas plasma treatment: The evaluation of three dogmas.* Free Radical Biology and Medicine **2021**, 10.1016/j.freeradbiomed.2021.02.035.

[77] Nicolson, G.L.; Ash, M.E. *Membrane lipid replacement for chronic illnesses, aging and cancer using oral glycerolphospholipid formulations with fructooligosaccharides to restore phospholipid function in cellular membranes, organelles, cells and tissues.* Biochimica et Biophysica Acta Biomembranes **2017**, 10.1016/j.bbamem.2017.04.013.

Index

Printed in the USA
CPSIA information can be obtained
at www.ICGtesting.com
LVHW081755041124
795688LV00005B/597

9 781032 356921